Communication with Extraterrestrial Intelligence

D0861335

SUNY scholarly conferences

Communication with Extraterrestrial Intelligence

Edited by
Douglas A. Vakoch

Published by State University of New York Press, Albany

© 2011 State University of New York

All rights reserved

Printed in the United States of America

No part of this book may be used or reproduced in any manner whatsoever without written permission. No part of this book may be stored in a retrieval system or transmitted in any form or by any means including electronic, electrostatic, magnetic tape, mechanical, photocopying, recording, or otherwise without the prior permission in writing of the publisher.

For information, contact State University of New York Press, Albany, NY
www.sunypress.edu

Production by Ryan Morris
Marketing by Michael Campochiaro

Library of Congress Cataloging-in-Publication Data

Communication with extraterrestrial intelligence (ceti) / edited by Douglas A. Vakoch.
         p. cm.
    Includes bibliographical references and index.
    ISBN 978-1-4384-3793-4 (hardcover : alk. paper)
    ISBN 978-1-4384-3794-1 (pbk. : alk. paper)
    1. Life on other planets.   2. Interstellar communication.   I. Vakoch, Douglas A.

    QB54.C653 2011
    576.8'39—dc22 2010046427

10 9 8 7 6 5 4 3 2 1

ACC LIBRARY SERVICES AUSTIN, TX

To Frank Drake

Honoring a half century leading the
Search for Extraterrestrial Intelligence

Contents

Contents

Contents

PART II
Active SETI: Should We Transmit?

Contents

PART III
Interstellar Message Construction:
Can We Make Ourselves Understood?

Preface

Just over fifty years ago, in April 1960, astronomer Frank Drake conducted Project Ozma, the first experiment to search for radio signals from extraterrestrial civilizations. Although we have not found evidence of intelligence beyond Earth in the intervening decades, our understanding of the cosmos has increased tremendously. As merely two examples, we now know that planets circle other stars in abundance and that life can survive in harsh environments ranging from the cores of nuclear reactors to the vacuum of outer space. It is fitting, then, that as we search for signs of extraterrestrial technologies we should do so in the broader context of astrobiology, the discipline that studies the origin, evolution, distribution, and future of life in the universe.

This book arose from the most recent biennial Astrobiology Science Conference (AbSciCon), organized by the National Aeronautics and Space Administration (NASA) and held the last week of April 2010. Chapters in this volume cover three broad themes in the Search for Extraterrestrial Intelligence (SETI). First, leading researchers examine the latest developments in observational SETI programs, as well as innovative proposals for new search strategies and novel approaches to signal processing. Next, both proponents and opponents of "Active SETI" debate whether humankind should be transmitting intentional signals to other possible civilizations, rather than only listening. Finally, constructive proposals for interstellar messages are juxtaposed with critiques that ask whether any meaningful exchange is possible with an independently evolved civilization, given the constraints of contact at interstellar distances, where a round-trip exchange could take centuries or millennia.

Those familiar with the history of interstellar communication will recall an earlier conference when reading the title of this book. The first major international conference on interstellar communication, jointly sponsored by the U.S. National Academy of Science and the USSR Academy of Sciences, was held in Soviet Armenia in 1971. The proceedings of that joint meeting were published as *Communication with Extraterrestrial Intelligence*

(CETI)—the preferred name for interstellar communication in the early 1970s (Sagan 1973). Both of these conferences recognized the importance of including perspectives from multiple fields in the physical sciences, social sciences, and humanities. Participants of the 1971 meeting covered a range of disciplines beyond the usual gathering of astronomers and engineers, and included representatives of fields such as anthropology, history, and linguistics. Similarly, the three "Search for Intelligent Life" sessions at AbSciCon 2010 encouraged cross-disciplinary dialogue by including scholars from all of these disciplines and more, as indicated in the following chapters.

The use of "communication" in the title of this book captures the potentially bidirectional nature of an exchange between civilizations, albeit over intergenerational timescales, as well as the interpretive challenges of designing and decoding the messages that are exchanged. The present volume covers a range of topics too often ignored by recent treatments of SETI, which have emphasized only the science and technology of listening for signals from other civilizations. This book complements the traditional focus on receiving signals by also analyzing how and why humankind might begin transmitting in earnest, as well as grappling with the complexities of making ourselves understood.

As we reflect on a half-century of SETI research, we are reminded of the expansion of search programs made possible by technological and conceptual advances. Project Ozma searched in a single radio channel; thanks to advances in computing, current searches can scan millions of frequencies per second, and with much greater sensitivity. So too have we broadened our search strategies, complementing the long-standing emphasis on detecting radio signals with new searches for brief yet powerful laser pulses. In this spirit of ongoing exploration, the contributors to this book advocate a diverse range of approaches to make SETI increasingly more powerful and effective, as we embark on the next half-century of searching for intelligence beyond Earth.

Douglas A. Vakoch
Mountain View and San Francisco, California

Reference

Sagan, Carl, ed. 1973. *Communication with Extraterrestrial Intelligence (CETI)*. Cambridge: MIT Press.

Foreword

AbSciCon—the Astrobiology Science Conference (http://www.lpi.usra.edu/meetings/abscicon2010/)—is like Woodstock for astrobiologists. Every two years, the rock stars of this field gather together to report on progress in their studies of the origin and evolution of life on Earth, planetary habitability, and the search for evidence of extraterrestrial life.

More than seven hundred astrobiologists came to AbSciCon 2010, traveling from every continent—including Africa and Antarctica, where astrobiology field research is ongoing—for this science "jam" in League City, Texas.

Recent research findings relevant to astrobiology and on the AbSciCon agenda include evidence of past and perhaps present liquid water on Mars as well as an ice-covered liquid water ocean on Europa, the discovery of hundreds of extrasolar planets, observations of plumes of water-ice particles erupting from Saturn's moon Enceladus, the possibility of prebiotic chemistry on and liquid water beneath the surface of Titan, and identification of new forms of microbial life in an ever-widening range of extreme Earth environments.

Enduring and widespread interest in the origin and evolution of life and the possibility of extraterrestrial life is—to resort to cliché—both a blessing and a curse. This broad, deep, cross-cultural engagement with the subject adds strength to the scientific rationale for astrobiology research and provides a great opportunity to foster science education with people of all ages.

That's the blessing part. Here's the curse part: the scientific definition of astrobiology is not necessarily the same as the public conception of astrobiology.

One aspect of the study of the origin and evolution of life in the universe that scientists have not yet found a way to adequately explain to nonexpert audiences is the vast chasm that stretches between our understanding of the emergence of life and the emergence of intelligence in life. Nonexperts have far less trouble than scientists do in condensing and simplifying the immense "spaces" of time and complexity that lead to prebiotic chemistry and then to life, to molecules and then to cells, and to microbial life and then to intelligent life. And it's not because they're ignorant. It's because they're not scientists. They're not trained to think like scientists. And they don't need to be.

The terms *astrobiology* and *SETI*—the search for extraterrestrial intelligence—are widely recognized inside and outside the scientific community. What these terms mean to people outside the community is something that we inside the scientific community might do well to understand, and explain, better than we do.

In the 1980s, the U.S. National Aeronautics and Space Administration (NASA) began a SETI research program culminating in the initiation of a search project in 1992, listening for radio signals originating from extraterrestrial technology. In 1993, Congress terminated the program, and NASA bowed out of the SETI enterprise. In the United States, since then, SETI work is privately funded. NASA's astrobiology program now funds research on the evolution of complex life, but not the evolution of intelligence.

In popular culture, nonetheless, boundaries between the scientific search for evidence of extraterrestrial life and SETI—and, for many out there on the fringes of credibility, even "UFOlogy"—are blurred or even erased altogether. Continuing interest in the subjects contributes to demand for popular depictions of them that are highly speculative and, sometimes, even a bit scary.

Adding to the complexity of the educational challenge facing astrobiologists is that, for experts and nonexperts as well, a wide range of opinions exists on exactly what "life" is, and what "intelligence" is.

Many pop culture takes on extraterrestrial life—or, more colloquially, aliens—appear to rest on the assumptions that life is common in the universe, and like us, and that the evolution of intelligence is easy.

About a week before the commencement of AbSciCon 2010, on April 26, the mass media widely reported remarks by renowned physicist Stephen Hawking that intelligent extraterrestrial life could be dangerous to Earthlings. (Hawking offered this view on a Discovery Channel television program called *Stephen Hawking's Universe*.)

The consensus at AbSciCon on Hawking's speculation was that while he is entitled to speculations, astrobiologists don't necessarily envision contact with extraterrestrial life as a risk. In fact, most astrobiologists are working hard to find evidence of extraterrestrial life.

How can the scientific community improve public understanding of these subjects?

Extraterrestrial intelligent life, and extraterrestrial technology, may not be recognizable to us, no matter how intelligent we may be. Recognizable or not, these phenomena may only exist so far away from Earth that scientists may never be able to detect, let alone verify, any signs of them. If scientists do find a signal, they must verify that it is coming from an extraterrestrial technological source, and this procedure is not simple. In addition, there will always be some who will argue that verification of a signal does not equal proof that extraterrestrial intelligence exists.

Foreword

The greatest challenge to SETI is the size of the space to be searched. SETI technology—radio and optical telescopes, signal detection equipment and signal processing systems—has greatly advanced since SETI got its start. Nonetheless, current technology can only search a small fraction of our own Milky Way galaxy. The universe is populated with billions of galaxies (estimates range from 125 billion to 500 billion). While scientists have greatly expanded our knowledge and understanding of the universe over fifty years of study, the universe is largely unexplored. (These facts do not, of course, argue that we should not continue exploring.)

While it's true that astrobiology has not yet yielded any evidence, or otherwise validated any claims, of the existence of extraterrestrial life, the solar system is a far less formidable search space than the universe at large. Astrobiologists are working on ways to detect extraterrestrial life in our solar system, should it exist, and Mars is a favorite search target. NASA's Mars Science Laboratory, scheduled for launch in 2011, is the agency's first astrobiology mission to Mars since Viking in the 1970s.

Detecting evidence of extraterrestrial life is a daunting task. Instruments must be extremely sensitive, and spacecraft must be prepared carefully to avoid contamination of results. Though some astrobiologists press for more in situ analysis of planetary materials, most agree that in situ analysis is far more limited than terrestrial laboratory analysis, and as yet no samples of planetary materials have been collected and returned to Earth. Astrobiologists are identifying biomarkers that would be signals for the possible presence of extraterrestrial life, as we know and as we don't know life.

As the volume and complexity of knowledge about the origins, evolution, and distribution of life in the universe grow, and as questions about the future of life in the universe continue to proliferate, astrobiologists will do well to remain mindful of public interest in their work and keep tending to the task of explaining their science.

Linda Billings, PhD
Member, AbSciCon 2010 Science Organizing Committee

Pamela Conrad, PhD
Co-Chair, AbSciCon 2010 Science Organizing Committee

Janet Siefert, PhD
Co-Chair, AbScicon 2010 Science Organizing Committee

Acknowledgments

Without the support of the Science Organizing Committee of Astrobiology Science Conference (AbSciCon) 2010, this book could not have happened. The Science Organizing Committee generously provided three sessions devoted to the search for intelligent life, in spite of an overabundance of excellent paper proposals on the range of topics comprising astrobiology and a paucity of slots in the conference schedule. I especially thank Linda Billings, Pan Conrad, Janet Siefert, and Mary Voytek. For logistical support at the Lunar and Planetary Institute, which hosted AbSciCon 2010, I am especially grateful to Elizabeth Wagganer. For her support in holding a post-conference workshop on interstellar message construction at the University of Houston—Clear Lake, I thank Alexandra McDermott.

To the contributors of chapters that appear in this volume, I especially appreciate the depth and innovation of the research they share here. These authors deserve special thanks for producing polished final versions of their papers so soon after the conference. I am pleased that many longtime colleagues involved in the Search for Extraterrestrial Intelligence (SETI) worldwide were able to participate in this project, along with other researchers relatively new to the field.

I am grateful to my colleagues at the SETI Institute, who have created over the past quarter-century a unique multidisciplinary environment for astrobiological research. I especially thank John Billingham, Tucker Bradford, Edna DeVore, Frank Drake, Andrew Fraknoi, John Gertz, Ly Ly, Chris Neller, Tom Pierson, Karen Randall, Pierre Schwob, Seth Shostak, and Jill Tarter. More recently, I warmly acknowledge the administration, faculty, staff, and students of the California Institute of Integral Studies (CIIS), especially for support from Katie McGovern, Joseph Subbiondo, and Judie Wexler. In addition, I thank Harry and Joyce Letaw, as well as Jamie Baswell, for their intellectual and financial contributions to promoting SETI.

For encouraging me to publish this volume through the State University of New York Press, I thank Gary Dunham. For overseeing this project from receipt of the manuscript to its printing and beyond—as well as for the prospect of giving a talk at his daughter's observatory—I thank James Peltz. For coordinating all aspects of the book's production with efficiency, accuracy, and good humor—all while on a tight timetable—I am indebted to Ryan Morris. For innovatively and capably marketing this work—through media traditional as well as new—I am grateful to Michael Campochiaro.

Finally and most importantly, to my wife Julie Bayless I am thankful in more ways than I can begin to express.

Part I

Latest Advances in the Search
for Extraterrestrial Intelligence (SETI)

Exoplanets, Extremophiles, and the Search for Extraterrestrial Intelligence

Jill C. Tarter

1.0. Introduction

We discovered the very first planetary worlds in orbit around a body other than the Sun in 1991 (Wolzczan and Frail 1992). They were small bodies (0.02, 4.3, and 3.9 times as massive as the Earth) and presented a puzzle because they orbit a neutron star (the remnant core of a more massive star that had previously exploded as a supernova) and it was not clear whether these bodies survived the explosion or reformed from the stellar debris. They still present a puzzle, but today we know of over five hundred other planetary bodies in orbit around hundreds of garden variety stars in the prime of their life cycle. Many of these planets are more massive than Jupiter, and some orbit closer to their stars than Mercury around the Sun. To date we have not found another planetary system that is an exact analog of the Earth (and the other planets of our solar system) orbiting a solar-type star, but we think that is because we have not yet had the right observing instruments. Those are on the way. In the next few years, we should know whether other Earth mass planets are plentiful or scarce.

At the same time that we have been developing the capabilities to detect distant Earths, we have also been finding that life on Earth occurs in places that earlier scientists would have considered too hostile to support life. Scientists were wrong, or at least didn't give microbes the respect they deserve. We now know that extremophiles can exist (and sometimes thrive) in the most astounding places: at the bottom of the ocean around hydrothermal vents, in ice, in pure salt, in boiling acid, and irradiated by massive doses of UV and X-rays. There do appear to be places on Earth that are too dry for even these (mostly microbial) extremophiles, or perhaps our sensors aren't yet sensitive enough to find them.

Since life as we know it is so extraordinarily hardy, might it exist today (or in the past) on any of the exoplanets that are being found? A group of scientists known as astrobiologists is trying to answer that question. This

3

chapter will discuss what appears to be possible in the near future, as well as the questions that will likely remain unanswered until new technologies enable new explorations in the more distant future. It might even turn out that our first indication of another inhabited world will be the signals deliberately generated by its inhabitants, as sought by scientists involved in the search for extraterrestrial intelligence, or SETI.

2.0. Exoplanets

In his book *Plurality of Worlds*, Steven J. Dick (1984) has chronicled the millennia of discourse about other inhabited worlds, based upon deeply held religious or philosophical belief systems. The popularity of the idea of extraterrestrial life has waxed and waned and, at its nadir, put proponents at mortal risk. Scientists at the beginning of the twenty-first century have a marvelous opportunity to shed light on this old question of habitable worlds through observation, experimentation, and interpretation, without recourse to belief systems and without risking their lives. A good place to keep track of the latest planet discoveries is the interactive catalog of the Extrasolar Planet Encyclopedia Web site <http://exoplanet.eu/catalog.php>. At a glance you will be able to see how many new planets have been announced in the time between when I reviewed the copyediting of this chapter, and when you are reading it. As of November 27, 2010, there are 504 planets orbiting 422 bodies (including those puzzling pulsar planets).

Exoplanets are primarily detected by indirect techniques: astrometry, radial velocity studies, transits, and gravitational micro-lensing. The first two of these detection methods measure the reflex motion of the star due to the mutual gravitational attraction between planet and star. The transit method measures the minute diminution of brightness that occurs periodically when a favorably aligned planet passes between its star and our telescope, blocking some of the star's light. The final method (gravitational micro-lensing) measures the brightening of a distant star when another (unseen) star and its orbiting planet align perfectly and the gravitational masses of the star and planet bend the light from the distant star causing it to appear brighter. Only very recently have we actually seen images of objects we believe to be planets in orbits around stellar hosts. The Hubble Space Telescope (HST) was used with a chronographic mask to block out the light from the star Fomalhaut. (As an analogy, think of holding up your hand to block out the light from a distant street light while you look for something faint in the area surrounding your hand.) In 2004 observations showed a small (single-pixel) bright object located in a large disk of dust, far from the star (115 astronomical units from the star, or 115 times as far from the star as the Earth is from the Sun). This object was confirmed as a planet Fomalhaut b when an HST 2006 observation showed that it had moved slightly along a believable orbital track. A series of rapid images of the star HR 8799 using ground-based telescopes allowed observers to remove the effects of atmospheric distortion and to image three giant-planet point-sources orbiting at distances of 24, 38, and 68 AU, far from the star.

The reflex motions of a star that are induced by the planetary orbit are greater and easier to observe if the star mass is small. What we haven't found thus far is a planet like the Earth orbiting its star at just the right distance so that its surface temperature might be conducive to permitting liquid water. This just-so region around any star is called the habitable zone. For the Sun, it stretches from just outside Venus's orbit to the orbit of Mars. For more massive stars the habitable zone is farther from the star, and it is closer in for lower mass stars. We think that such terrestrial analogs exist, but we have not had the capability to detect them until now, with the launch of spacecraft capable of making precise measurements of stellar brightness to search for transits. The Kepler spacecraft is expected to detect a handful of Earth-sized planets within the next few years. If it doesn't, we will have to revise our thinking about the way stars and protoplanetary disks of gas and dust actually form planets.

Most of the planets discovered are giants, and many have been surprising, and many are in orbits that are highly inclined to the equatorial plane of the star, unlike those of our own solar. "Hot Jupiters" in short period orbits very close to their host stars, and an abundance of high eccentricity (noncircular) orbits were surprising discoveries, though we are beginning to have reasonable explanations based on interactions with viscous protoplanetary disks and a version of cosmic billiards. Initially it was assumed that the presence of hot Jupiters would doom any terrestrial planets within the habitable zone, but recent studies by Raymond et al. (2006) argue that it is possible to have wet, terrestrial planets, even though a Jupiter-mass planet has migrated through their orbital radii on its way toward the star. The near-term future, with a suite of new instruments, holds promise for the detection of other Earths, large moons of gas-giant planets, and other potentially habitable cosmic real estate. The next obvious question is "Will they be inhabited?"

3.0. Extremophiles and Weird Life

Astrobiology is the science that deals with the origin, evolution, distribution, and future of life in the universe. It has been successful in bringing together scientific specialists from many different disciplines to tackle these big-picture questions. Many of my colleagues at the SETI Institute are astrobiologists studying organisms living in extreme conditions (by human standards) in an attempt to better understand the origin of life on Earth and the potential habitable real estate for life beyond Earth.

In the past few decades we have expanded the range of conditions recognized as suitable for life. Life is no longer confined between the boiling and freezing points of water. Hyperthermophiles live at high temperatures (and sometimes also high pressures), the current record holder being archean microbial *Strain 121* that thrives at 121 °C metabolizing iron, but it can survive up to 130 °C (Kashefi and Lovley 2003). At the other extreme, the psychrophilic bacterium *Psychromonas ingrahamii* survives and reproduces (very slowly) at −12 °C in the ice off Point Barrow, Alaska (Breezee et al. 2004). Macroscopic ice worms occupy and move through the Alaskan glaciers as well as the methane ice seeps on the floor of the Gulf of Mexico using natural antifreeze to protect their cellular structures (Fisher et al. 2000). Sunlight, once argued to be the source of energy for all life, is completely absent miles beneath the surface of the ocean, around the deep hydrothermal vents, where a rich and diverse community of organisms thrives in the dark, at enormous pressures. Small blind shrimp there have developed IR-sensing eye spots to navigate the vent environs or to travel from one vent to another using their thermal signatures (Pelli and Chamerlain 1989). Some chemical process (or processes) within the vents also produces minute quantities of visible light that are harvested by green sulfur bacteria for photosynthesis even though many hours go by between photons (Beatty et al. 2005).

Humans increasingly protect themselves from exposure to UV radiation as the protective layer of atmospheric ozone thins, since our DNA lacks sufficient repair mechanisms to survive intense radiation environments. Yet organisms inhabiting the highest freshwater lakes on Earth, in the caldera of the Lincancabur volcano overlooking the Atacama desert, have adapted to the huge UV load their altitude and evaporating environment present (Cabrol et al. 2005). Colleagues from the SETI Institute free-dive in these lakes each austral spring to catalog these organisms and study the DNA repair mechanisms they have elaborated. If life once occupied Mars, similar mechanisms might have been employed by organisms seeking to survive the dual stresses of increased radiation and desiccation due to loss of planetary atmosphere. Even more spectacular, *Deinococcus radiodurans* can withstand millions of rads of hard radiation because its DNA repair mechanisms are so effective. This skill probably evolved, not because it encountered naturally high radiation environments, but because desiccation causes the same sorts of breaks in DNA linkage; a sufficiently robust repair mechanism can endow survival independent of the damage source (Mattimore and Battista 1996). As a result, this microbe is the focus of many bioremediation programs to deal with radioactive materials, and is being sought in the Atacama to demarcate areas just too dry for life.

Neutral pH was once thought essential for life, but we now know of many organisms capable of living in either extremely acidic or highly alkaline environments. Cyanobacteria and fish can survive at pH ~4, but the acidophile red alga *Cyanidium caldarium* and the green alga *Dunaliella acidophila* can live at pH below 1 (Rothschild and Mancinelli 2001). In the ground waters of industrial slag heaps, extremely alkaline-tolerant microbes have been found thriving at a pH of 12.8 (Roadcap et al. 2006). While salt has been used historically to preserve food from decay due to bacterial action, halophilic archean microbes have been found living within pure NaCl crystals (Rothschild et al. 1994).

Astronomers may find it extremely unpleasant to live beyond the "just right" bounds of our current terrestrial environment, but clearly life has a greater tolerance and no lack of innovative ways of making a living. Within the past few years, we have begun to accept the concept of the "deep hot biosphere," and to acknowledge that perhaps ten times as much biomass is resident in the crust beneath our feet, as compared to the surface biomass with which we have long been familiar (Gold 1998). All these biological adaptations must inform our searches for life elsewhere in the universe.

Mindful of the adaptability of life to extreme environments, we should reconsider our own solar system (where we may have some hope of systematic in situ sampling), and we should expand our inspection to any bodies capable of providing raw materials and energy sources that might be exploited by biology of any sort. Thus, in addition to the terrestrial planets Venus, Mars, and Earth (which were all biologically connected during an earlier epoch of planet building and bombardment), we should consider the large icy satellites of Jupiter and Saturn, namely Europa, Ganymede, Callisto, and Enceladus, where liquid, briny, water oceans are thought to exist beneath their icy outer crusts. Titan might be a world hosting biology on its surface without liquid water as a solvent, although the presence of a subsurface water ice slush is now fairly certain. Perhaps results from the Venus Express mission of the European Space Agency (ESA) will shed light on whether the chemical disequilibria previously noted in the Venusean atmosphere require explanations involving Grinspoon's revival of the Sagan and Salpeter "sinkers, floaters, hunters and scavengers" (Schulze-Makuch et al. 2004). For the foreseeable future NASA will continue a "follow the water" strategy, returning to Mars with robots and humans to look for signs of extinct life from a wetter, warmer epoch or even for subsurface extant life. The seasonal and inhomogeneous appearance of trace amounts of methane in the tenuous Martian atmosphere (Mumma et al. 2005) might be explained by the presence of methanogens in subsurface liquid aquifers, or by the more prosaic, geological transformation of olivine rock, but that too requires flowing water. Subsequent missions may venture to Europa to verify the existence of a massive water ocean and to examine the tantalizing discolorations near the surface cracks that could be the end products of organic molecule irradiation (Dalton et al., 2003), and then, later still, return to make a sterile penetration of the ice and search for life in the water below.

The resilience and diversity of extremophiles have now also focused attention on life as we don't yet know it. Life on Earth (even extremophiles) seems completely connected; life as we know it appears to have had a single common ancestor. But what about life as we don't yet know it? Might it exist on Earth today in extreme environments, and remain undetected because of our instrumental biases toward carbon-based organisms? Might it exist on other bodies in our solar system, and in the planetary systems of other stars, perfectly suited to those local environmental conditions? What are the limits of organic life in planetary systems? It is a heady question that, if answered, may reveal just how crowded our Earth and the cosmos could be with alien biology. In 2007, the National Academy of Sciences released a report titled *The Limits of Organic Life in Planetary Systems* (Space Studies Board 2007), that is, life with an alternate biochemistry—what they called "weird life." How would we recognize life based on different biosolvents, different nucleotides, different metabolic pathways? What instruments should we develop to aid human and robotic explorers undertaking a search for other forms of life?

It seems impossible to avoid one particular trap of being twenty-first-century humans. In seeking life, or its technological by-products, we cannot search for what we cannot conceive, and it is also impossible to guarantee that we will correctly interpret what we find. This conundrum has been shared by all past explorers. Those who were successful pushed ahead, with the tools at their disposal, or tools they invented. As my colleague Seth Shostak is fond of saying "Columbus didn't wait for a 747 to cross the Atlantic"; neither should astrobiologists.

4.0. Biosignatures

Moving out beyond the solar system, the focus will be on exoplanets, but here it will not be possible to consider in situ sample collections, at least not for a long time; remote observations will have to suffice. The first task will be the detection of, and subsequent imaging of, a terrestrial-mass planet within the stellar habitable zone. Just how exactly like the Earth does another environment have to be in order to host life? Ward and Brownlee (2000) have argued that an exact duplicate of the terrestrial environment—its history, large moon, and giant-planet shields—are required for anything bigger than microbial life. However, Darling (2001) reviews the arguments and concludes that other astronomers might exist on many worlds. In fact, the evidence is consistent with life, including intelligent life, existing on many worlds or exclusively only on Earth; there is as yet *no evidence*. We have an example of "one." We cannot know from tracing the detailed history of that single example what the branching ratio might be for the experiment of life; how many other ways might things have gone but didn't, at what rate, with what end result? The number "two" will be all-important in answering this question—as in the second example of an independent origin of life.

The next step will be to attempt to conduct a chemical assay of the atmosphere of any terrestrial planet imaged in orbit around nearby stars. Transiting hot Jupiters have already permitted the first analysis of chemical constituents of exoplanet atmospheres. Observations of HD 209458b reveal sodium, hydrogen, oxygen, and carbon in an extended atmosphere and/or escaping from the planet (Vidal-Madjar et al. 2004). The Terrestrial Planet Finder (TPF) (previously studied by NASA and now on indefinite hold) and the Darwin constellation under development by ESA may eventually launch during the first half of the twenty-first century, perhaps combined as an international mission. Telescopes on these spacecraft are intended to suppress the light from a central star, using either an occulting coronagraph at visible wavelengths or interferometric nulling in the IR, thereby spatially resolving and directly detecting reflected starlight from any terrestrial exoplanets orbiting within the stellar habitable zone; enormous precision is required to separate the very faint reflected light from the planet from the very much brighter star close by. Once an image has been formed, very long observations will attempt to collect sufficient light to reveal absorption lines in the exoplanet spectrum due to trace atmospheric constituents that might be clues to the presence of biology on the planetary surface.

As difficult a technical challenge as implementing these spacecraft will be, perhaps an even greater challenge will be deciding what spectral signature does or does not constitute a reliable biomarker. Using the present Earth as an example, chemical disequilibrium is one very promising sign. The coexistence of the very reactive molecular oxygen and methane gases in our own modern atmosphere is the direct result of photosynthetic cyanobacteria and plants as well as the fermenting bacteria within termites, ruminants, and rice paddies (Lovelock 1965; Lovelock and Margulis 1974). But the paleoearth would have presented a very different picture during the billions of years when life was present, but had not yet participated in the elevation of atmospheric O_2 levels; for that world, we must ask about the undeniable biosignatures of methanogens. Additionally, one must ask if there are any abiotic processes that can yield the same result. For exoplanets, the harsh realities of the remote observational circumstances are further challenges; it does not now seem feasible to observe simultaneously the visible bands of oxygen and the thermal IR signature of methane in a single instrument. The broad absorption feature of ozone (at λ 9.3 μm) in the atmosphere of an exoplanet is currently the favored biosignature for complex life as we know it, if the host star is sun-like (Léger 1999; Kasting and Catling 2003).

When the primary is an M dwarf, other molecules such as N_2O, and CH_3Cl may be detectable along with ozone (Segura et al. 2005). While there are abiotic sources of oxygen, and therefore ozone, an ongoing biological source appears to be required for substantial atmospheric ozone concentrations on a geologically active planet. Any future announcements of atmospheric ozone detection from an exoplanet and a potential linkage to biota are likely to be accompanied by a long list of caveats, which the media will ignore. However, spectral absorption features alone are unlikely to distinguish between a future detection of alien microbes or mathematicians. To find the latter we need to search for technosignatures; we need to do SETI.

5.0. Technosignatures: The Search for Extraterrestrial Intelligence

"Are we alone?" is really a loaded question. The detection of life on another world (extant or extinct) is a real possibility within the lifetimes of the people living today. That is a thrilling possibility. Detection of irrefutable biosignatures will provide the pivotal "number two" in the record of life in the universe, but at a very deep level, humans want to know whether other intelligent creatures also view the cosmos and wonder how they came to be. Like the term *life*, there is really no acceptable definition of "intelligence." Nevertheless, we may be able to remotely deduce its existence over interstellar distances. If we can find technosignatures—evidence of some technology that modifies its environment in ways that are detectable—then we will be permitted to infer the existence, at least at some time, of intelligent technologists. As with biosignatures, it is not possible to enumerate all the potential technosignatures of technology as we don't yet know it, but we can define systematic search strategies for equivalents of some twenty-first-century terrestrial technologies. The science fiction author Arthur Clarke has suggested that "any sufficiently advanced technology will be indistinguishable from magic"; if they occur at all, detections of advanced technologies will probably occur as the accidental result of our detailed studies of the natural universe.

SETI is that subspecialty of astrobiology that currently conducts systematic explorations for other technology as we know it; primarily it searches for electromagnetic radiation—radio or optical signals. SETI pre-dates the current field of astrobiology, beginning life as a valid field of exploratory science in 1959 with the publication of the first paper on this subject in a refereed journal (Cocconi and Morrison 1959). This first paper advocated a search for radio signals, but the suggestion of searching for pulsed optical laser signals followed shortly thereafter in 1961 (Schwartz and Townes 1961). SETI provides another plausible avenue for discovering habitable worlds by attempting to detect the actions of technological inhabitants. More than one hundred SETI searches can be found in the literature (Tarter 2001). (To see the search strategies that have been used, look at <https://observations.seti.org>.) The list begins with Project Ozma in 1960 the first search for radio signals from nearby sun-like stars (Drake 1960). While these searches may seem like a large effort, the sum total of all these investigations has covered only a minute fraction of the search parameter space. Imagine looking at a single glass of water scooped from the ocean, and examining it to see whether there are any fish in the ocean. That's a pretty good analogy to how much of the cosmos we've been able to search so far—one glass from an entire ocean.

Signals might be generated for the benefit of the transmitting civilization or to deliberately attract the attention of another civilization. While it may be possible for us to detect unintentional leakage radiation from another technology, deliberate signals, transmitted to be detectable, are the most likely to be found. Furthermore, any detectable signals will have originated from a technology far older than our own. If technology tends to be a long-lived phenomenon among galactic civilizations, then statistics favor the detection of signals from a technology during its old age. If technology, in general, is a short-lived phenomenon, then it will be undetectable because the chance that two short-lived technological civilizations would not only be close to one another, but also overlap in time during the thirteen billion year history of the Milky Way is vanishingly small. For this reason, Philip Morrison has called SETI the archeology of the future. The finite speed of light guarantees that any detected signal will tell us about the transmitter's past, but the detection of any signal tells us that it is possible for us to have a long technological future (Mallove 1990). This is one of the things that makes SETI important to me, and why I feel that I have the best job in the world.

Although we've been doing SETI for more than fifty years now, most of the time, most SETI searches are off the air. Historically, the searches have been conducted on telescopes constructed for other scientific observing programs, so there has been little telescope time available for SETI (the piggyback SETI@home project is a notable counterexample to this). Today the situation is changing as new instruments intended for dedicated SETI use are commissioned and beginning to look at the sky. This is an exciting time because our tools may finally be getting to be commensurate with the magnitude of our task.

Deliberate signals, intended to be detected, might be engineered in one of two ways: they could appear to be "almost astrophysical," or they could appear to be "obviously technological," The distinct benefit of the former scheme is that such signals are very likely to be captured as a young technology (such as ours) begins to deploy multiple sensors to study the universe around it. Eventually, some graduate student searching through observational databases might discover something peculiar about one of the entries, and thus discover that the signal is actually engineered. Our rapidly improving astronomical observing capability and our curiosity about the cosmos should ensure that we eventually discover any such "almost astrophysical" signals. On the other hand, the detection of "obviously technological" signals will require construction of specific instrumentation not available from astronomical observing programs, because the characteristics of the extraterrestrial signals are precisely those that we expect nature to be unable to produce. Today we search for narrowband radio signals—a single channel on the radio dial. Natural radio emission occurs at many frequencies, but technology can compress a lot of power into a single frequency. The computerized signal detection algorithms in use can detect narrowband signals that are continuously on, or that pulse on and off, and they can be constant in frequency or change frequency during the observation due to accelerations between transmitter and receiver. Natural, background radio noise from galactic synchrotron emission rises rapidly at frequencies below 1 GHz (1 billion Hz), while the noise from atmospheric water vapor and oxygen contributes above 10 GHz; radio SETI searches have a goal of systematically exploring the naturally quiet Terrestrial Microwave Window from 1–10 GHz. Since the signals may be as narrow as 1 Hz, it means we need to search through nine billion channels on the radio dial. While nature is quiet at these frequencies, our cell phones, satellites, garage door openers, and microwave ovens generate lots of signals that cause interference. Radio telescopes are located far from population centers, and astronomers must use a great deal of computational effort to work around this interference.

At optical frequencies, signals exhibiting extreme time compression (short, broadband laser pulses) are searched for with photon counters having nanosecond rise times, a regime with no known sources of astrophysical background (Howard and Horowitz 2001). Because interstellar dust begins to absorb optical pulses over distances beyond ~1000 light years, it is desirable to extend the optical SETI search into the infrared so that more of the galaxy becomes accessible. This will happen when, and if, the requisite fast photon counters become available and affordable in the IR. Other modulation schemes employed by current terrestrial communications technologies can produce signals whose statistical properties differ from the Gaussian noise of astrophysical emitters, but they are harder to recognize than the simple artifacts now being sought. As Moore's Law delivers more affordable computing, SETI programs are beginning to search for more complex signals (Harp et al. 2011).

At any frequency, there are two basic search strategies that can be implemented: move quickly across the sky (or a portion thereof) to cover as much of the spatial dimension of the cosmos as possible, or select individual directions deemed to have a higher a priori probability of harboring a technological civilization and make targeted observations in those directions for longer periods of time. The former strategy minimizes the assumptions about the source of the signal, but in general, the surveys will achieve poorer sensitivity as the result the smaller telescopes and shorter dwell times typical of this strategy. By developing a list of plausible targets, it is possible to achieve significantly better sensitivity through integration and signal processing gain for a wider range of signal types. However, if the target list is constructed under the wrong assumptions, detection probabilities are lowered rather than improved. In both cases it is desirable, but seldom affordable, to accomplish the task of signal detection and recognition in real-time, or near-real-time, so that immediate follow-up of candidate signals is enabled before they can vary with time, and opportunities for distinguishing between terrestrial and extraterrestrial technologies can be exploited.

Dan Werthimer and a group at UC Berkeley's Space Science Lab has operated a series of increasingly capable SERENDIP detectors operating at radio observatories in a commensal (or piggyback) mode to achieve maximum access to the sky for conducting random, SETI sky surveys. The SETI observers don't control the telescope pointing, but they eventually end up surveying most of the available sky from any site. Recently, SERENDIP V began taking data at Arecibo, the world's largest radio telescope, working with the multibeam ALFA receiver to search seven directions on the sky simultaneously and analyze 300 MHz of the spectrum in the vicinity of the 21 cm Hydrogen line. A few percent of the data collected by ALFA and SERENDIP V, at the precisely 1420 MHz (21 cm), is analyzed on the pioneering SETI@home global distributed computing platform (http://setiathome.ssl.berkeley.edu/). If you are interested, you can download a screen saver that works as a background program on your own computer to search for SETI signals in a small piece of recorded data automatically shipped to you from UC Berkeley, and you can join more than three million people who are helping the search. A commensal SETI search at Parkes Observatory in NSW, based on earlier SERENDIP technologies, is now being renovated, and may provide another opportunity for participation in SETI.

In Northern California, the SETI Institute and the University of California Berkeley Radio Astronomy Lab have partnered to build the Allen Telescope Array (ATA) at the Hat Creek Radio Observatory for the purpose of simultaneously surveying the radio sky for signals of astrophysical and technological origin (DeBoer et al. 2004). Ultimately, the array will consist of 350 antennas, each 6.1m in diameter, extending over a maximum baseline of 900 m. Currently it is operating as the ATA-42, with the first forty-two dishes spread over 300 m. The ATA provides simultaneous access to any frequency between 500 MHz and 11.2 GHz, a system temperature ~50° K, and four separately tunable intermediate frequency channels feeding a suite of signal processing backends. The backend instrumentation can produce wide-angle radio images of the sky with ~20000 resolution pixels and 1024 spectral channels per pixel, and at the same time, study three point sources of interest within its large field of view using phased up beams at two different frequencies. This new approach to commensally sharing the sky allows SETI and traditional radio astronomical science to both utilize the telescope nearly full time. Unlike, the SERENDIP systems at Arecibo, the SETI signal detection takes place in near-real-time. The ~250,000 stars in our current catalog of "habstars" provides a few stellar targets in every array field of view at lower frequencies and enables efficient commensal observing. Another catalog of about 3500 target directions is enabling a survey of twenty square degrees surrounding the galactic center, within which are located some ten billion distant stars. Whereas the targeted searches of nearby "habcat" stars would detect transmitters as strong as the current radar signals we generate to study our ionosphere, at the distance of the galactic center this survey would detect a transmitter with a power equivalent to twenty thousand Arecibo planetary radars.

The ATA is the first attempt to manufacture a radio telescope by taking advantage of cost discounts from economies of scale, consumer off-the-shelf components (primarily from the telecommunications industry), and inexpensive commercial manufacturing technologies. Like ASKAP, now being built in Western Australia, and MeerKAT under construction in South Africa, the ATA is one of the pathfinders for the Square Kilometre Array, a project to build an international observatory with one hundred times the collecting area of the full 350-dish ATA. The ATA is very much a Moore's Law telescope. Whereas all the data from 500 MHz to 11.2 GHz is brought to the central processing center as analog signals, only a small portion of that data is currently digitized and sent to the astronomical and SETI processors. In the future, as digitizers and processors get cheaper, the array will improve its capability by investigating more and more of the spectrum simultaneously. The ATA will continue to allow us to increase the speed with which we explore the cosmos.

Moore's Law improvements have helped optical SETI (OSETI) programs every bit as much as they have enabled the ATA. At Harvard, Paul Horowitz and his students are using a new, dedicated, OSETI telescope for a survey of the 60 percent of the northern sky visible from the Oak Ridge Observatory in Massachusetts (Horowitz et al. 2006). The telescope is housed in a building enthusiastically constructed with student labor. The 72" primary and 36" secondary mirrors have been manufactured inexpensively by fusing glass over a spherical form and then polishing, because the system does not require image quality optics. The detection system is based on eight pairs of 64-pixel Hamamatsu fast photodiodes, and custom electronics for real-time detection. This new telescope searches for powerful transmitters from a large collection of stars by conducting meridian transit scans of the sky in 1.6° × 0.2° strips (with a dwell time, due to the Earth's rotation, of about one minute). The sky visible from that site can be scanned in approximately 150 clear nights. The survey sensitivity should be adequate to detect laser pulses from the analog of a current Helios-class laser being transmitted through a 10 m telescope up to a distance of 1000 light years. In Australia, at the Campbelltown Rotary Observatory at the University of Western Sydney, an OSETI program has been in operation for several years conducting a targeted search of nearby stars (Bhathal 2011).

Looking toward the future, when the ATA is fully built out, ASKAP and MeerKAT become operational, and then eventually the Square Kilometre Array comes on, we are hoping that some of the students attending the Astrobiology Science Conference will be inspired to take over from us. At the SETI Institute, we plan for success and actively try to educate the next generation of scientists. For example, we have developed a yearlong integrated science curriculum for ninth graders, called *Voyages Through Time*.

In addition to Phillip Morrison's hopeful characterization of SETI as the archeology of the future, I think that SETI is extraordinarily important because it provides an opportunity to change the perspective of every person on this planet. The successful detection of a signal, or even the serious discussion of that possibility, would have the effect of holding up a mirror to the Earth. In this mirror we, all of us, would be forced to see ourselves as Earthlings, all the same when compared to the detected extraterrestrials. SETI can help to trivialize the differences among humans that we find so divisive today. This is why, when I got to make a wish to change the world as part of my TED prize (http://www.tedprize.org/jill-tarter/) (Tarter et al. 2011), I said, "I wish that you would empower Earthlings everywhere to become active participants in the ultimate search for cosmic company."

SETI might succeed in my lifetime, in your lifetimes, or never. There is no satisfactory way to make an estimate. The wisest summary still remains the last sentence in the original 1959 *Nature* journal article: "The probability of success is difficult to estimate, but if we never search the chance of success is zero" (Cocconi and Morrison 1959, 846). So I invite you to stay tuned because some of us are determined to keep searching, and we could use your help.

Note

This chapter is an adaptation of J. C. Tarter, "Extremophiles and Exoplanets: Expanding the Potentially Habitable Real Estate in the Galaxy," in *ISS2009: Genes to Galaxies* (Sydney: Science Foundation for Physics, 2009).

Works Cited

Beatty, J. T., J. Overmann, M. T. Lince, A. K. Manske, A. S. Lang, R. E. Blankenship, C. L. Van Dover, T. A. Martinson, F. G. Plumley. 2005. An obligately photosynthetic bacterial anaerobe from a deep-sea hydrothermal vent. *Proceedings of the National Academy of Sciences* 102 (26): 9306–10.

Bhathal, R. 2011. The OZ OSETI project. In *Communication with extraterrestrial intelligence (CETI)*, ed. D. A. Vakoch. Albany, NY: State University of New York Press.

Breezee, J., N. Cady, and J. T. Staley. 2004. Subfreezing growth of the sea ice bacterium "Psychromonas ingrahamii." *Microb. Ecol.* 300:4.

Cabrol, N. A., E. A. Grin, L. Prufert-Bebout, L. Rothschild, and A. N. Hock. 2005. Field and diving exploration of the highest lakes on Earth: Analogy of environment and habitats with early Mars and life adaptation strategies to UV [Abstract]. 2005 Biennial Meeting of the NASA Astrobiology Institute, Boulder, CO. *Astrobiology* 5: 305–306.

Cocconi, G., and P. Morrison. 1959. Searching for interstellar communication. *Nature* 184: 844–46.

Dalton, J. B., R. Mogul, H. K. Kagawa, S. L. Chan, and C. S. Jamieson. 2003. Near-infrared detection of potential evidence for microscopic organisms on Europa. *Astrobiology* 3: 505–529.

Darling, D. 2001. *Life everywhere: The maverick science of astrobiology.* New York: Basic Books.

DeBoer, D., W. J. Welch, J. D. Dreher, J. C. Tarter, L. Blitz, M. D. Davis, M. Fleming, D. Bock, G. Bower, J. Lugten, G. Girmay-Keleta, L. D'Addario, G. Harp, R. Ackermann, S. Weinreb, G. Engargiola, D. Thornton, and N. Wadefalk. 2004. The Allen Telescope Array. *Ground-based Telescopes*, ed. J. M. Oschmann, Proc. SPIE 5489: 1021–28.

Dick, S. J. 1984. *Plurality of worlds: The extraterrestrial life debate from Democritus to Kant.* Cambridge: Cambridge University Press.

Drake, Frank. 1960. How can we detect radio transmissions from distant planetary systems. *Sky and Telescope* 19: 140–43.

Fisher, C. R., I. R. MacDonald, R. Sassen, C. M. Young, S. Macko, lS. Hourdez, R. Carney, S. Joy, and E. McMullin. 2000 Methane ice worms: *Hesiocaeca methanicola* colonizing fossil fuel reserves. *Naturwissenschaften* 87: 184–87.

Gold, T. 1998. *The deep hot biosphere.* New York: Springer.

Harp, G. R., R. F. Ackermann, S. K. Blair, J. Arbunich, P. R. Backus, J. C. Tarter, and the ATA Team. 2011. A new class of SETI beacons that contain information. In *Communication with extraterrestrial intelligence*

(CETI), ed. D. A. Vakoch. Albany, NY: State University of New York Press.

Horowitz, P., C. Coldwell, A. Howard, D. Latham, B. Stefanik, J. Wolff, and J. Zajac. 2006. Targeted and all-sky search for nanosecond optical pulses at Harvard-Smithsonian, preprint, http://seti.harvard.edu/oseti/oseti.pdf.

Howard, Andrew W., and Paul Horowitz. 2001. Is there RFI in pulsed optical SETI? *The Search for Extraterrestrial Intelligence (SETI) in the Optical Spectrum III*, eds. Stuart A. Kingsley and Ragbir Bhathal, Proc. SPIE 4273: 153–60.

Kashefi, K., and D. R. Lovley. 2003. Extending the upper temperature limit for life. *Science* 301: 934.

Kasting, J. F., and Catling, D. 2003. Evolution of a habitable planet. *Annual Review of Astronomy and Astrophysics* 41: 429–63.

Léger, A., M. Ollivier, K. Altwegg, N. J. Woolf. 1999. Is the presence of H_2O and O_3 in an exoplanet a reliable signature of a biological activity? *Astronomy and Astrophysics* 341: 304–11.

Lovelock, J. 1965. A physical basis for life detection experiments. *Nature* 207: 568–70.

Lovelock, J. E., and L. Margulis. 1974. Atmospheric homeostasis by and for the biosphere: The Gaia Hypothesis. *Tellus* 26: 2–9.

Mallove, E. F. 1990. Extraterrestrial search marks 30th anniversary: An interview with Prof. Philip Morrison. *MIT Tech Talk*, November 7.

Mattimore, V., and J. R. Battista. 1996. Radioresistance of *Deinococcus radiodurans*: Functions necessary to survive ionizing radiation are also necessary to survive prolonged dessication. *Journal of Bacteriology* 178 (3): 633–37.

Mumma, M. J., R. E. Novak, T. Hewagama, G. L. Villanueva, B. P. Bonev, M. A. DiSanti, M. D. Smith, and N. Dello Russo. 2005. Absolute abundance of methane and water on Mars: Spatial maps. *BAAS* 37: 669.

Pelli, D. G., S. C. Chamerlain. 1989. The visibility of 350 °C black-body radiation by the shrimp *Rimicaris exoculata* and man. *Nature* 337: 460–61.

Raymond, S. N., A. M. Mandell, and S. Sigurdsson. 2006. Exotic Earths: Forming habitable worlds with giant planet migration. *Science* 313: 1413–16.

Roadcap, G. S., R. A. Sanford, J. Qusheng, J. R. Pardinas, and C. M. Bethke. 2006. Extremely alkaline (pH>12) ground water hosts diverse microbial community. *Ground Water* 44: 511.

Rothschild, L. J. and R. L. Mancinelli. 2001. Life in extreme environments. *Nature* 409: 1092–1101.

Rothschild, L. J., L. J. Giver, M. R. White, and R. L. Mancinelli. 1994. Metabolic activity of microorganisms in gypsum-halite crusts. *J. Phycol.* 30: 431–38.

Schulze-Makuch, D., D. H. Grinspoon, O. Abbas, L. N. Irwin, and M. A. Bullock. 2004. Sulfur-based survival strategy for putative phototrophic life in the Venusian atmosphere. *Astrobiology* 4: 11–18.

Schwartz, R., and C. Townes. 1961. Interstellar and interplanetary communication by optical masers. *Nature* 190 (4772): 205–208.

Segura, A., J. F. Kasting, V. Meadows, M. Cohen, J. Scalo, D. Crisp, R. A. H. Butler, and G. Tinetti. 2005. Biosignatures from Earth-like planets around M dwarfs. *Astrobiology* 5: 706–25.

Space Studies Board. 2007. *The limits of organic life in planetary systems.* Washington, DC: The National Academies Press.

Tarter, J. C. 2001. The search for extraterrestrial intelligence (SETI). *Annual Review of Astronomy and Astrophysics* 39: 511–48.

———, A. Agrawal, R. Ackermann, S. K. Blair, M. T. Bradford, D. M. Cooper, G. Harp, J. Jordan, T. Kilsdonk, K. E. Smolek, K. Randall, R. Reid, John Ross, G. S. Shostak, and D. A. Vakoch. 2011. Getting the world actively involved in SETI searches. In *Communication with extraterrestrial intelligence (CETI)*, ed. D. A. Vakoch. Albany, NY: State University of New York Press.

Vidal-Madjar, A., J.-M. Désert, A. Lecavelier des Etangs, G. Hébrard, G. E. Ballester, D. Ehrenreich, R. Ferlet, J. C. McConnell, M. Mayor, and C. D. Parkinson. 2004. Detection of oxygen and carbon in the hydrodynamically escaping atmosphere of the extrasolar planet HD209458b. *Astrophys. J.* 604: L69.

Ward, P., and D. Brownlee. 2000. *Rare Earth: Why complex life is uncommon in the universe.* New York: Springer.

Wolszczan, A., and D. A. Frail. 1992. A planetary system around the millisecond pulsar PSR 1257+12. *Nature* 355: 145.

Current and Nascent SETI Instruments in the Radio and Optical

*Andrew Siemion, Henry Chen, Jeff Cobb, Jim Cordes,
Terry Filiba, Adam Fries, Andrew Howard, Josh von Korff,
Eric Korpela, Matt Lebofsky, William Mallard, Peter McMahon,
Aaron Parsons, Laura Spitler, Mark Wagner, Dan Werthimer*

1.0. Background

By far the most common type of SETI experiments are searches for narrowband continuous-wave radio signals originating from astronomical sources. These searches are based on a number of fundamental principles, many first described by Frank Drake in 1960. Paramount among them is the fact that sufficiently narrow signals are easily distinguishable from astrophysical phenomena. The spectrally narrowest known astrophysical sources of electromagnetic emission are masers, with a minimum frequency spread of about one kHz. Additional support for the possible preference for narrowband interstellar transmissions by ETI include the immunity of narrowband signals to astrophysical dispersion and consideration of similarities to our own terrestrial radio communication systems. Further encouragement is provided by the existence of a computationally efficient matched filter for searching for narrowband signals, the Fast Fourier Transform. More recently, searches have begun targeting other signal types, such as broadband dispersed radio pulses (Siemion et al. 2009).

Optical SETI, the name usually ascribed collectively to SETI operating at optical wavelengths, was first proposed in 1961 by Schwartz and Townes shortly after the development of the laser. Searches have been conducted for both pulsed emission (e.g., Howard et al. 2004) and continuous narrowband lasers (e.g., Reines and Marcy 2002). Pulsed optical SETI rests on the observation that humanity could build a pulsed optical transmitter (using, for example, a U.S. National Ignition Facility-like laser and a Keck Telescope-like optical beam former) that would be detectable at interstellar distances. When detected, the nanosecond-long pulses would be a factor of ~1000 brighter than the host star of the transmitter during their brief flashes (Howard et al.,

2004). Such nanosecond-scale optical pulses are not known to occur naturally from any astronomical source (Howard and Horowitz 2001).

1.1. Extant SETI Searches

Our group is involved in a variety of ongoing searches for signatures of extraterrestrial intelligence, spanning the electromagnetic spectrum from radio to optical wavelengths. The most publicly well known of these is our distributed computing effort, SETI@home (Anderson et al. 2002). Launched in 1999, SETI@home has engaged more than five million people in 226 countries in a commensal sky survey for narrowband and pulsed radio signals near 1420 MHz using the Arecibo radio telescope. SETI@home is currently operating over a 2.5 MHz band on the seven beam Arecibo L-band Feed Array (ALFA). Participants in the project are generating the collective equivalent of 200 TeraFLOPs/sec and have performed more than 1.4×10^{22} FLOPs to date.

Another of our radio SETI projects, the Search for Extra-Terrestrial Radio Emissions from Nearby Developed Intelligent Populations (SERENDIP) (Werthimer et al. 1995), is now in its fifth generation and is currently being conducted in a collaboration between UC Berkeley and Cornell University (Table 2.1). In June 2009 we commissioned SERENDIP V.v, the newest iteration of the three-decade-old SERENDIP program.[1] This project utilizes a high performance field programmable gate array (FPGA)–based spectrometer attached to the Arecibo ALFA receiver to perform a high sensitivity sky survey for narrowband signals in a 300 MHz band surrounding 1420 MHz. The SERENDIP V.v spectrometer analyzes time-multiplexed signals from all seven dual-polarization ALFA beams, commensally with other telescope users,

Table 2.1.

Program	Bandwidth (MHz)	Resolution (Hz)	Number of Channels	Date	Location
SERENDIP I	0.1	1000	100	1979–1982	Hat Creek, Goldstone
SERENDIP II	0.065	1	64K	1986–1990	Green Bank, Arecibo
SERENDIP III	12	0.6	4M	1992–1996	Arecibo
SERENDIP IV	100	0.6	168M	1998–2006	Arecibo
SERENDIP V.v	300	1.5	2G	2009–Present	Arecibo

SERENDIP past and present. SERENDIP V.v is multiplexed (128M channels / 200 MHz bandwidth instantaneous).

effectively observing two billion channels across seven 3 arcminute pixels. A copy of this instrument is currently deployed by the Jet Propulsion Laboratory on a 34 m Deep Space Network (DSN) dish, DSS-13, in Barstow, California.

Our optical pulse search (Lampton, 2000) is based at UC Berkeley's thirty-inch automated telescope at Leuschner Observatory in Lafayette, California. The detector system consists of a custom-built photometer, employing three photomultiplier tubes (PMTs) fed by an optical beamsplitter to detect the concurrent (within ~1 ns) arrival of incoming photons across a wavelength range 300 nm $< \lambda <$ 650 nm. This "coincidence" detection technique improves detection sensitivity by reducing the false alarm rate from spurious and infrequent pulses observed in individual PMTs. PMT signals are fed to three high speed amplifiers, three fast discriminators, and a coincidence detector (See Figure 2.1, where detections are measured by a relatively slow [1 MHz] Industry Standard Architecture [ISA] counter card.) The photometer features a digitally adjustable threshold level to set the false alarm rate for a particular sky/star brightness.

During a typical observation, the telescope is centered on a star and detection thresholds are adjusted so that the false alarm rate is sufficiently low. Currently, we record three types of events: single events, when an individual PMT output is greater than the voltage threshold originally set; double events, when any two of the PMTs output exceeds the threshold in the same nanosecond-scale time period; and triple events, when all three PMTs concurrently exceed threshold. Voltage thresholds are set so that false triple events are very rare and false double events occur only a few times in an ~5 minute observation. A duplicate of this instrument is in place at Lick Observatory near San Jose, California (Stone et al. 2005).

1.2. New Instruments

Historically, the level of technology and engineering expertise required to implement a SETI instrument was quite high. As a result, SETI programs have been limited to just a handful of institutions. Our group is developing two new instruments possessing several advantages over previous generations of SETI instrumentation—an Optical SETI Fast Photometer for optical SETI and the Heterogeneous Radio SETI Spectrometer for observations in the radio. Both are constructed from widely available modular components with relatively simple interconnects. The designer logs into the instrument using Linux and programs in C, obviating the need for cumbersome interfaces (e.g., JTAG) and languages (e.g., VHDL/Verilog). Further, these instruments are scalable and easily upgradable by adding additional copies of commercially available parts (compared with the money and time-consuming upgrades of previous instruments that involved complete redesigns of PC boards and

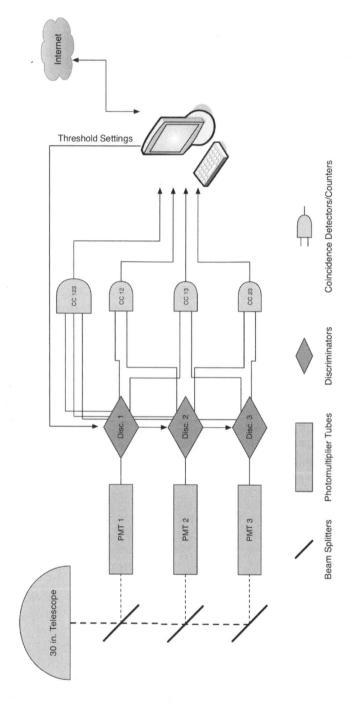

Figure 2.1. Schematic diagram for the existing analog electronics and ISA digital counter card used to read out the Lick and Leuschner SETI photometers. An optical telescope feeds three photomultiplier tubes using optical beamsplitters. Analog electronics threshold the signals from three PMTs and coincidence detectors trigger low-speed counters. While this system is effective, all detailed data about the event trigger (the digitized light profile) is lost.

ASICs). Collectively, these advances will enable much wider participation in SETI science.

Our next generation Optical SETI Fast Photometer (OSFP) is based on the same front-end optics and photodetectors as the original Berkeley OSETI instrument, but adds a flexible digital back-end based on the Center for Astronomy Signal Processing and Electronics Research (CASPER, see below) DSP instrument design system. The programmable FPGA-based digital back-end will allow us to improve sensitivity by implementing sophisticated real-time detection algorithms, capture large swaths of raw sampled voltages for diagnostics or centroiding and perform efficient rejection of interference based on pulse profiles.

Our newest radio SETI instrument, the Heterogeneous Radio SETI Spectrometer (HRSS), is also CASPER-based. HRSS will take advantage of the wide bandwidth capabilities of a high speed analog-to-digital converter (ADC) paired with an FPGA to digitize, packetize, and transmit coarse channelized spectral regions to flexible, off-the-shelf CPUs and graphics processing units (GPUs) for fine spectroscopy and RFI rejection. This architecture will not only provide for economical entry into cutting edge SETI research (see Table 2.2), its use of standard C programming on CPUs and GPUs will enable the DSP instrument internals to be accessible for students with only modest instrumentation experience. The HRSS architecture is highly scalable and inexpensive, paving the way for future spectrometers with very high bandwidth (many GHz) covering many beams simultaneously.

The complete instrument system for both HRSS and OSFP, including digitization and packetization hardware, digital signal processing (DSP) algorithms, and control software, will be made publicly available for students and researchers worldwide.

1.3. Open Source Hardware Infrastructure

All of the instruments discussed here take advantage of the open source, modular DSP instrumentation framework developed by the Center for Astronomy Signal Processing and Electronics Research (CASPER) (Parsons et al. 2006). This international collaboration seeks to shorten the astronomy instrument development cycle by designing modular, upgradeable hardware and a generalized, scalable architecture for combining this hardware into a signal-processing instrument. Employing FPGAs, FPGA-based chip-independent signal-processing libraries, and packetized data routed through commercially available switches, CASPER instrument architectures look like a Beowulf cluster, with reconfigurable, modular computing hardware in place of CPU compute nodes. Thus, a small number of easily replaceable and upgradeable hardware modules may be connected with as many identical modules as necessary to meet the computational requirements of an application, known colloquially as "computing by the yard." Such an architecture can provide orders of magnitude reduction in overall cost and design time and will closely track the early adoption of state of the art IC fabrication by FPGA vendors.

The Berkeley Emulation Engine (BEE2) system was CASPER's first attempt at providing a scalable, modular, economic solution for high-performance DSP applications (Chang et al. 2005). The BEE2 system consists of three hardware modules: the main BEE2 processing board, a high speed ADC board for data digitization and an iBOB board primarily responsible for packetizing ADC data onto the Ethernet protocol. Communication between hardware modules takes place over standard 10 Gbit Ethernet (10 GbE) protocol, allowing for the integration of commercial switches and processors.

The next generation Virtex-5-based "ROACH" board (Reconfigurable Open Architecture for Computing Hardware) replaces, but interoperates with, both the BEE2 and IBOB boards. ROACH includes a single Xilinx Virtex-5 FPGA (SX95T, LX110T, LX155T), four 10 GbE-CX4 ports, two ADC ports, up to 8 GB of DDR2 memory, 72 Mbit of QDR, and an independent control and monitoring PowerPC processor. ROACH remains compatible with all current and next generation ADC boards.

All CASPER boards may be programmed via a set of open source libraries for the Simulink/Xilinx System Generator FPGA programming language (Parsons et al. 2008). These libraries abstract chip-specific components to provide high-level interfaces targeting a wide variety of devices. Signal-processing blocks in these libraries, such as polyphase filterbanks, Fast Fourier Transforms, digital down converters, and vector accumulators, are parameterized to scale up and down to arbitrary sizes, and to have selectable bit widths, latencies, and scaling.

2.0. SERENDIP V.v

Over the last thirty years SERENDIP spectrometer development has closely tracked the Moore's Law growth in the electronics industry, with new spectrometers processing ever-larger bandwidths while achieving finer spectral resolution. SERENDIP V.v is the most powerful spectrometer yet built as part of the SERENDIP project. SERENDIP V.v was installed at Arecibo Observatory in June 2009 and operates commensally with other experiments on the ALFA multibeam receiver. Currently, the spectrometer multiplexes beam-polarizations through a single-beam 200MHz digital signal processing chain via a computer-controlled RF switch.

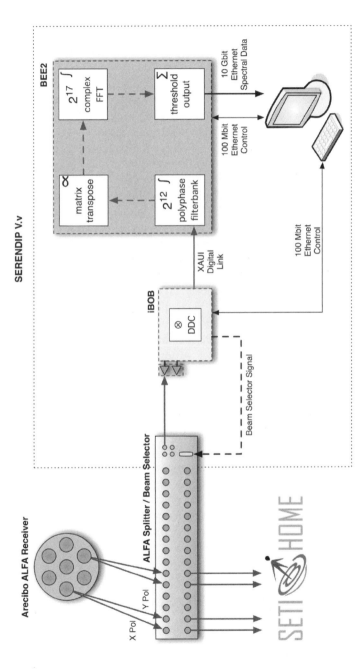

Figure 2.2. SERENDIP *V.v* instrument architecture. Analog signals from the ALFA receiver, mixed down to IF, are fed to a computer-controlled switch. One copy of the input is relayed to the SETI@home data recorder and a time-multiplexed beam is sent to the SERENDIP *V.v* spectrometer. The spectrometer samples the incoming IF signal at 800 Msps, digitally down converts the data to a complex baseband representation, performs a two-stage channelization (yielding ~1 Hz spectral resolution) and outputs over-threshold frequency channels to a host PC.

The SERENDIP V.v system architecture and dataflow are shown in Figure 2.2. ALFA signals for all fourteen beam-polarizations are fed into an RF switch, with a single output fed into a high speed ADC sampling at 800 Msps. An iBOB board mixes the sampled signal down to baseband, decimates to a 200 MHz bandwidth and transmits the serialized data stream to a BEE2 via a high speed digital link. Processing on the BEE2 is split into four stages, each of which occupies a separate FPGA on the board. The data stream is (1) coarse channelized via a 4096pt polyphase filter bank (PFB), (2) matrix transposed by a "corner turner" to facilitate a second stage of channelization, (3) fine channelized using a conventional 32768pt Fast Fourier Transform (FFT) and finally, (4) "thresholded," in which each fine frequency "bin" (1.49 Hz wide) is compared against a scaled coarse-bin average to pick out fine bins of interest. Local averages are calculated per PFB channel by averaging the same data being fed to the FFT in parallel with the transform. This way, the total power in each PFB bin can be accumulated while the FFT is being computed (via Parseval's theorem).

The thresholding process triggers "hits" for fine/FFT bins that are greater than or equal to the threshold power. For practical reasons, the number of hits reported per coarse/PFB bin is capped via a software-adjustable setting, usually set to report fine bins between 15–30 times the average power. The reported hits are assembled into UDP packets on board the BEE2 and transmitted to a host PC. The host PC combines spectrometer data with meta-information, such as local oscillator settings and pointing information, and writes the complete science data stream to disk. To date, SERENDIP V.v has commensally observed for approximately nine hundred hours. Analysis efforts are underway, in parallel, at both UC Berkeley and Cornell.

While both SERENDIP V.v and SETI@home operate simultaneously and commensally on the same RF signal, SERENDIP V.v differs in the key respect that the computationally intensive Fourier Transform is performed internally, rather than through distributed computing. This forces SERENDIP spectrometers to use a much simpler search algorithm than SETI@home employs. However, since the SERENDIP spectrometer is collocated with the telescope, it has access to a much larger bandwidth. SERENDIP and SETI@home are thus complementary, in that together they can look with both a panoramic gaze across many MHz and with microscopic precision near the 21 cm "watering hole."

3.0. Heterogeneous Radio SETI Spectrometer

The HRSS instrument system bridges our previous radio SETI programs by connecting open source FPGA-based signal processing hardware and software to an easily programmable GPU-equipped multicore CPU back-end, thus

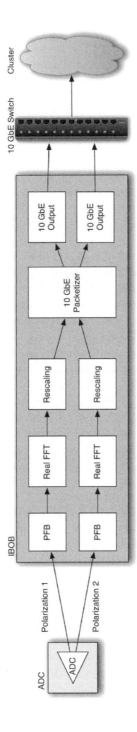

Figure 2.3. The Packetized Astronomy Signal Processor Block Diagram: the data flow inside the FPGA. Two data streams from the ADC are channelized by the PFBs and FFTs. After rescaling and bit selection (18 bits down to 8 bits), the data are packetized and sent out over two 10GbE links to a server or 10GbE switch.

achieving an economical student-friendly SETI instrument. The low-cost, scalable architecture used in HRSS will enable more widespread deployment than previous instruments, potentially increasing both the sky and frequency coverage of the radio SETI search space. With previous instruments, difficulty in programming the hardware precluded implementing intricate algorithms directly into the real-time data flow. The flexibility of the CPU/GPU back-end of HRSS will readily enable arbitrarily sophisticated algorithms in the real-time processing pipeline, including dynamic interference rejection and immediate follow-up.

The prototype for HRSS is the existing Packetized Astronomy Signal Processor (PASP) (McMahon 2008), based on the CASPER iBOB. This reconfigurable FPGA design channelizes two signals, each of 400 MHz bandwidth (digitizing at 800 Msps), packetizes, and distributes the channels to different IP addresses using a runtime programmable schema. Figure 2.3 shows a block diagram of the PASP instrument. Two signals (e.g., two polarizations) are fed into an iBOB using a dual ADC board. Each polarization is sent through a PFB, which channelizes the streams. Individual channels are buffered into packets and sent out over 10 GbE links to a cluster of servers via a 10 GbE switch or a single back-end server directly connected to the iBOB.

The PASP design is highly reconfigurable. The number of channels, number of IP addresses, and the packet size can all be easily adjusted in the instrument's Simulink design. This design can support a variety of back-end processing options simply by adjusting these three parameters. The number of channels adjusts the size of the sub-band for each processing element. Dividing the 400 MHz band into sixteen channels creates large 25 MHz sub-bands, which may require a faster server, but this can be balanced by increasing the number of channels and thereby reducing the size of the sub-bands and processing demand. The number of IP addresses also controls the bandwidth each back-end server receives. In a sixteen-channel design with only eight IP addresses, each IP will receive two channels. In a server with multiple processing elements (e.g., multiple CPU cores or GPUs), these channels can be processed in parallel. HRSS will largely consist of a port of the PASP design to the new CASPER ROACH board, taking advantage of the larger FPGA, enabling a larger bandwidth and improved interface. The iBOB can be difficult to interface with, requiring a JTAG connection to reprogram the board and a very limited shell program to interact with the FPGA. In contrast, ROACH provides a full Linux OS.

Additional software running on connected CPUs/GPUs will finely channelize the sub-bands and identify possible events for further processing. This software will initially be developed in ANSI-C to allow maximum portability. Once the C-based system is fully prototyped, we will optimize for

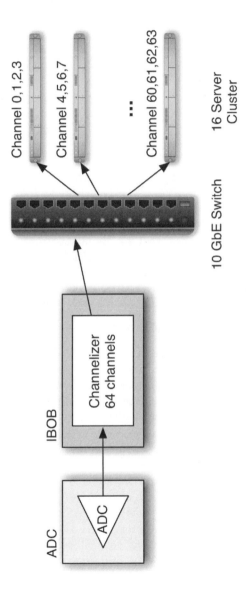

Figure 2.4. The Heterogeneous Radio SETI Spectrometer Block Diagram: an example configuration of the HRSS instrument. The iBOB channelizes ADC data into 64 channels and sends it over a 10GbE switch to a cluster of 16 servers. The number of channels and servers can be reconfigured based on available server processing power.

GPU hardware and specialized languages to extract more processing power from servers with graphics capabilities. We are investigating both OpenCL and CUDA as target languages. CUDA will provide excellent performance, but can only be compiled for NVIDIA GPUs. OpenCL is designed to compile for generic CPU and GPU platforms, but it may not provide the performance efficiency of an architecture-specific language such as CUDA.

Figure 2.4 shows an example configuration of HRSS with a cluster of servers on the back end. It has a PASP configured for sixty-four channels and sixteen IPs, a 10 GbE switch, and a cluster of back-end servers. The reconfigurability of the PASP design makes the required size and computing power of the back-end processing cluster highly elastic, scaling from a single server to a cluster of high-powered servers.

As shown in Table 2.2, HRSS is extremely cost effective compared to other SETI spectrometers with ~1 Hz spectral resolution. HRSS is less expensive than SERENDIP V.v, primarily because it uses a newer single-FPGA ROACH board paired with commodity computing hardware instead of an iBOB with a 5-FPGA BEE2 board. The HRSS architecture can be easily scaled up to processing a 1.5 GHz dual polarization signal on a single ROACH board using currently available dual 3 Gsps ADCs and multiple fine channelization nodes. The cost of each additional 125 MHz/ dual polarization module is about a factor of three less than the first module.

4.0. Optical SETI Fast Photometer

The forthcoming Optical SETI Fast Photometer (OSFP) is based on the same front-end optics and photodetectors as the original Berkeley OSETI instrument, but adds a flexible digital back-end based on the CASPER DSP instrument design system. This instrument will significantly improve our sensitivity to pulsed optical signals, and lower some of the barriers to wider engagement in optical SETI searches.

The digital back-end for the instrument, Figure 2.5, will be constructed from modular CASPER components; direct sampling PMT outputs with two dual 8 bit, 1500 Msps ADC boards and using a single ROACH board for DSP. This board features a variety of interfaces for connection to a control computer, accommodating a variety of experiment parameters. For high-threshold, low-event rate searches, the ROACH's 100 Mbit Ethernet should be sufficient for data acquisition. For low thresholds, or characterization of instrument PMTs, the ROACH's 10GbE interfaces can be used for transferring many events and/or large swaths of raw sampled voltages.

The programmable FPGA-based digital back-end will allow us to improve sensitivity by implementing sophisticated real-time detection algorithms. In our existing system at Leuschner observatory, the detection algorithm is very

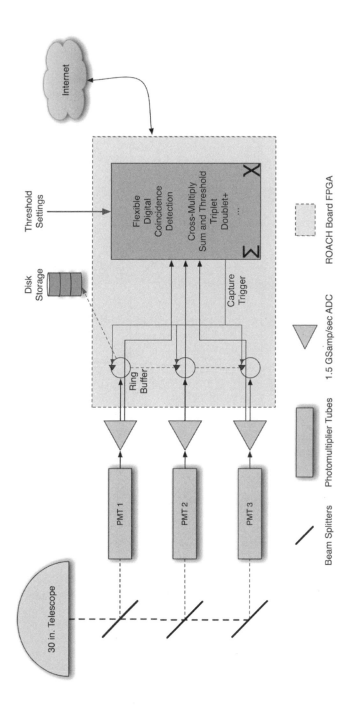

Figure 2.5. Schematic diagram for the OSFP digital backend. An optical telescope feeds three photomultiplier tubes (PMTs) using optical beamsplitters. The PMT outputs are digitized directly using fast 1.5 Gsps ADCs and fed into the ROACH Virtex-5 FPGA for processing. Voltage samples are fed into a deep 4 Gb DRAM ring buffer in parallel with a programmable event trigger that results in the readout of the ring buffer and capture of raw event data, including pulse profiles and possible information content. The digital logic for the instrument is fully reconfigurable and compatible with the CASPER open-source instrument design tool flow.

Table 2.2.

SETI Spectrometer	Bandwidth (MHz)	Beams	Pol's	Cost	Normalized Cost per MHz/beam/pol
SERENDIP V.v UCB, deployed at Arecibo & JPL	200	1	1	$40K	$200
HRSS First 125 MHz dual pol bands	125	1	2	$9K	$40
HRSS additional 125 MHz dual pol bands (after the first 125 MHz)	125	1	2	$3K	$15

HRSS costs compared to other SETI spectrometers. Costs do not include labor.

simple—all three PMT signals must be above a programmable threshold to trigger an event. With OSFP, one can implement more sophisticated detection algorithms. For example, a multistage trigger could be implemented that requires the sum of the three digitized PMT outputs to exceed a threshold as well as the requirement that the signal levels in the three streams be similar to each other. The ability to perform significant computations on the data streams in real-time is a crucial aspect of this design. We envision searching for multiple signal types simultaneously, including weak pulse trains with repetition times from ns to ms and violations of Poisson statistics in photon arrival times (indicating a non-astrophysical source). False positive signals can also be efficiently rejected based on pulse profiles (a capability sorely lacking in the current threshold-based instrument).

The large amount of DRAM available on the ROACH board will enable buffering of raw PMT waveforms and triggered write-to-disk based on high-confidence events. Such a capability will enable detailed analysis of an event, including precise determination of pulse arrival times using centroiding. Upon detection of a coincidence event, a user-adjustable section of the corresponding waveform, along with microsecond time-tagging provided by a GPS 1 pulse per second (PPS) system, will be packetized and transmitted to a host computer over one of the ROACH's Ethernet interfaces. A parallel, streaming DSP design will enable the instrument to operate at 100 percent duty with reasonable waveform buffers. Should a significant event be detected, the software system will automatically alert the observer to the possible signal detection, and will optionally automatically cause the telescope to continue observing the same sky coordinates where the telescope was pointed when the reported flash arrived.

In anticipation of the real-time computing capabilities of OSFP, we have performed preliminary simulations of several pulse detection algorithms. These simulations model the entire front-end of the system, including the optical beamsplitter, PMTs, and ADCs. In initial work, it appears the optimal algorithm involves thresholding the cross-correlation of each pair of PMT waveforms, but we continue to evaluate tradeoffs in sensitivity and false alarm rate. Future simulations will allow us to improve our algorithms by incorporating more elaborate detection criteria.

Use of CASPER hardware and gateware for this instrument guarantees an upgrade path when faster ADCs become available, eventually allowing full Nyquist sampling of the PMT bandwidth. All optics and detector components for the existing front-end are available off the shelf from Hamamatsu and Edmunds Industrial Optics. The entire assembly can be constructed without special tools, and complete instructions and parts lists are available at http://seti.berkeley.edu/opticalseti.

Acknowledgments

The Berkeley SETI projects are funded by grants from NASA and the National Science Foundation, and by donations from the friends of SETI@ home. We acknowledge generous donations of technical equipment and tools from Xilinx, Fujitsu, Hewlett Packard and Sun Microsystems.

Note

1. SERENDIP experiment numbering proceeds from IV (four) to V.v (five point five) to accommodate the ambiguous naming of the SERENDIP V computing board, which was in fact never used for searches for extraterrestrial intelligence.

Works Cited

Anderson, D., J. Cobb, E. Korpela, M. Lebofsky, and D. Werthimer. 2002. SETI@home: An experiment in public-resource computing. *Communications of the ACM* 45 (11): 56–61.

Backer, D., P. Demorest, R. Ramachandran, I. Stairs, R. Ferdman, D. Nice, and I. Cognard. 2005. Realtime astronomy signal processing in PC clusters. Paper presented at the International Union of Radio Science General Assembly XXVIII, India.

Chang, C., J. Wawrzynek, and R. Brodersen. 2005. BEE2: A high-end reconfigurable computing system. *IEEE Design and Test of Computers* 22 (2): 114–25.

Drake, F. 1960. How can we detect radio transmissions from distant planetary systems. *Sky and Telescope* 19: 140–43.

Howard, A. W., and P. Horowitz. 2001. Is there RFI in pulsed optical SETI? *The Search for Extraterrestrial Intelligence (SETI) in the Optical Spectrum III*, ed. Stuart A. Kingsley and Ragbir Bhathal, Proc. SPIE 4273: 153–60.

Howard, A. W., P. Horowitz, D. T. Wilkinson, C. M. Coldwell, E. J. Groth, N. Jarosik, D. W. Latham, R. P. Stefanik, A. J. Willman, J. Wolff, and J. M. Zajac. 2004. Search for nanosecond optical pulses from nearby solar-type stars. *Astrophysical Journal* 613: 1270–84.

Lampton, M. 2000. Optical SETI: The next search frontier. In *Bioastronomy '99—A New Era in Bioastronomy*, Astronomical Society of the Pacific Conference Series 213: 565–72.

McMahon, P. 2008. Chapter 5: A load balanced spectrometer for coherent dedispersion applications. In *Adventures in radio astronomy instrumentation and signal processing*, 81–90. MSc Thesis, University of Cape Town, South Africa.

Parsons, A., D. Backer, C. Chang, D. Chapman, H. Chen, P. Crescini, P-Y. Droz, C. de Jesus, D. MacMahon, K. Meder, J. Mock, V. Nagpal, B. Nikolic, A. Parsa, B. Richards, A. Siemion, J. Wawrznyek, D. Werthimer, and M. Wright. 2006. PetaOp/Second FPGA signal processing for SETI and radio astronomy. Paper presented at the Asilomar Conference on Signals and Systems, Pacific Grove, California.

Parsons, A., D. Backer, A. Siemion, H. Chen, D. Werthimer, P-Y. Droz, T. Filiba, J. Manley, P. McMahon, A. Parsa, D. MacMahon, and M. Wright. 2008. A scalable correlator architecture based on modular FPGA hardware, reusable gateware, and data packetization. *The Publications of the Astronomical Society of the Pacific* 120 (873): 1207–21.

Reines, A. E., and G. W. Marcy. 2002. Optical search for extraterrestrial intelligence: A spectroscopic search for laser emission from nearby stars. *The Publications of the Astronomical Society of the Pacific* 114 (794): 416–26.

Schwartz, R., and C. Townes. 1961. Interstellar and interplanetary communication by optical masers. *Nature* 190 (4772): 205–208. D. Anderson, G. Bower, J. Cobb, G. Foster, M. Lebofsky, J. van Leeuwen, and M. Wagner. 2010. New SETI sky surveys for radio pulses. *Acta Astronautica*, arXiv:0811.3046.

Stone, R. P. S., S. A. Wright, F. Drake, M. Munoz, R. Treffers, and D. Werthimer. 2005. Lick observatory optical SETI: Targeted search and new directions. *Astrobiology* 5: 604–11.

Werthimer, D., D. Ng, S. Bowyer, and C. Donnelly. 1995. The Berkeley SETI program: SERENDIP III and IV instrumentation. In *Progress in the Search for Extraterrestrial Life*, ed. Seth Shostak, Astronomical Society of the Pacific Conference Series 4: 293–302.

Candidate Identification and Interference Removal in SETI@home

Eric J. Korpela, Jeff Cobb, Matt Lebofsky,
Andrew Siemion, Joshua Von Korff, Robert C. Bankay,
Dan Werthimer, David Anderson

1.0. Introduction

SETI@home, a search for signals from extraterrestrial intelligence, has been recording data at the Arecibo radio telescope since 1999. These data are sent via the Internet to the personal computers of volunteers who have donated their computers' idle time toward this search. The SETI@home client software, which runs on these computers, corrects the data for a wide variety of possible accelerations of the transmitter or receiver ranging from −100 Hz/s to 100 Hz/s. At each possible Doppler drift rate, the software performs a sensitive analysis to detect four types of potential signals: (1) narrowband continuous wave signals, (2) narrowband signals that match the Gaussian profile expected as an extraterrestrial signal drifts through the telescope field of view, (3) repeating pulses found using a fast folding algorithm, and (4) signals representing a series of three signals at constant frequency, evenly spaced in time (Korpela, Werthimer, Anderson, Cobb, and Lebofsky 2001).

To date, SETI@home volunteers have detected more than 4.2 billion potential signals. (http://setiathome.berkeley.edu/sci_status.html) While essentially all of these potential signals are due to random noise processes, radio frequency interference (RFI), or interference processes in the SETI@home instrumentation, it is possible that a true extraterrestrial transmission exists within this database. Herein we describe the process of interference removal being implemented in the SETI@home post-processing pipeline, as well as those methods being used to identify candidates worthy of further investigation.

2.0. Candidate Identification

Several properties make a candidate worthy of reobservation. Primarily, a good candidate should be persistent in its position in the sky. If we detect a frequency from a certain sky position, and detect an identical frequency from a point on the sky many degrees away, there are two possibilities: an extraterrestrial civilization has multiple beacons separated by hundreds of light years, all of which are Doppler corrected for the motions of the planet Earth, or we've detected a source of terrestrial interference. The latter is, of course, far more probable.

A good candidate should be persistent in time. For example the "Wow!" signal (Gray and Marvel 2001) had extremely high power, and it had the appropriate Gaussian profile for a point source drifting through the telescope's field of view, but despite repeated attempts at follow-up detections it has never been seen again. That makes it unlikely that the "Wow!" signal was a high duty cycle extraterrestrial beacon.

A good candidate should be persistent in frequency. When examined again it should appear at a similar frequency (but perhaps not identical due to uncorrected Doppler effects). Allowing too large a frequency difference makes it more likely that random noise events or unrelated interference could be considered to be part of a candidate.

The SETI@home candidate identification ranks groups of signals by their persistence in time, their spatial proximity, their dissimilarity to signals generated by random noise processes, their dissimilarity to known interference sources, and their proximity to interesting celestial objects (nearby or solar-type stars, known planetary systems, etc.) It assigns a score based upon the probability that the set of signals seen from a point in the sky would occur due to random noise processes, with lower scores being better.

Early in the project, candidate identification was an arduous process that was undertaken at intervals ranging from six months to more than a year. Because this process would access every signal in the SETI@home database several times, it was very I/O intensive and would require months to complete. To remove this shortcoming, we have designed a Near-Time Persistency Checker (NTPCkr).

The SETI@home pipeline keeps track of incoming potential signal locations by pixelating the sky in an equal area pixelization scheme. When a signal comes in, the corresponding sky pixel is marked as "hot" and given a time-stamp. Since a given area of sky tends to be observed several times in a short period, this pixel is allowed to "cool" for several weeks. At this point, if no further signals for that pixel are received, it is marked as ready for analysis.

The NTPCkr examines the signals within that pixel and adjacent pixels to determine a candidate score based upon the above criteria. It is our goal that the score represent the probability that the set of potential signals associated with the candidate could arise due to random noise processes. The existing candidates are ranked in order of this score from lowest (least noise-like) to highest (most noise-like).

3.0. Interference Removal

In the past, it has been our practice to perform interference removal on the entire set of potential signals detected by our instruments. Again, this method requires that the entire database be examined multiple times, which is inefficient.

Because narrowband correlations are very unlikely to occur due to random noise processes, candidate groups containing interference are ranked very highly on our candidate lists. Therefore, we now run interference rejection on candidate groups in order of their ranking. A candidate containing a lot of interference will have a good (low) score because it is not noise-like. The interference removal process will remove many of the non-noise-like signals, resulting in a candidate that is more noise-like, and thereby increasing (worsening) the score.

The interference removal techniques we use are independent and, because of the random access nature of the database, can be run in any order. After interference rejection, the candidate position is again marked as ready for analysis by the NTPCkr.

3.1. Radar Removal

By far, the most common source of interference in the SETI@home data set is radar stations on the island of Puerto Rico. Although these stations do not transmit within the 1.4GHz band received by the ALFA receiver used by SETI@home, signals from the radars do leak into the band, appearing as short duration, high intensity signals that change frequency rapidly with a large frequency component near the receiver central frequency. This component typically breaks up into multiple stable harmonics when seen in the recorded data. Fortunately, the radars are periodic, transmitting pulses of a few microseconds duration every few milliseconds. The pulse patterns and periods are known or can be measured. The Arecibo Observatory has build a radar blanking signal that is synchronized with the strongest radar and can be recorded with the data. However, this signal only removes the strongest radar and if the period or phase of that radar changes, it can take some time for the blanking signal to become resynchronized.

Therefore, we have built a software equivalent. This software radar blanker examines the data for radar pulses fitting the pattern of one of several known radars, determines the repetition period for that pattern and generates a signal indicating at what time the radar pulses should be present. Before distributing data to our volunteers, we replace these sections of data with a computer-generated noise-like signal. This typically results in a sensitivity loss of about 1.2 dB for strong narrowband signals with durations longer than the interpulse period. This loss is acceptable considering the alternative of filling the signal database with unwanted radar signals.

Our remaining interference mitigation methods are applied to the results returned by our volunteers after they have been inserted into our science database.

3.2. Zone Interference Removal

Zone Interference Removal removes signals that are contained within a "zone," which is a region of parameter space known to contain a large number of invalid signals. The parameters that define a zone can include a range of radio detection frequency, base-band frequency, period (for pulsed signals), detection time, the identity of the receiver, and the version of software used for various stages of the analysis process.

The top panel of Figure 3.1 shows the frequency distribution of 378,362,077 potential pulsed signals detected by SETI@home between July 5, 2006, and September 16, 2009. The vertical bands that are present indicate frequencies that are overrepresented and are probable RFI frequencies. We use a statistical analysis to determine which frequencies appear too frequently on differing sky positions to be due to noise processes. Those frequencies define the exclusion zones. Pulses determined to be within these zones (6.6 percent of the total) are shown in the middle panel. The lower panel shows the distribution of pulses that remain after those within zones have been removed.

The RFI frequency zones are typically quite narrow. We have identified 35,000 frequencies, covering less than 1 percent of our band which are subject to frequent interference. These zones contain between 5 and 20 percent of the detected signals depending upon signal type. As our software matures, our zone definitions are changing to better match interference characteristics. Signals determined to be within the zones are marked as interference and are excluded from future candidate scoring computations. This analysis can be done on other parameters (for example: pulse period or Doppler drift rate) to design RFI exclusion zones for those parameters as well.

3.3. Short-Term Fixed-Frequency Interference Removal

Some sources of interference are present at constant frequencies for periods of time ranging from hours to days, but not for sufficiently long to define

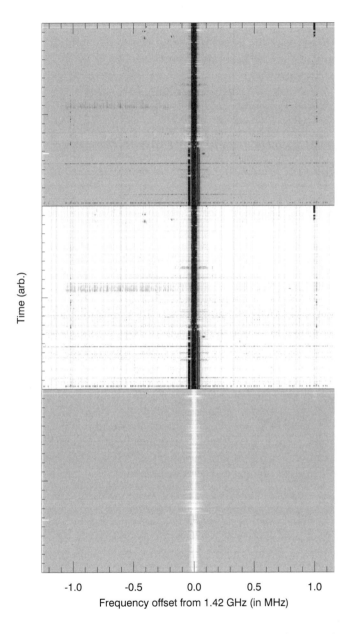

Figure 3.1. These plots show the frequency distribution of pulses detected by SETI@home. The upper panel shows all pulses. The middle panel shows pulses determined to be due to persistent interference sources. The lower panel shows the pulse frequency distribution after the interference has been removed. Note that some interference remains.

a zone. Because celestial objects stay in our field of view for seconds to minutes, we can use this property to remove these sources of interference. By examining a time range around a potential signal we can calculate the probability of coincidence with another signal with similar frequency but seen at a different sky position. If this probability falls below a threshold (~10^{-4}) we conclude that the signals are due to an interference source.

3.4. Removal of Interference that Drifts in Frequency

Some sources drift in frequency, even over short periods of time. For these methods we use the octant-excess drifting interference detection and removal method described by Cobb, Lebofsky, Werthimer, Bowyer, and Lampton (2000). Adjacent signals in time and frequency, but at different sky positions, are allocated into octants of frequency-time space surrounding the signal being examined. A significant statistical excess in an octant and the octant 180 degrees opposite indicates the presence of an RFI source drifting in frequency.

Again, a probability computation is used to determine the likelihood that this excess is due to random noise, and if this computation falls below a threshold, the signal being examined is marked as being due to interference.

3.5. Crowdsourced Interference Removal

The final stage of candidate identification requires examination of the top candidates by eye to detect forms of interference that might get past the first three layers of RFI removal. Because of the small amount of manpower available in the form of SETI@home staff members, we intend to develop a "crowdsourced" candidate investigation method. Similar to Stardust@home, it will use fabricated candidates, some containing RFI and others that are RFI clean, to train volunteers in identifying RFI and ranking candidates.

The lists of best candidates will be available online. Volunteers can then submit an opinion whether each of the signals making up the candidate is due to RFI. These votes will be used (in conjunction with the volunteer training scores) to modify the candidate score, which will alter the rankings.

Acknowledgments

The SETI@home and Astropulse projects are funded by grants from NASA and the National Science Foundation, and by donations from the friends of SETI@home. Observations are made at the NAIC Arecibo Observatory, a facility of the NSF, administered by Cornell University.

Works Cited

Cobb, J., M. Lebofsky, D. Werthimer, S. Bowyer, and M. Lampton. 2000. SERENDIP IV: Data acquisition, reduction, and analysis. In *Bioastronomy 99: A New Era in the Search for Life*, ASP Conference Series, 213: 485–89.

Gray, R. H., and K. B. Marvel. 2001. A VLA search for the Ohio State "Wow." *The Astrophysical Journal*, 546: 1171–77.

Korpela, E. J., D. Werthimer, D. Anderson, J. Cobb, and M. Lebofsky. 2000. SETI@home—Massively distributed computing for SETI, *Computing in Science and Engineering* 3 (1): 78–83.

A New Class of SETI Beacons
That Contain Information

*Gerald R. Harp, Robert F. Ackermann, Samantha K. Blair,
Jack Arbunich, Peter R. Backus, Jill C. Tarter,
and the ATA Team*

1.0. Introduction

When designing a signal for interstellar transmission to an unknown but technologically competent species, we must consider how that species (e.g,. humans) might discover the signal *as distinct from the galactic background radiation*. Note that SETI is different from most earth-based communication problems because (1) we don't know, a priori, how much uninteresting naturally generated power is arriving from any random point on the sky and (2) we must invent a process that separates the SETI signal from the background noise (which includes both the galactic background and the receiver noise).

At the same time, most SETI researchers suspect that an extra-solar civilization will wish to communicate nontrivial information to humans. Until now, the primary focus of radio SETI observations has been narrowband signals or strong broadband pulses. These signals can be used only for "beacons," since they convey no "message" beyond a single symbol of information. It is usually suggested that the message information will be communicated in an entirely different signal mode (or with extremely low symbol rate consistent with the narrowband criterion).

We introduce a new signal type where the beacon and message are encoded in one and the same signal. The proof of principle described here opens the road for invention of more SETI beacons that have this property. The point of this chapter is to show that a signal that is both easily discovered and contains a message is possible and takes no more than a factor of two times the computational power of a narrowband SETI search. Another advantage is that our proposal searches an "orthogonal" space to narrowband or pulsed SETI, opening an uncharted territory for exploration. One disadvantage of the proposed signal type is that with the algorithms presented here, it gives lower signal power to background power ratio than narrowband SETI for equal power transmitters.

We propose to recover signals using autocorrelation spectroscopy. While a more generalized concept was first suggested in 1965 (Drake 1965), to the best of our knowledge no prior published work on radio SETI searches using autocorrelation spectroscopy exists. We begin this chapter with an extended introduction to put this work into perspective, followed by a description of the technique, and finishing with preliminary observations taken with the Allen Telescope Array (ATA) that prove the concept of the technique and detection algorithm.

1.1. Some Constraints Due to the Galactic Background Radiation

Starting with the extragalactic background, it is a natural law of the universe that almost all galactic radiation arises from sources with relatively large bandwidths (between 500 Hz masers [Cohen 1987] and 10^{19} Hz gamma ray bursts). In the radio frequency range natural signals have time-varying amplitude within a narrow bandwidth that is not distinguishable from Gaussian white noise (Figure 4.1). This is true even in the case of narrow spectral lines or masers; in the frequencies of emission the signals are noise-like. In the time domain, there are pulsars with time variations as small as 1 ms (Backer 1982). But these unusual sources have recognizable characteristics that allow SETI researchers to identify them. For this reason we endeavor to exclude noise-like signals (in the absence of further information)[1] from our search space as being most probably associated with natural sources.

Most galactic radiation is indistinguishable from Gaussian random noise because it obtains from a large number of unresolved independent radiators situated very far from the telescope. For the radiation to be visible across vast distances, a great number of similarly radiating sources is required since those sources radiate incoherently and usually have separations greater than the light crossing time between radiators (though there are exceptional cases). According to Rice's mean value theorem, radiators may have any spectral character (or frequency occupation) and will tend to Gaussian random noise in a narrow bandwidth as the number of sources $\to \infty$. Such natural radiation is compounded with the receiver noise, which is also nearly Gaussian white noise in character.

1.2. Information Content

Nyquist's and Shannon's theorems (Nyquist 1928; Shannon 1949) give us some insight. For a given signal, the maximal received information rate (symbol rate) is roughly equal the bandwidth of the signal detector. For example, a continuous sine wave has zero symbol rate (symbols per second) if the signal

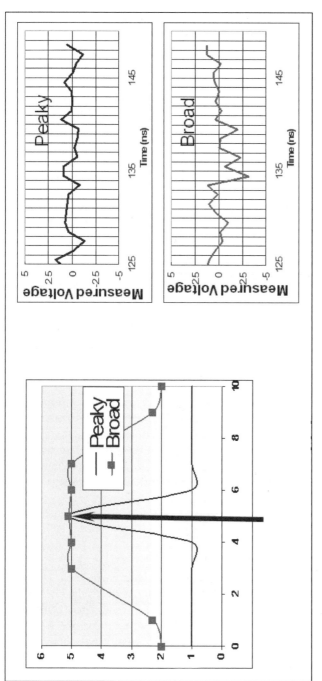

Figure 4.1. Left: Highly schematic representation of two fictitious radio signals represented in the frequency domain: A broadband source (Broad) and a relatively narrowband source such as spectral line radiation (Peaky). The gains of the two features are normalized and measured in a narrow bandwidth centered at the black arrow. The voltage signals from these sources might look like the two graphs on the right, which show Gaussian random noise for the electric field amplitude as it arrives at the telescope. This is an important quality of natural signals and successful SETI observations must focus on non-Gaussian random signals.

appears in only one frequency bin of the detector.[2] Similarly, when singular wideband pulses arrive, we obtain only one bit of information. Comparatively, many wideband signal types allow information recovery with a symbol rate up to its received frequency bandwidth.

Generally, SETI transmitters can never take full advantage of the transmitted bandwidth because (1) a maximally compressed information signal appears as Gaussian white noise, hence is *indistinguishable* from natural radiation, and (2) a corollary to Shannon's theorem is that a maximal symbol rate is obtained when there is no redundancy between one sample and the next, making it impossible to decode (without the key). We can find/decode beacons only if they contain substantial redundancy, that is, the signal must strike a balance between redundancy (which makes it noticeably artificial), and detectability/decodeability. Here we are assuming that we have no prior knowledge of the transmitted signal. If we know (by some means) or guess a part of the encoded information, then this requirement is relaxed. An example of this is narrowband pulsed SETI, where we "guess" that signal form is essentially a sine wave, but carries information in very slow (few Hz) pulses.

An example of extreme redundancy is the sine wave or single-pulse signal, which contains only one bit of information but is easily noticeable against the galactic background.

1.3. Some Constraints Due to the Interstellar Medium (ISM)

To reach us, the signal must traverse the space between transmitter and receiver. This space is filled with dust, neutral gas, and ionized gas. Between 1–10 GHz, the most important source of signal distortion is the free electrons in the ionized gas or plasma (mostly ionized hydrogen). While traversing plasma, electromagnetic (EM) photons acquire an effective rest mass equal to $\hbar\omega_p$, and travel more slowly than the speed of light. The plasma angular frequency ω_p is given by (Jackson 1975),

$$\omega_p = \frac{4\pi\rho_e e^2}{m_e} \tag{1.1}$$

where ρ_e is the free electron density (electrons per cubic meter). The constants e, and m_e are the electron charge and electron mass, respectively. The average interstellar medium has about one electron per cubic centimeter (Cordes and Lazio 1992; Cordes, Lazio, and Sagan 1997) leading to plasma frequency ω_p ~0.1 rad/s ~0.16 Hz ~4 $\times$ 10^{-17} eV. If the ISM plasma were homogenous with this value, EM frequencies below this cutoff would not propagate. Higher frequency EM waves would propagate, but more slowly than the speed of light, with higher frequencies travelling faster (Fitzpatrick 2006). In the radio frequency range 1–10 GHz, for a transmitter located a distance L from the receiver, the light travel time τ for the signal is given by

$$\tau \cong \frac{L}{c}\left(1 + \frac{\omega_p^2}{2\omega^2}\right).$$ (1.2)

Another derivation of this delay can be found in Thompson (2001). With this expression it is straightforward to simulate signal distortion in a uniform ISM plasma. This is done in Figure 4.2 where a single pulse with length 50 nanoseconds (left) is dispersed according to Equation. After traversing only the short distance (4 LY) the pulse is been broadened by a factor of 40 and arrives at the detector with 30x lower peak power and reducing the signal to background radiation level by a factor of 6-30.

First-order dispersion correction (de-dispersion) can be done even when the plasma is not homogeneous as long as the electron column density or "dispersion measure," DM = $\int_0^L \rho_e \, dL$ between transmitter and receiver is known. De-dispersion recovers a narrow pulse from the dispersed pulse on the righthand side of Figure 4.2, and works equally well with other kinds of wideband signals. Unfortunately, the DM of received signals is generally not known since the interstellar medium is clumpy on all length scales and often we do not know the distance to the source. The result is that for recovery of any wideband signal, such as the pulse above, a search over a large number of test values for DM is required (unless you have prior knowledge, such as when the pulse was sent),[3] which makes searches for wideband signals computationally inefficient as compared with narrowband signals.

For any but the nearest transmitters, it is also necessary to understand the nonlinear scattering of signals caused by the inhomogeneous electron density in the ISM. Scattering defects are variously known as pulse broadening, signal fading, and scintillation. Cordes et al. (1997) conclude that "scintillations are very likely to allow initial detections of narrowband signals, while making redetections extremely improbable. . . . This conclusion holds for relatively distant sources but does not apply to radio SETI toward nearby stars (~100 pc)." The reason re-detections are improbable is that even continuous signals from distant sources will fade in and out at the receiver location (analogous to listening to a radio station at long distance). Scintillation is negligibly small at short distances (<100 pc) and lower frequencies (<3 GHz). Scintillation grows monotonically larger at greater distances and higher frequencies. For example, at 1 GHz and distances >500 pc, 100 percent fading (in and out) is expected. See references for more detail.

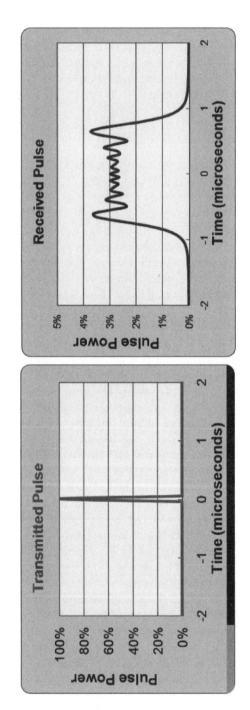

Figure 4.2. Left: A 50 ns pulse as observed near the transmitter located the same distance from Earth as the nearest star (Proxima Centauri). Right: The dispersed pulse as received by a detector on Earth. The received pulse is not quite symmetrical about Time = 0 because dispersion is in nonlinear in frequency.

1.4. Conventional SETI and Pulse Searches

Because of dispersion and scattering, conventional SETI searches have focused on continuous or slowly (few Hz) pulsed signals that are nearly monochromatic in a given reference frame. Such signals propagate through the ISM with little or no corruption (apart from fading) since a multipath sum of a monochromatic signal cannot change its frequency. At the SETI Institute, observers look for "drifting" signals with bandwidth of <10 Hz and Doppler drift rate up to 1 Hz/second. It is important to search over Doppler drift space because our Earth is accelerating as it rotates about its axis, the Sun, and galactic center. The same can be said for the transmitter.[4] Such acceleration causes compression/dilation of the signal in the time domain, leading to time-dependent frequency changes for the monochromatic signal. The Doppler drift of a narrowband signal is proportional to its frequency for a given relative acceleration. This variation is partially mitigated by the acceptance of signals with bandwidth up to 10 Hz. For the purposes of this chapter, we shall not consider the frequency variation of drift rate and consider drift rates up to 1 Hz/s as being an acceptable range to allow most SETI transmitters to be observed at any frequency.

In "coherent" searches over short time periods (e.g., one second) Doppler drift is negligible and a simple Fourier Transform[5] can be applied to detect a narrowband signal. This detection algorithm is considered fast and scales as $N_S \ln(N_S)$ where N_S is the number of samples in the measurement. To its detriment, a search for pulses such as that in Figure 4.2 would require the same searches over frequency (and if necessary drift rate), but additionally requires a search over pulse start time and DM;[6] therefore, it is substantially less efficient than a search for narrowband beacons.[7]

1.5. Persistence

With current telescope and computing capabilities it is impossible to reliably determine the direction of arrival when an artificial signal enters the telescope. The telescopes are indeed pointed in specific directions, yet bright sources can leak into the receiver through sidelobes of the telescope. In optical photography, this phenomenon is known as lens flare and can be caused by the Sun or a bright source at the edge or even outside of the image field.[8] The streaks and artifacts of lens flare have analogs in radio imaging, corrupting the radio signal measured from a single point (or pixel) on the sky.

With sufficient computational power plus many more dishes or omnidirectional antennas, it might be possible to determine with high confidence the direction of arrival of a signal with only one observation. The state of the art, however, requires telescopes to re-observe the same signal over and over, with changes in pointing, focus, and then corroboration from other observatories before a signal can be identified as having extraplanetary origin. Practically speaking, signals must be persistent over days or weeks (ideally, forever). If signals are not persistent, then we cannot prove their alien origin and for safety we classify them as human-made. Given the signal fading problems described above, current radio SETI searches are limited to relatively close by, or enormously powerful transmitters. The authors hope that this limitation may one day be overcome with a clever algorithm (sooner) or by brute force (later).

Interesting signals without persistence are observed *thousands* of times each day at the SETI Institute. Figure 4.3 for example shows a result obtained in a narrowband SETI search near the PiHI frequency (the number π times the HI observing line of 1420.4 MHz). This (extremely powerful) ~10 second pulse of narrowband radiation appeared in one fifty second observation period

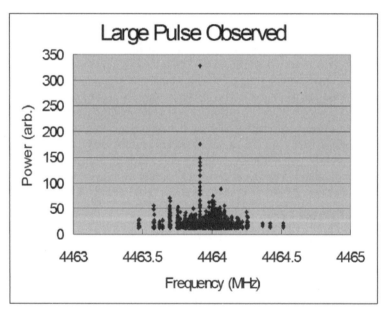

Figure 4.3. A pulse with maximum power $>300\sigma$ above the noise background was observed on a nearby star (~100 LY, (J2000 RA, Dec) = (32.211809°, 22.441734°)) in the HabCat Catalog (Turnbull, 2003). This pulse is interesting since it appears to arrive from the direction of a potentially habitable star and because it appears very close (within the expected Doppler shift tolerance caused by relative motion) to the "magic" PiHI frequency of 4462.3 MHz, this signal appeared in only one observation and never thereafter. Given the proximity of this source, we do not expect substantial fading in the ISM; hence the signal is really not present, most of the time.

but was never re-observed. This pulse has interesting features: it is observed at a magic frequency in the direction of a nearby and potentially habitable star. Yet we cannot be sure this signal was created intentionally or unintentionally by some transmitter on Earth. Hence, after multiple observations over two weeks and no re-detection, we gave up (although this direction is added to a catalogue of directions to re-observe as time permits).

1.6. Dimensionality of Search

In a crude way, we can talk about computational efficiency in terms of the number of physical parameters (or dimensions) that must be searched in order to complete a survey for a subtype of ET signals. We have already discussed four physical dimensions of search for "frequency compact"[9] signals: time of arrival, emitted center frequency, Doppler drift (relative acceleration between transmitter and receiver), and DM. To this list we add two more: signal bandwidth and symbol dilation. Symbol dilation[10] is most easily understood in the context of a binary encoded signal which consists entirely of 0's and 1's. ET is free to choose the time duration of a single bit and interval between bits, from very slow (milliHertz or less) up to the Nyquist limit for the transmitter bandwidth.

For the most general signal type, it is necessary to search over all six of these dimensions, but special signal types have been discovered that "project out" some search dimensions and lower the computation time. For example, continuous narrowband signals require only a two-dimensional search space (frequency and Doppler drift). Extremely narrow (e.g., ps-ns) pulsed signals require a four-dimensional search space (time of arrival, frequency, Doppler drift, DM).

To help with this discussion we present a list of the number of different values each parameter must take for the six search parameters mentioned above as constrained by the parameters of the ATA and our current processing capabilities (1–10 GHz, observation length ~100s, narrowest frequency channel bandwidth ~1 Hz, maximum frequency bandwidth = 100 MHz) in Table 4.1.

This analysis is oversimplified since not all search dimensions are comparably constrained. In a narrowband frequency search the number of different acceleration values (~200 for a 100 second measurement at 1 GHz with 1 Hz resolution) that must be probed for a reasonably complete SETI search is much less than the number of different frequencies (9 billion 1 Hz bins in the range 1–10 GHz). Yet we can say that a pulse search is always computationally less efficient than a narrowband search since pulse searching requires the same search variables as for narrowband, plus time of arrival and DM. As mentioned below, certain algorithms can take, for example, two search dimensions and reduce them to one dimension as far as computation

Table 4.1

Search Parameter	Approx number of parameter values
Time of arrival	100 s @ 10 ns sampling = 10^{11}
Center frequency	10 GHz at 1 Hz sampling = 10^{10}
Frequency Bandwidth	Up to 100 MHz = 10^8
Dispersion Measure (DM)	Of order $N_S = 10^{10}$
Relative Acceleration (Doppler rate)	Up to 10 Hz/s at 10 GHz = 2000
Symbol dilation Autocorrelation Signal Duration (Repetition rate, "Gold Code," see section 2.)	Of order $N_S = 10^{10}$ 100s @ 1 μs steps = 10^8

List of the relative sizes of parameter search spaces for the 6 observation variables defined above. Recall that N_S is the number of samples in the measurement. However, a computation time cannot be obtained from a simple product of the numbers on the righthand side, see text for details.

is concerned. This argument was first pointed out to the authors by David Messerschmitt.

2.0. A New Beacon Proposal

There is a substantial historical bias toward narrowband searches in radio SETI. In the Project Cyclops report (Oliver and Billingham 1973), this seminal SETI document reads, "Beacons . . . will surely be highly monochromatic." The later book *SETI 2020* proposed for searches including narrowband signals and high bandwidth pulsed signals. About wideband signal varieties such as frequency-shift encoding and spread spectrum, *SETI 2020* reads, "[S]uch a scheme would be so computationally intensive that our searching for stars and frequencies would [be very inefficient]." In this context, we now introduce a specialized set of wideband signals that do not suffer this negative consequence envisioned in *SETI 2020*.

A key goal of our work is to minimize the number of dimensions over which we search. Above we noted that narrowband searches are more effective than pulse searches since the latter require searches over start time and DM while the former do not. Is there a wideband signal that can be recovered with a computational cost similar to a narrowband search?

Consider this: The transmitter sends an *arbitrary* signal with arbitrary length and arbitrary bandwidth, communicating nonzero symbol rate. After a short delay (<1s), an adjacent transmitter (or the same one; see below) sends a second copy of the signal. The two signals are superimposed at the receiver.

We assume that transmission began sometime far in the past and continues until sometime far in the future, and that the bandwidth of the transmission equal to or larger than that of our detector. We call this a twice-sent signal. Because the time scale for ISM fluctuations is on the order of seconds to hours (Cordes and Lazio 1992; Walker 1998), we can take advantage of the fact that both signals suffer *identical* dispersion and scattering by the ISM, provided the delay between transmissions is less than one second. We can discover such signals with an efficient autocorrelation algorithm.

The autocorrelation spectrum of a signal $A(t)$ as a function of signal delay τ is computed from

$$A(\tau) = \int_0^{N_s} S(t) \ S(t - \tau) \ dt, \tag{1.3}$$

where $S(t)$ is the received signal as a function of time, and the time interval between samples is scaled to unity.[11] We assume that the detector emits a regularly sampled voltage and N_s is the number of samples. $A(t)$, also known as the delay spectrum of the above described signal, will show a strong peak where $\tau = \tau_0$ the delay between the transmission of the signals mentioned above.

As an example, consider Figure 4.4 (left) where one message is sent twice with a delay of seven samples. In this simulation we do not consider the effects of noise. Calculating the autocorrelation of the received signal using gives the result on the right. On the right there is a strong peak (beacon) with power equal to half the total transmitted power of the composite signal. Using this scheme, the transmitting civilization has the opportunity to send us a great deal of information such as the complete works of Shakespeare or the complete embodiment of their society's knowledge. The message never has to repeat in order to discover this beacon; we require only that the transmitter stays on.

More than two copies of the signal could be sent with delays of $\tau_0, 2\tau_0,$ $3\tau_0,$ etc., creating multiple copies of the beacon at regularly spaced delay values. Multiple-copy encoding is explicitly used for space communications so as to provide robust error correction during information retrieval. For multiply-sent signals the analysis is modified to search for a series of peaks all having the same delay separation.

However, a multiply-sent signal has more redundancy hence reduced information-carrying capacity,[12] so there is a trade-off between redundancy and information content. For beacon discovery, there is no signal to background advantage for multiple copy signals (though there is an advantage for information retrieval due to increased redundancy). Furthermore, we must guess (or search) over the number of delay copies, which makes the beacon search more complex. If the number of delay copies is not two, as in the

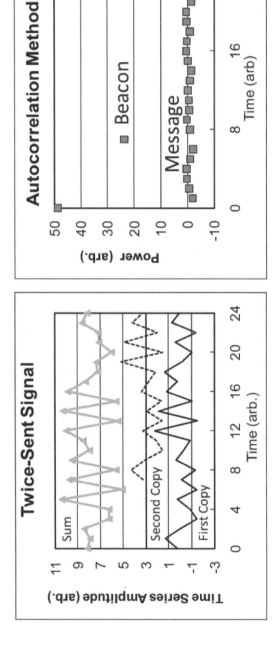
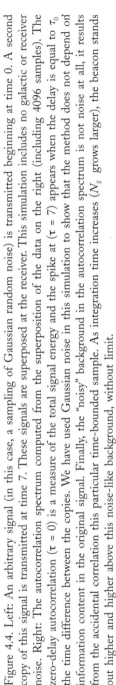

Figure 4.4. Left: An arbitrary signal (in this case, a sampling of Gaussian random noise) is transmitted beginning at time 0. A second copy of this signal is transmitted at time 7. These signals are superposed at the receiver. This simulation includes no galactic or receiver noise. Right: The autocorrelation spectrum computed from the superposition of the data on the right (including 4096 samples). The zero-delay autocorrelation ($\tau = 0$) is a measure of the total signal energy and the spike at ($\tau = 7$) appears when the delay is equal to τ_0 the time difference between the copies. We have used Gaussian noise in this simulation to show that the method does not depend on information content in the original signal. Finally, the "noisy" background in the autocorrelation spectrum is not noise at all, it results from the accidental correlation this particular time-bounded sample. As integration time increases (N_s grows larger), the beacon stands out higher and higher above this noise-like background, without limit.

twice-sent signal, then we suggest that the next most likely value is infinity. Having an infinitude of repeats means that the beginning (preamble) of the signal is always available in any snippet of data. However, unless the time duration of the message is finite and less than the detection time, no part of the data would be recoverable since there would be fewer sampled symbols than the number of symbols representing the signal. For this reason we favor the twice-sent signal type over any other number and we believe this chapter marks the first suggestion of this approach.

Another alternative use of multiply-sent signals is an encoding scheme with amplitude-shift, frequency-shift, or phase-shift keying where a snippet of information is repeated over and over without overlap, but with regular breaks in amplitude, frequency, or phase. For example, in binary phase-shift keying the phase of the transmitted signal is changed (or not changed) at regular intervals to indicate the transition between two different symbols of information (a 0 or a 1). This scheme is used in GPS communications, and in Section 4 we present real observations that demonstrate how autocorrelation uncovers signals of this type. Other types of encoding that use repeating signals can also be discovered using autocorrelation. An important point to our proposal is that it is not necessary to know what code snippet is being repeated to make the discovery.[13]

Besides encoding information, twice- or multiply-sent signals have other advantages. No search over Doppler drift, time of arrival, signal dilation, or symbol rate is required. Finally, even high levels of distortion caused by the intergalactic, interstellar, interplanetary media or earth's ionosphere and troposphere can be tolerated since short time scale correlation is highly resistant to signal distortion. This means that we could detect (but perhaps not decode) signals from much farther away than the limit for narrowband signals set by interstellar scattering mentioned in Section 2.3. Again, as long as both copies of the signal are distorted in the same way, they will correlate. Distortion may be a problem for signal recovery, however, as described below.

As a comparison with the narrowband and pulse searches described above, a search for autocorrelation peaks in a twice-sent signal requires the production of an autocorrelation spectrum and then a thresholding step. Since we look for strongest autocorrelation peak apart from $\tau = 0$, the size of the search space is N_S. As we shall see below, autocorrelation uses the same fast algorithms as for a narrowband SETI search and can be accomplished in equal time with twice as much computational power.

2.1. Signal, Background Radiation, and Noise

The transmitted signal has the form $S_{\text{trans}}(t) = s(t) + s(t - \tau_0)$ where τ_0 is the delay introduced between the two copies of the signal by the transmitting civilization. In a realistic example, noise is introduced by the galactic background radiation and by the receiver itself, and we detect N_S samples of the received signal $S(t)$:

$$S(t) = s(t) + s(t - \tau_0) + N(t). \qquad (1.4)$$

$N(t)$ is the sum of galactic background and receiver noise, and can be assumed to be Gaussian white noise to a good approximation in most cases. We define the measured variances

$$\langle s^2 \rangle = \frac{1}{N_S} \int_0^{N_S} |s(t)|^2 \, dt$$

$$\qquad (1.5)$$

$$\langle N^2 \rangle = \frac{1}{N_S} \int_0^{N_S} |N(t)|^2 \, dt$$

Performing the autocorrelation transform and admitting the possibility that the computations are performed using a complex representation of the measured signal, we obtain

$$AC(\tau) = \int_0^{N_S} S^*(t - \tau) \, dt =$$

$$(a) \int_0^{N_S} [s(t - \tau_0) \, s^*(t - \tau)] \, dt =$$

$$(b) \int_0^{N_S} [s(t)s^*(t - \tau) + s(t)s^*(t - \tau_0 - \tau) + s(t - \tau_0) \, s^*(t - \tau_0 - \tau)] \, dt = \qquad (1.6)$$

$$(c) \int_0^{N_S} [N(t)s^*(t - \tau) + N(t)s^*(t - \tau_0 - \tau) + s(t) \, N^*(t - \tau) + s(t - \tau_0) \, N^*(t - \tau)] \, dt$$

$$(d) + \int_0^{N_S} N^*(t - \tau) \, dt$$

The righthand side is divided into four parts, each of which contributes in its own way to $AC(\tau)$. Beginning with (a), this is our "SETI signal" or beacon term, and this term reaches a maximum value of $N_S\langle s^2 \rangle$ when $\tau = \tau_0$.

For other values of delay τ, (a) as statistical properties similar to the three terms in (b). We wish to estimate the behavior of these terms as a function of N_S, but this is impossible without detailed knowledge of $s(t)$, so we make some reasonable guesses. If we assume that the transmitting civilization sends us a signal "dense" with information, then it may have statistical properties similar to Gaussian white noise. As mentioned above, the signal must contain some redundancy to permit decoding, but suppose that every so often they send a preamble which gives us a "key" to decode the rest of the information. Then most of the time, all the terms (b) (and term (a) when $\tau \neq \tau_0$) will grow as $\sqrt{N_s}\langle s^2 \rangle$ as $N_s \to \infty$. While we cannot rely on this exact scaling behavior, we can rely on the aliens to make reasonable choices about how the information is encoded. Indeed, encoding schemes where all the terms in (b) grow less slowly than $\sqrt{N_s}$ may be possible.

Before continuing, we consider a special encoding scheme where a snippet of data is sent over and over many times with a constant repeat rate. In this case, all the terms in (a) and (b) will take their turns integrating up coherently and we observe a comb of autocorrelation spikes much that first suggested by Drake (1965). See Section 4 for a real-world example of this kind of beacon.

We identify the terms (c) and (d) as "noise" terms since they all contain an integral over a product where one of the terms is Gaussian white noise. For now we shall make the assumption that $s(t) = N(t)$ since the transmitter is far from the receiver and swamped by the galactic background. In this case, (d) $>>$ (c) $>>$ (b), and we may neglect the terms in (c) (and (b) for twice-sent signals and the noise signal is dominated by (d). Since the noise is Gaussian white distributed, the leading background noise term as $N_s \to \infty$ is $\dfrac{\sqrt{N_s}}{2} \langle N^2 \rangle$.[14]

With this analysis and the assumptions mentioned above we can estimate the signal to background ratio $SBR(\tau = \tau_0)$ (the ratio of the beacon peak to a typical background point in the autocorrelation spectrum) to be

$$\lim_{N_s \to \infty} SBR(\tau = \tau_0) = 2 \sqrt{N_s} \frac{\langle s^2 \rangle}{\langle N^2 \rangle}. \tag{1.7}$$

To summarize, AC will contain a strong peak for $\tau = \tau_0$ that is $\sqrt{N_s}$ greater than the typical AC value where $\tau \neq \tau_0$. This is our main mathematical result. In a comparison with conventional narrowband SETI, the factor of 2 on the righthand side is canceled out by the fact that the transmitted power in $s(t)$ is only half of the total transmitted power for a twice-sent signal.

Speaking of a conventional SETI, over short time periods the scaling with N_s is the same as for a "matched filter" detection algorithm. Matched filters are known to be optimal linear filters for maximizing signal to noise ratio (here, signal to background ratio). It is straightforward to perform a computation for the signal to background ratio in a narrowband search to be

$$SNR_{NB} = N_s \frac{\langle s^2 \rangle}{\langle N^2 \rangle}. \tag{1.7}$$

Thus, the scaling law for twice-sent signals as a function of number of samples N_s is not as favorable as for conventional narrowband SETI. However, in narrowband SETI the maximum integration interval is one second since we are open to Doppler drift rates as high as 1 Hz/second. Beyond a one second interval, a search over Doppler drift correction is required, which increases the complexity of the algorithm. There is no limit to the coherent integration time in the autocorrelation method.

As an aside, recent research on alternative "matched filtering" approaches to discovery of message-bearing SETI signals (Messerschmitt 2010) show that optimal *SBR* scaling with N_s (Equation 1.8) can be obtained at the cost of substantially greater signal processing. Specifically, one must search over start time, symbol rate, Doppler drift, and DM.[15] In this approach, one guesses at least part of the message content (e.g., first one hundred bits of the binary representation of π) and matches to that. In this chapter we cannot do justice to this developing field and leave its exposition to the future.

2.2. Implementation of Autocorrelation Spectroscopy

The implementation of autocorrelation spectroscopy is a trivial extension of the processing required for conventional narrowband SETI as shown in Figure 4.5. In narrowband searches, the measured signal is Fourier transformed (FT) with a filter bank based on the fast Fourier transform. The resultant frequency power spectrum (PS) is formed by squaring the results of the FT and then performing a threshold operation (upward-pointing arrow in Figure 4.5). These signals that pass through the threshold detector are declared "interesting" and then followed up on, until the direction of arrival can be confirmed.

To perform the autocorrelations required for this proposal, we take the same power spectrum values and perform an inverse FT (or FT^{-1}) and pass the result through a thresholding filter (or a comb filter for multiple-copy versions of the proposed signal type, see Section 4.1). A search for autocorrelated signals can be carried out simultaneously with a conventional SETI search using the same FT engine (with a trivial sign inversion) and the same thresholding detector software. We propose that in future narrowband SETI search systems, autocorrelation spectroscopy should be added for a total cost increase of less than a factor of 2.

3.0. Signal Recovery

Once the beacon signal has been identified, all the arguments about search spaces fade since the transformation of a single known ET signal is negligible

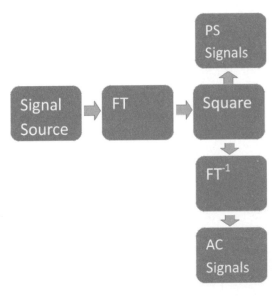

Figure 4.5. A block diagram showing the processing steps to recover narrowband SETI signals using the power spectrum (PS), and its relationship with the processing steps to recover twice-sent beacons using autocorrelation (AC). The figure indicates that the first three steps of processing are identical. For twice-sent signal detection, the thresholding step takes place after a second (inverse) Fourier Transform. Since the Fourier transform step is usually the most computationally intensive step, AC spectroscopy takes about twice the compute power and can be accomplished in the same time using the same computational programs as narrowband SETI.

compared to the massive computations required to find the signal in the first place (see Table 4.1).

Here we discuss some ideas for recovery of the message in a twice-sent beacon. The superimposed twice-sent beacon resulting from the signals in Figure 4.4, left, can be described as a message (first signal) convolved with a pair of delta functions separated by τ_0 in time. This convolution can be represented as a multiplication in the frequency domain (convolution theorem). In practice, we perform a discrete Fourier transform (DFT) on both the message and a second "filter" of equal length which contains only zeros except for two Dirac delta functions separated by τ_0 and having the value $1/\sqrt{2}$. (We use this normalization so that the total power in the filter sums to unity.) The Fourier transforms of both signal and filter are multiplied bin by bin, and the product undergoes an inverse DFT. The resulting signal is the superposed twice-sent signal and this could be fed directly into a single transmitter for communication with Earth (i.e., two different transmitters are not necessary, though they may be convenient).

Some information is lost in this convolution. The Fourier transform of a double delta function $F_{\tau_0}(f)$ is described by

$$F_{\tau_0}(f) = z'[1 + \exp(i\,2\pi\,\tau_0 f)] = z\,\cos(2\pi\,\tau_0 f) \tag{1.9}$$

where z is a complex number with magnitude $\sqrt{2}$. $F_{\tau_0}(f)$ takes the value 0 when $\tau_0 f = \dfrac{1}{2}, \dfrac{3}{2}, \dots, n + \dfrac{1}{2}$, so whatever information was present at these frequencies is destroyed by the convolution. Of course, the transmitting civilization knows this, and they are likely to use only frequency bands where the cosine function is high, perhaps in areas where $F_{\tau_0}(f) > \dfrac{1}{\sqrt{2}}$.

Note that these zeros could be entirely avoided by sending the second signal with different amplitude than the first. If the first signal is transmitted with amplitude 1 and the second with amplitude b, then the areas of destructive interference in frequency space would take the value $(1-|b|)^2$. Though not as dramatic, this still has negative consequences on signal recoverability in the presence of background radiation and noise. Choosing $b \neq 1$ also reduces the beacon *SNB* from its maximum value at $b = 1$.

In this section, we have not yet discussed the impact of background noise on message recovery. The effects of noise are draconian. We first define the Fourier conjugates of the message $s(t)$ and noise $N(t)$ to be $s(f)$ and $N(f)$, respectively. Unless $s(f) > N(f)$ for a given frequency, it is impossible to recover the message information for that frequency. However, if we have identified a SETI beacon using autocorrelation, then not only do we know τ_0 but we know we are dealing with a SETI beacon. If necessary, humanity will build a sufficiently sensitive radio telescope to achieve $s(f) \gg N(f)$. This might not be necessary—recall that an autocorrelation search at radio frequencies has yet to be performed (or at least published). To promote the discussion we simply assume $s(f) \gg N(f)$ from here on.

We define the signal to background ratio in the frequency domain $SBR(f)$ as[16]

$$SBR(f) = \frac{\langle s(f) \rangle}{\langle N(f) \rangle}\,\cos(2\pi\,\tau_0 f) \tag{1.10}$$

Since we know the positions of the zeros of $s(f)$ we can estimate $\langle N(f) \rangle$. Using this information and the peak positions of $s(f)$ we can estimate $\langle s(f) \rangle$. Now we define a dimensionless threshold P (~10) which determines the reliability we wish to achieve in the message recovery. We obtain poor estimates of the message anywhere $SBR(f) < P$, so we must discard this data (e.g., set to zero or something similar). The result $s'(f)$ is then divided by $F_{\tau_0}(f)$ to remove the distortion caused by the double delta function. Finally, an estimate of the original message $s'(t)$ is obtained from an inverse Fourier transform of $s'(f)$. Alternatively, the civilization may choose to encode their message in the frequency domain, where our message estimate is $s'(f)$. In either case, we may have a very good estimate of the message, provided that the transmitting civilization has encoded their message to avoid the zeros of $s(f)$. Notice that for any $N < \infty$ there will be finite probability of an error in the signal recovery. This defect might be further mitigated if the transmitting civilization chooses a message with inherent redundancy that allows error correction in signal recovery.

There is more that could be said about the fascinating topic of message recovery and how to optimize it. Here we present only the simplest possible approach as a demonstration of the feasibility and limits of message recovery for the twice-sent beacon.

4.0. Proofs of Principle Using the Allen Telescope Array

As follow-through on this proposal as a search for interstellar beacons we have begun observations of known point-source emitters such as pulsars, masers, galaxies, and human-made satellites. For now we choose directions where there are known strong point-source emitters, with the idea that some of the received radiation may have embedded autocorrelation information that is not apparent in ordinary radio astronomical, narrowband, or narrow pulsed SETI observations. Hence, we have hope of detecting autocorrelation power even in a relatively short measurement of duration ms to minutes.

For an astronomical example we show a measurement of methanol maser emission in the W3OH molecular cloud region in Figure 4.6. On the left we show the part of the frequency power spectrum where methanol maser emission is found. On the right we show the autocorrelation spectrum. The autocorrelation contains bumps and oscillations associated with the rather narrow 100 kHz frequency bandwidth of the maser signal. In Figure 4.6 we plot time delays of up to 150 μs since this is the most interesting region. However, the transmitting civilization must use a value of τ_0 much greater than the delay range of the maser spectrum so that it will stand out from the background. We did not discover any interesting features for larger values of delay, up to fractions of a second in these prototype experiments. In future reports we will show results from longer time integrations and more sources.

The concept behind measuring masers is that an extraterrestrial civilization might "pump" such masers to cause amplitude variations in the maser power.

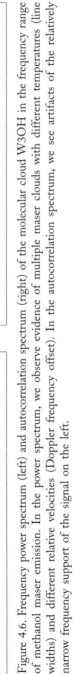

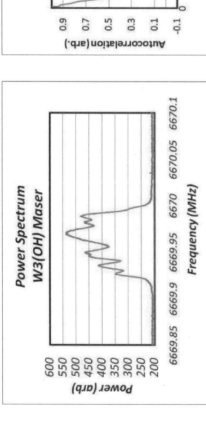

Figure 4.6. Frequency power spectrum (left) and autocorrelation spectrum (right) of the molecular cloud W3OH in the frequency range of methanol maser emission. In the power spectrum, we observe evidence of multiple maser clouds with different temperatures (line widths) and different relative velocities (Doppler frequency offset). In the autocorrelation spectrum, we see artifacts of the relatively narrow frequency support of the signal on the left.

Tommy Gold proposed this idea in the late 1960s and early 1970s, and this idea was shown to be feasible by Joel Weisburg (2005), who found a maser being modulated by pumping from a nearby pulsar.

In a second example, we display in Figure 4.7 the power and autocorrelation spectra from a GPS satellite. GPS communication uses binary phase shift key encoding where each bit of information is represented by twenty copies of a 1023 bit "gold code" sent sequentially with a repeat rate of 1 ms. Although the artificial nature of this signal is hardly apparent in the power spectrum (and over longer time periods would be even smoother), the artificial nature of the received signal is obvious in the autocorrelation spectrum on the right. This is precisely the kind of AC spectrum expected for multiple-time-sent SETI signals and indicates intelligent origin, in this case human origin.

4.1. Generalizations

We have already discussed many generalizations of the twice-sent beacon, including multiply-sent versions and repeating code schemes (e.g., binary phase shift keying). One can consider giving the twice-sent signals different amplitudes. This would decrease the detectability of the beacon, but would allow, at least in principle, full recovery of all message data as a function of frequency or time. More complex schemes can be constructed with multiply-sent signals copies having different amplitudes, polarization,[17] prime number delay ratios, etc. Such approaches may decrease detectability and/or maximum message symbol rate, but may improve signal recovery.

In another direction, twice-sent beacons could be redundant in the frequency domain rather than in the time domain, just as narrowband signals are the frequency domain equivalent of single pulses. However, redundancy in the frequency domain implies that the two signal copies will not be identical upon arrival due to differential ISM-related scattering (and the delay will be changed by dispersion, which could cause trouble for signal recovery). We expect even better ideas based on our simple suggestion will be forthcoming both from our group and other groups performing research in this field.

5.0. Conclusions

We propose a new class of beacon signals that contain rich information content while standing out substantially from the galactic background radiation. As beacons, they are highly resistant to distortions during the voyage from transmitter to receiver. These signals are detectable with a simple and efficient autocorrelation algorithm. Although autocorrelation techniques for signal detection have been speculated upon in the past, here we provide

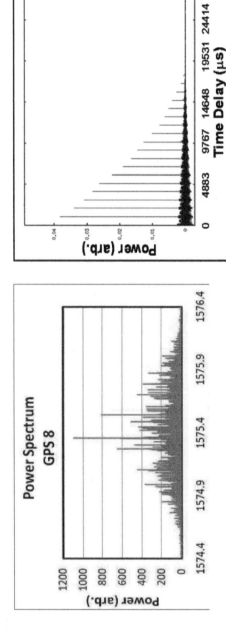

Figure 4.7. Power spectrum and autocorrelation spectrum from a GPS satellite known as PRN08. The broad frequency support on the left looks fairly noise-like within an envelope of ~2 MHz. Hence a narrowband SETI search would be confounded by this artificial signal. However, the autocorrelation spectrum shows a series of peaks separated by 1 ms which is clear evidence of this signal's artificial origin.

observational evidence with a GPS satellite that demonstrates the feasibility for detecting such beacons and identifying them as having intelligent origin.

We describe an implementation of an autocorrelation detector that can run simultaneously on the same data stream in a narrowband SETI search and uses the same computational blocks. Thus, both narrowband SETI and autocorrelation SETI can be run together in real-time with less than a factor of 2 augmentation of compute resources. Finally, we present an introduction to the challenges faced by humans once these signals are detected. Solutions and suggestions about how to analyze data for signal recovery are discussed. Because these signals contain substantial information, they are a more straightforward method to actually communicate (one way) useful information from the extra-solar civilization to human kind.

Conventional narrowband SETI and pulse searches are promising ways to look for extra-solar civilizations. At the same time, alternative searches such as the one proposed here are worth investigation and merit the relatively small additional effort to carry them out in parallel.

Acknowledgments

The first phase of the ATA was funded through generous grants from the Paul G. Allen Family Foundation. UC Berkeley, the SETI Institute, the National Science Foundation (Grant No. 0540599), Sun Microsystems, Xilinx, Nathan Myhrvold, Greg Papadopoulos, and other corporations and individual donors contributed additional funding. This work was also supported by the NSF through award AST-083826 and by NASA award NNG05GM93G. The authors gratefully acknowledge David Messerschmitt and Ian Morrison for fruitful conversations and constructive criticisms of the poster associated with this work.

Notes

1. Notice that some artificial signals may be "noise-like," i.e. spread spectrum signals. A spread-spectrum algorithm could distinguish these signals from real background radiation and noise.
2. Note that the signal might contain information discoverable by a different detector with, e.g., smaller bin size. However, for a given detector, once the detection is made no signals encoded in a single detector bin may be recovered.
3. Since a pulse search requires a search over both start time and DM, it is possible to order the searches by (1) performing a search over start times followed by an easy estimate of DM, or (2) by performing the DM search followed by an easy time of arrival estimate. It may be

possible to optimize the computation in pulse searches by choosing the appropriate order.

4. This search can be obviated by the assumption of a "global" reference frame such as the one associated with the galactic center.

5. In the numerical domain where signals are sampled regularly with a common time interval, an FFT-based poly-phase filter bank is generally used to approximate the Fourier Transform.

6. See Note 3.

7. Our analysis attempts to leave out any "value judgments" placed on the relative success of a narrowband or narrow-pulse search. The authors believe that until we have discovered at least one transmitting civilization, our opinions about the likelihood of one signal type being more probable than another are simply opinions.

8. For photographs illustrating lens flare caused by the sun, see <http://photosbycarla.blogspot.com/2008/08/lens-flare-tutorial.html>. In radio astronomy imaging, similar effects occur not only from the sun but from strong satellite transmitters, and other human-made transmitters in the air or on the ground.

9. By frequency compact we mean that in whatever bandwidth is chosen for a signal, the frequency occupancy over that bandwidth is on the order of unity. We will not consider signals with very sparse frequency support in this chapter.

10. The major contributing factor to symbol dilation searches is the fact that we don't have a timing reference from the transmitting civilization. However, another important factor is dilation caused by the relative velocity of transmitter and receiver. The symbol dilation can sometimes be recovered from autocorrelation. However, the relative velocity changes with time, so not only do we have to divine the symbol rate, but its time derivative.

11. For the sake of clarity, we use integral notation even though the numerical computations are performed as summations.

12. Here we are focusing on multiply-sent signals where all copies of the signal have the same amplitude. In this case, there are frequency ranges where the signals interfere destructively and can lead to zeros in the received frequency spectrum. If different copies are sent with different amplitudes, 90° phase offset (electric field is a real-valued function), polarization, etc. then it is possible to arrange the carrying capacity to be not reduced, but the signal to background ratio is still impacted. From the viewpoint of data extraction in a noisy environment, partial destructive interference still limits the information carrying capacity, hence the tradeoff remains.

13. In another approach, we are aware that David Messerschmitt has suggested that one *can* guess the code snippet. It may be, for example, the first one hundred digits of the binary representation of the number π or e. After choosing the code snippet, signal encoding proceeds as usual. We will not pursue this interesting suggestion in this chapter.

14. Here we use the property that $N(t)$ and $N(t - \tau)$ when $\tau \neq 0$ are independent Gaussian random numbers.

15. David Messerschmitt has developed an algorithm where the start time and DM searches can be combined into, essentially, a one-dimensional search. This kind of clever algorithm may possibly be expanded to further reduce the computation load in matched filter searches.

16. Here we take *SNB* to be the quantity that is used during the thresholding process, where we look for peaks that stand out far above the background level. This threshold is set by choosing a specific value of *SNB*. To calculate the probability of false alarm rates, a different quantity (approximately the square of *SNB*) may be required.

17. For polarization, it is best to choose circularly polarized polarizations since the ISM plasma is also influenced, in general, by an unknown magnetic field along the direction of travel. This introduces (another) delay between the two signal copies.

Works Cited

Backer, D. C., S. R. Kulkarni, C. Heiles, M. M. Davis, and W. M. Goss. 1982. *Nature* 300: 615.

Cohen, R. J., G. Downs, R. Emmerson, M. Grimm, S. Gulkis, G. Stevens, and J. C. Tarter. 1987. Narrow polarized components in the OH 1612 MHz maser emission from supergiant OH-IR sources. *Monthly Notices of the Royal Astronomical Society* 225: 491–98.

Cordes, J. M., and T. J. W. Lazio. 1992. NE2001.I. A new model for the galactic distribution of free electrons and its fluctuations, 2010, from http://arxiv.org/abs/astro-ph/0207156.

Cordes, J. M., T. J. W. Lazio, and C. Sagan. 1997. *Astrophysical Journal* 487: 782.

Drake, F. D., ed. 1965. *Current aspects of exobiology: The radio search for intelligent extraterrestrial life.* Oxford: Pergamon Press.

Fitzpatrick, R. 2006. Classical electromagnetism: An intermediate level course, 2009, from http://farside.ph.utexas.edu/teaching/em/lectures/node100.html.

Jackson, J. D. 1975. *Classical electrodynamics.* New York: John Wiley and Sons.

Messerschmitt, D. 2010. Unpublished manuscript.

Narayan, R. 1986. Maximum entropy image restoration in astronomy. *Annual Review of Astronomy and Astrophysics* 24: 127–30.

Nyquist, H. 1928. Certain topics in telegraph transmission theory. *Trans. AIEE 47*, 617–44.

Oliver, B. M., and J. Billingham, eds. 1973. *Project Cyclops: A design study of a system for detecting extraterrestrial intelligent life.* NASA CR 114445. Moffett Field, CA: NASA Ames Research Center.

Shannon, C. E. 1949. Communication in the presence of noise. *Proceedings of the Institute of Radio Engineers* 37 (1): 10–21.

Thompson, A. R., J. M. Moran, G. W. Swenson. 2001. *Interferometry and synthesis in radio astronomy.* New York: Wiley-Interscience.

Turnbull, M., and J. C. Tarter. 2003. Target selection for SETI. I. A catalog of nearby habitable stellar systems. *Astrophysical Journal Supplement*: 181–98.

Walker, M. A. 1998. Interstellar scintillation of compact extragalactic radio sources. *Monthly Notices of the Royal Astronomical Society* 204: 307–11.

Weisberg, J. M., S. Johnston, B. Koribalski, and S. Stanimirovic. 2005. Discovery of pulsed OH maser emission stimulated by a pulsar. *Science* 309: 106–10.

Getting the World Actively Involved in SETI Searches

Jill C. Tarter, Avinash Agrawal, Rob Ackermann,
Samantha K. Blair, M. Tucker Bradford, Danese M. Cooper,
Gerald R. Harp, Jane Jordan, Tom Kilsdonk,
Kenneth E. Smolek, Karen Randall, Rob Reid,
John Ross, G. Seth Shostak, Douglas A. Vakoch

1.0. Introduction

In February 2009, the TED organization (Technology, Entertainment, and Design) and its parent body, the Sapling Foundation (Anderson n.d.) committed to helping enable a particular wish to change the world made by J. C. Tarter: *"I wish that you would empower Earthlings everywhere to become active participants in the ultimate search for cosmic company."* The motivation behind this wish is the extraordinary power that SETI has to encourage individuals to reconsider their place in, and intimate connection with, the universe, and to adopt a more cosmic perspective, to internalize the commonality of all human Earthlings, and ultimately to trivialize the differences among them. Of all human pursuits, SETI is one that ought to be global by its very nature; detection of evidence of another distant technology will change everything for all of us. This wish was made in light of the current state of digital and social networking technologies, which may now be able to achieve a new level of globalization. This paper describes a year of progress in concretizing the various aspects of the wish, and in learning to take full advantage of a nontraditional source of support for astrobiology and scientific research.

2.0. The World Is Ready to Get Involved

A decade ago, the introduction of the innovative distributed computing effort called SETI@home (Anderson et al. 2001) and its rapid adoption by millions of computer users around the world, demonstrated that SETI is something that ordinary people want to be involved in. The popularity of this program put distributed computing on the map and eventually gave rise to the active community of citizen scientists who participate in dozens of scientific research programs today. There is appetite for more involvement; for ways in which these citizen scientists can contribute more than service computing by actively improving the projects in which they participate.

3.0. A Wish in Three Pieces

There are three different categories of individuals that we want to engage with us using the newly commissioned Allen Telescope Array (ATA) (Welch et al. 2009) at the Hat Creek Radio Observatory in Northern California in order to improve the speed, sensitivity, and scope of our searches for engineered signals, and thereby improve our chances for a successful detection. The open source community of software developers will be able to work with our existing published code base; taking from it what pieces they may find useful for other purposes and adding features that will improve its efficiency and ease of use. There are also opportunities to develop more visibility into the real-time search and provide the world with a way of checking in on our progress. The community of software and communications engineers and students with technical understanding of digital signal processing will be able to help us expand the types of signals that our searches can recognize. Existing algorithms are well suited to a class of signals characterized by extreme frequency compression, but are not a good match to complex signals of higher dimensionality. (See contributions by Blair et al. [2011] and Harp et al. [2011] in this volume). The cost of computing is now becoming sufficiently low, and hosted storage (for raw time-series data from the ATA) along with cloud computing resources are being made available in support of this wish, so that we can challenge these DSP-savvy individuals to produce clever, more sensitive signal-detection algorithms that are capable of being implemented within real-time observing programs on the telescope. The third group of individuals is everybody else: the crowd. These are the people whose help we want most, because this is how the world will change. We are planning to use a combination of social networking and gaming technologies to build a vibrant, passionate, and strongly connected community (or a tribe, in the terminology of Godin [2008]). This tribe helps us by participating in the real-time search using their eyes as pattern-recognition tools to augment the extant set of implemented algorithms and to decide whether the next observation should follow up on a signal they have discovered, They help themselves by connecting to one another and using the SETI framework to better understand their human sameness, when contrasted with an independently evolved technological civilization elsewhere.

4.0. The Allen Telescope Array

The Allen Telescope Array (ATA) is the first example of a large-number-of-small-dishes (LNSD) array designed to be highly effective for commensal surveys of conventional radio astronomy projects and search for extraterrestrial intelligence (SETI) targets at centimeter wavelengths. The ATA will consist of 350 6m-diameter dishes when completed, which will provide an outstanding survey speed and sensitivity. In addition, the many antennas and baseline pairs provide a rich sampling of the interferometer uv plane, so that a single pointing snapshot of the array of 350 antennas yields an image in a single field with about 15,000 independent pixels. This number, the ratio of antenna beam width to array pattern beam width, is much smaller than the number of baselines and shows the large redundancy of the array. The goal is good image quality and high brightness sensitivity. Other important features of the ATA include continuous frequency coverage over 0.5–10 GHz and four simultaneously available 600 MHz bands at the back-end, which can be tuned to different frequencies anywhere within the overall band—thus, as many as four independent observing projects could be conducted simultaneously. Within these bands there are both 100 MHz spectral-imaging correlators for making radio maps of the large field of view, and beamformers for phasing up the array to focus on single pixels. The correlators have 1024 channels with adjustable overall bandwidths, which permit high spectral resolution. Up to thirty-two separate beams may be formed to feed either SETI signal detectors or other processors, for example, pulsar-timing systems. The ATA is a joint project of the SETI Institute in Mountain View, California, and the Radio Astronomy Laboratory of the University of California, Berkeley. The initial design grew out of planning meetings at the SETI Institute summarized in the volume *SETI 2020* (Ekers, Cullers, Billingham, and Scheffer 2002). The design goals were:

1. continuous frequency coverage over as wide a band as possible in the range 0.5–10 GHz for both SETI and conventional radio astronomy;

2. an array cost improvement approaching a factor of ten over current array construction practices;

3. large sky coverage for surveys;

4. a collecting area as large as one hectare for a point source sensitivity competitive with other instruments;

5. interference mitigation capability for both satellite and ground-based interference sources;

6. both imaging correlator and beamformer capability with rapid data reduction facilities.

The performance design goals have been achieved. However, the achieved cost factor is closer to a factor of five than ten due to the fact that full funding for the array was not available at the start of the project, and this same lack of funding has limited the current array size to forty-two rather than 350 dishes.

The ATA is located at the Hat Creek Radio Observatory of the University of California, Berkeley, in northern California. Figure 5.1 shows the forty-two-element array in operation today, and an artist's conception of the anticipated 350-element array at the observatory.

5.0. setiQuest

To galvanize and support this new community, we have launched setiQuest. org, an interactive site designed and supported by Last Exit LLC. At this site interested participants can register to become involved in three different threads of the TED wish, and join forums where we are beginning to post tutorials about the basics of SETI signal detection as it has been practiced to date. This site is also the vehicle for developing the various tools that we will need to support the community interactions.

The existing SETI signal detection code for the Allen Telescope Array is referred to as SonATA (SETI on the ATA). It receives several hundred seconds of time series data (complex voltages) from three phased array beamformers that select three target stars, or grid positions in a galactic plane survey, to be observed simultaneously as part of a two-stage pipeline. In the first stage, these data are coarsely channelized by polyphase filters, and then Fourier Transformed to fine frequency bins of 1 Hz resolution. During the second stage, the data are normalized and a set of efficient algorithms analyzes the two-dimensional frequency/time domain for patterns indicative of narrowband continuous, or slowly pulsing, signals that may be changing frequency due to the relative acceleration between a transmitter and the ATA receivers. While this analysis is proceeding, more data at another frequency are being collected and binned for analysis. Any detected patterns that are judged statistically significant are classified as potential candidate signals. These candidates are compared with a database of known interference, and candidate signals from different beams are intercompared, If the same signal is observed in more than one beam, it is classified as interference that has entered the sidelobes of the array. If no candidate signals persist after this interference-rejection process, then the pipeline continues uninterrupted, analyzing the data acquired at the next frequency. If candidate signals persist, the pipeline operation is broken, and automated follow-up observations are

Figure 5.1. (left) Aerial image of operational ATA-42 array at the Hat Creek Radio Observatory in northern California. (right) Artist's concept of the full 350-dish ATA as it will appear when fully funded.

initiated to ascertain the nature of the candidates. The SonATA code represents an instantiation in commodity servers of legacy code that has been developed at the SETI Institute over more than a decade and three generations of full- to partial-custom signal detection hardware systems. SonATA is currently being cleaned up for publication as open source code. We expect that this complex code (approximately 350,000 lines) will be published in functional tranches starting in mid-2010, along with test datasets.

Figure 5.2 illustrates the detection of a narrowband carrier signal from the distant Voyager 1 spacecraft at the edge of the solar system. While your eye may have difficulty finding the signal in this two-dimensional waterfall plot of frequency versus time, (a) the SETI signal detection code makes a credible 6σ detection (b). This is a good illustration of the fact that once the signal type is defined, an algorithm can do a better job of detection than can the human observer.

A full day of raw, time-series observational data from the three beamformers at the ATA is more than 100 TB. In order to follow up on candidate signals immediately, and to avoid massive storage requirements, signal detection is done in near-real-time and all data are discarded except those data in the vicinity of detected candidate signals. Current SonATA algorithms search for narrowband continuous and pulsed signals. They have some sensitivity to strong complex signals, but miss or inconsistently represent weak broadband signals with any complexity. Narrowband signals have been favored by SETI programs since the initial Project Ozma observations (Drake 1960), because a circularly polarized sinusoid will propagate across the Milky Way Galaxy with little or no absorption or distortion. Indeed, narrowband signals remain attractive, but given our own telecommunications technology, complex signals of higher dimensionality (particularly broadbandwidth) and greater information content should also be considered now that computational capacity is becoming more affordable. For this reason setiQuest.org has begun to provide access to raw time-series data from the ATA stored in the cloud, with new datasets appearing weekly. Individuals with the skills and tools to analyze these data are being encouraged to download files and explore them for additional signal content. The next step in this process will be to construct a user-friendly interface to both the datasets and the computational resources available in the cloud so that community members can compile and debug their analysis tools locally, but actually process the data remotely. Farther in the future, we hope to encourage larger community participation by holding a contest to determine the most effective algorithms for detecting a wide variety of signals present in a prepared dataset. The winning algorithm will then be made available to the open source code-developing community and we will work to make the algorithm fast, and efficient enough to be implemented on available computing resources at the ATA and become part of the ongoing searches.

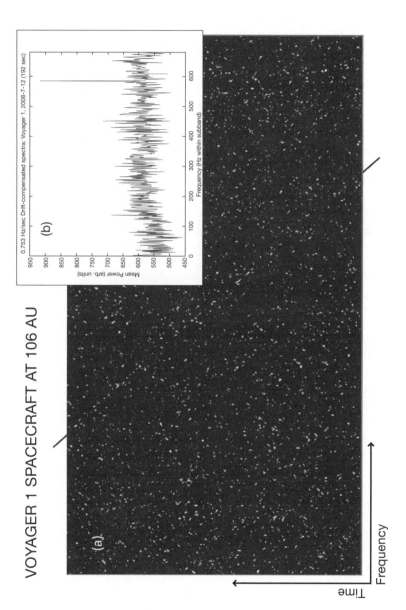

Figure 5.2. (a) waterfall plot of 8.4 GHz carrier wave from *Voyager 1* spacecraft at a distance of 106 AU. Lines extending beyond frame are a guide for visualizing the signal. (b) statistical detection of drifting signal with SETI algorithm.

Thus far we've been describing how to improve our searches for signals we can imagine or describe. What about unexpected, anomalous patterns in the data? How might we find what we aren't looking for? The human brain and eyes and ears are often extremely sensitive to patterns that differ from noise—indeed, they have a potential for detecting patterns where none exist. After many years of SETI observing at the ATA and other radio telescopes, we know that there are frequency bands that contain a very large number of signals of many types. In these bands there are in fact so many signals detected as candidates by our existing algorithms that it is impossible to classify all of them and complete the interference rejection processes before the next pipeline stage ends. Rather than partially complete analyses to an upredictable level, our SETI search procedures declare these bands unobserved. Perhaps the new algorithms for broadband signals that we will be developing with the community will be able to deal with these unobserved bands in a timely fashion, we don't really know because we haven't examined them in detail. Raw data from these frequency bands will become part of the datasets stored in the cloud for more leisurely and detailed analyses. But we intend to also try to harness the talents and creativity of citizen scientists around the globe to tackle this problem. First, if we want individuals to "look" or "listen" to data, we need data visualization and signal processing experts to help us display the information in meaningful ways. Figure 5.1 is a frequency versus time waterfall plot of a small piece of the received data, it makes sense for seeing narrowband signals, but broadband signals with significant time structure could easily look like noise in this display. What other sets of orthogonal parameters should we be considering as two-dimensional displays? How do we compress the hundreds of MHz of raw data bandwidth from the ATA down to the few KHz of audio bandwidth the ear can sense without loosing information? We need immediate help from the tech-savvy setiQuest community to solve these technical challenges before we can recruit an army of citizen scientists. Once we have an assortment of two-dimensional visual displays and audio stream we will emulate the extremely successful Galaxy Zoo project (Lintott et al. 2008) and its successors. Data will be presented to the general population of citizen scientists, candidates they identify will be presented to a second tier of citizen scientists, and candidates that are vetted will be presented to a third cadre of citizen scientists who will compare them to a rogue's gallery of known interference. Anything that survives this entire process can trigger a reobservation. At first this process will all take place in the cloud in non-real-time. Eventually, as we learn what works, what doesn't, and what motivates our participants, we would like to move this into near-real-time operations on the ATA. Citizen scientists may actually cause our observing program to be interrupted to follow up on signals that coded algorithms might have missed. Will it work? Will anyone be interested? We'll let you know.

Figure 5.3 summarizes the various threads and opportunities of the setiQuest community. We would like individuals to participate as open source code developers, signal detection algorithm developers, citizen scientists, and volunteers to help us build the infrastructure needed to enable all of setiQuest.

Acknowledgments

This work has been supported in part by NSF grant AST-0540599, and by generous donations from the Paul G. Allen Family Foundation, Nathan Myhrvold, Greg Papadopoulos, Xilinx Corporation, Dell Inc., Intel Inc., Google Inc., Amazon Web Services, Github, Infosys Inc., and many other individual and corporate sponsors. The 2009 TED Prize is supported by the Sapling Foundation who also provided support for Rob Reid, Last Exit, and Avinash Agrawal the Director, Open Innovation at the SETI Institute.

Works Cited

Anderson, C. No date. http://www.ted.com/pages/view/id/42.
Anderson, D., D. Werthimer, J. Cobb, E. Korpela, M. Lebofsky, D. Gedye, and W. T. Sullivan. 2001. SETI@home: Internet distributed computing for SETI. *Bioastronomy 99: A new era in the search for life*, ASP Conference Series, Vol. 213, ed. G. Lemarchand and K. Meech, 511–17, San Francisco: Astronomical Society of the Pacific.

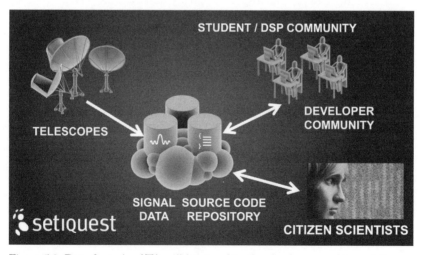

Figure 5.3. Data from the ATA will be stored in the cloud or served in real time to allow a global community of citizen scientists, open source code developers, and digital signal processing experts, who can develop new detection algorithms, to participate in improving our SETI searches and change the world.

Blair, S. K., D. G. Messerschmitt, J. C. Tarter, and G. R. Harp. 2011. The effects of the ionized interstellar medium on broadband signals of extraterrestrial origin. In *Communication with extraterrestrial intelligence (CETI)*, ed. D. A. Vakoch. Albany: State University of New York Press.

Drake, F. D. 1960. How can we detect radio transmissions from distant planetary systems? *Sky and Telescope* 19 (3): 140–43.

Ekers, R. D., D. K. Cullers, J. Billingham, and L. K. Scheffer, eds. 2002. *SETI 2020: A roadmap for the search for extraterrestrial intelligence.* Mountain View, CA: SETI Press.

Godin, S. 2008. *Tribes: We need you to lead us.* New York: Portfolio.

Harp, G. R., R. F. Ackermann, S. K. Blair, J. Arbunich, P. R. Backus, J. C. Tarter, and the ATA Team. 2011. A new class of SETI beacons that contain information. In *Communication with extraterrestrial intelligence (CETI)*, ed. D. A. Vakoch. Albany, NY: State University of New York Press.

Lintott, C. J., K. Schawinski, A. Slosar, K. Land, S. Bamford, D. Thomas, M. J. Raddick, R. C. Nichol, A. Szalay, D. Andreescu, P. Murray, and J. Vandenberg. 2008. Galaxy Zoo: Morphologies derived from visual inspection of galaxies from the Sloan Digital Sky Survey. *Monthly Notices of the Royal Astronomical Society* 389 (3): 1179–89.

Welch, J., D. Backer, L. Blitz, D. C.-J. Bock, G. C. Bower, C. Cheng, S. Croft, M. Dexter, G. Engargiola, E. Fields, J. Forster, C. Gutierrez-Kraybill, C. Heiles, T. Helfer, S. Jorgensen, G. Keating, J. Lugten, D. MacMahon, O. Milgrome, D. Thornton, L. Urry, J. van Leeuwen, D. Werthimer, P. H. Williams, M. Wright, J. Tarter, R. Ackermann, S. Atkinson, P. Backus, W. Barott, T. Bradford, M. Davis, D. DeBoer, J. Dreher, G. Harp, J. Jordan, T. Kilsdonk, T. Pierson, K. Randall, J. Ross, S. Shostak, M. Fleming, C. Cork, A. Vitouchkine, N. Wadefalk, and S. Weinreb. 2009. The Allen Telescope Array: The first widefield, panchromatic, snapshot radio camera for radio astronomy and SETI. *Proceedings of the IEEE, Special Issue: Advances in Radio Telescopes* 97: 1438–47.

The Effects of the Ionized Interstellar Medium on Broadband Signals of Extraterrestrial Origin

Samantha K. Blair, David G. Messerschmitt,
Jill C. Tarter, Gerald R. Harp

1.0. Introduction

The search for extraterrestrial intelligence is an ongoing endeavor. SETI has been searching for signals from other civilizations in our galaxy for five decades. Frank Drake spearheaded the effort by conducting the first search for extraterrestrial intelligence, Project Ozma, in the early 1960s using the NRAO facility in Green Bank, West Virginia. From this groundbreaking project has spawned a number of different searches for signals from extraterrestrials in both the radio and visible parts of the electromagnetic spectrum.

SETI Institute has traditionally searched for narrowband signals at low radio frequencies. The idea behind searching for very narrowband signals is that naturally occurring phenomena in space emit radiation that is "spread out" in frequency due to the natural emission processes so any signal that has very narrow linewidth would likely be artificially rendered, and also such signals can be efficiently detected by power spectrum estimates without detailed knowledge of the waveform. An example of a narrowband signal of technological origin is that of a human-made satellite in Earth's orbit. This search strategy certainly has merit and is founded on sound reasoning, but has thus far yielded no actual "detections." In light of this fact, it is of interest to extend the search to new classes of signals. The direction that modern radio and wireless communication has taken suggest including signals that are broadband and have much greater bandwidth than necessary (called spread spectrum).

The challenges posed in detecting these types of signals beg several questions. First, what are the advantages, if any, of using spread spectrum signals as a means to communicate with other civilizations? Secondly, what design challenges would the transmitting civilization face and how would those challenges constrain the type of signal that they could transmit? A third question is what impairments might the signal encounter in interstellar space as it propagates to Earth. This third question may actually help to elucidate an answer to the second question posed above. Answers to these questions can form a strategy for detecting these kinds of signals. The goal of this chapter is to begin to address some of the issues raised by these questions, and particularly question three, the impairments imposed by the ionized interstellar medium (IISM). Section 2 will address the motivation behind the choice of signal in extraterrestrial communication. In section 3, implications for SETI will be presented. The effects of the IISM on broadband signal propagation are discussed in section 4. The scattering model, along with the approach we take to understand potential impairments, is presented in section 5. Section 6 will conclude and identify needed future work.

2.0. Choice of Signal

The most fundamental limitation of detection of extraterrestrial radio signals of intelligent origin is white Gaussian noise (WGN) of natural origin in conjunction with large propagation losses over interstellar distances. Maximum detection sensitivity under these circumstances depends on knowledge of the signal, and thus SETI searches have assumed specific signal waveforms such as sinusoidal or pulse shapes. Optimum detection of a known signal in WGN consists of a correlation of reception and signal, and the resulting detection sensitivity depends on (1) the total signal energy and (2) noise power spectral density. This is true not only of "simple" signals such as pulses and sinusoids, but also arbitrarily complicated signal waveforms. Thus, while we need to know the shape of the signal waveform to build an optimum detector, that specific shape is irrelevant to the resulting detection sensitivity.

A signal of interstellar origin is, presumably, engineered to match a set of goals (e.g., maximum propagation distance to receiver), assumed resources available to the transmitter (e.g., available power) and receiver (e.g., collection area and computational facilities), and impairments (e.g., WGN and dispersion in the ionized interstellar medium (IISM)). If so, one of those impairments may be radio-frequency interference (RFI) at the receiver location, a growing challenge for our own SETI observations that may be appreciated by an advanced civilization. When a noise-optimum correlator is used, the shape of a known signal waveform does have considerable influence on susceptibility to different types of RFI.

A passband signal can be represented by a complex-valued baseband equivalent. The degrees of freedom (DOF) of the signal is defined as the number of complex numbers required to completely specify the signal waveform. For example, if a signal has bandwidth W and time duration T (this can only be approximated since a band-limited waveform is not strictly time-limited) then its DOF is WT, as can be observed in a Fourier series or sampling theorem expansion. If the signal waveform's energy is spread evenly over its DOFs (even for different basis functions), then it displays a robust immunity to interference; that is, interference is uniformly suppressed over a wide variety of interference waveforms. That uniform suppression also increases with large DOF. Such a signal waveform has characteristics resembling a burst of band-limited white noise. For example, a noise-like signal is relatively immune to pulse-like interference (since the signal is spread out in time) and carrier-like interference (since the signal is also spread out in frequency). For this reason, military communications (which suffer overt jamming) and modern wireless multiuser communication systems (which are dominated by self- and external interference) have increasingly utilized noise-like signal designs. This is called spread-spectrum signaling (Peterson, Ziemer, and Borth 1995).

3.0. Implications for SETI

Since noise-like signal waveforms are commonplace in terrestrial wireless designs, might they also be advantageous for interstellar communication? Like a pulse or sinusoid, a noise-like signal can be modulated by discrete data to convey information content. For example, transmitting one of two uncorrelated noise-like waveforms conveys one bit of information without a major effect on initial acquisition of the presence of a signal. As a basis for interstellar communication, this design has two major advantages. First, while a sinusoidal signal is highly susceptible to a sinusoidal interference (and likewise for pulse-like signals and pulse-like interference), a noise-like signal is immune to all kinds of interference, thus abetting observations in an interference-rich environment. Immunity is improved as WT increases, and bandwidth in particular is a plentiful resource in interstellar signal propagation. Second, by providing specific guidance on the shape of a signal waveform, design for interference provides an implicit form of coordination between transmitter and receiver. However, noise-like signal waveforms also introduce challenges. Having a large DOF also implies the need for the receiver to guess a large number of signal parameters. This can be addressed, as on earth, by algorithmic generation of the signal, giving the receiver the simpler task of guessing an algorithm. For example, the binary expansion of an irrational number such as π makes a good pseudorandom signal waveform generator (Tu and Fischbach 2005). Presumably, a SETI search for such signals would examine a set of alternative pseudorandom algorithms.

Another challenge is the frequency-dispersive and time-varying impairments arising in the IISM. Sinusoidal signals are relatively immune to dispersion and pulses are relatively immune to time-varying effects, but noise-like signals can be distorted by both types of propagation impairments. This is compounded by our inopportunity to accurately estimate these impairments prior to acquisition (although they may be estimated during an ongoing communication). Fortunately, the propagation impairments can be more easily and accurately estimated when the transmit signal waveform is known or assumed as compared to situations where the signal statistics are known but the waveform is not (such as natural sources like pulsars).

In summary, an advanced civilization seeking to attract our attention, unless it has at its disposal unlimited transmit power, is likely to engineer a signal for which the detection sensitivity is maximized. If its scientific and technological development is similar to our own, it presumably has comparable or more advanced knowledge of the IISM, and is also sensitive to the increasing RFI, but in view of large separation in both time and distance lacks knowledge of the specifics. If so, this may influence the signal design to assist with RFI immunity, and do this in the most robust manner possible, which is accomplished as described above. The choice of W and T is likely to be influenced by both knowledge of the ISM and (in the case of T) the strategy for a targeted search of different stars.

All these considerations suggest that we need to understand, for engineered as opposed to signals of natural origin (such as pulsars), the propagation impairments of the ISM and how they interact with other environmental factors (such as transmitter and receiver motion). Even the impact of the ISM on narrowband signals has not been fully characterized in light of modern astronomical observations of pulsars. In considering a wider class of possible signals, of particular interest is the frequency coherence (largest W for which frequency dispersive effects will be insignificant) and time coherence (largest T for which time-varying effects will be insignificant). Also of interest are statistical models of the IISM that allow detection strategies to incorporate knowledge of what specific phenomena are more likely than others. This understanding will shed further light on the characteristics of engineered signals that we should be searching for in our SETI observations.

4.0. Propagation Through the Ionized Interstellar Medium

The interstellar medium (ISM) is comprised of the gas and dust that reside in the region between the stars. The ISM is diverse and complex despite the fact that it is dominated by atomic and molecular hydrogen gas. Some of the space is filled with molecular clouds that range from Giant Molecular clouds that are tens of parsecs in size to small dense clumps that are fractions of a parsec in size. There are also ionized regions, the IISM, where the medium is a plasma of ionized hydrogen that can effect propagating radiation through electromagnetic interactions. These ionized regions include the Warm Ionized Medium, HII regions, and the Hot Ionized Medium, and have temperatures ranging from 8000 K to 10^7 K. It is these regions that have the greatest impact on signals being transmitted by extraterrestrial civilizations, as is evidenced by the dispersion of pulsar signals traveling through these areas.

Pulsar studies over the last several decades have made significant progress in understanding the dominant effects of the IISM on propagating signals: dispersion and scattering. Dispersion manifests itself as a frequency-dependent delay due to the interaction of electromagnetic radiation incident upon an ionized region of the ISM. The time delay is related to the dispersion measure, a quantity that is a function of the free electron density along the line of sight (LOS). At the frequencies used for radio communication through the ionosphere propagation through the ISM can be modeled by the cold plasma dispersion relation (Ramo, Whinnery, and Van Duzer 1993). Pulsar astronomers define the dispersion measure DM for convenience in applying this model

$$DM = \int_0^z n_e\, dz \tag{1}$$

$$t_{delay} = C \left(\frac{DM}{v^2} \right), \tag{2}$$

where DM is the dispersion measure, n_e is the electron density, dz is along the LOS (0 to z, distance to the source), v is the frequency, and C is a constant related to the mass and charge of an electron and the speed of light (Cordes and Sullivan 1995). Dispersion effects can be reversed at the receiver by de-dispersing the signal, and this is much easier for engineered signals than for signals of natural origin.

Scattering is a much more complicated phenomenon and results from the interaction of radio frequency radiation with turbulent IISM. Rickett (1990) describes the various effects of scattering on pulsar signals. The scale upon which scattering is based is the Fresnel scale which is defined as:

$$rf = \left(z\, \frac{\lambda}{2}\, \pi \right)^{\frac{1}{2}}, \tag{3}$$

where z is the distance to the source and λ is the wavelength. The phenomenon is quite different in two regimes: weak and strong (Rickett 1990). In weak scattering the intensity fluctuations in the trajectory of receiver motion due to a point source are relatively small, whereas in strong scattering the intensity fluctuations are relatively large. When scattering is caused by turbulence, the phase fluctuations are always large but may be very slowly varying. As radio waves are incident upon turbulent plasma, the electron density fluctuations cause the incident waves to be scattered into an "angular spectrum" of waves. Depending upon the fraction of the waves that are deviated, the scattering is weak or strong. In weak scattering, intensity fluctuations are cause by interference of scattered waves with the unscattered core, whereas in strong scattering intensity fluctuations are caused by interference of scattered waves with each other (Coles, Rickett, Gao, Hobbs, and Verbiest 2010). The strength of scattering can be calculated (loosely speaking) by integrating the turbulence spectrum along the propagation path. For this purpose, a parameter called the scattering measure (SM) is useful (Rickett 1990). If the phase spectrum of the ISM has the form

$$P(f) = \frac{C_n^2}{\kappa^{\frac{11}{3}}},$$
(4)

then SM is given by

$$SM = \int_0^z C_N^2 \, dz,$$
(5)

where C_N^2 is a measure of turbulence in the plasma, dz is the same as in (1), and κ is the wavenumber. For cases of strong scattering, two distinct processes contribute to the intensity variations: diffraction and refraction. The two contributions differ in both the scale (distance) and the time over which they are manifested. The spatial scales are related to the Fresnel scale by the following relationship:

$$r_f^2 = r_{diff} r_{ref}.$$
(6)

Diffraction results in shorter length and time scale intensity variations and refraction in longer length and time scales. In weak scattering, both diffractive and refractive scales approach the Fresnel scale, so the two regimes are indistinguishable (Rickett 1990). The parameter m_b^2 is used as a measure of the strength of scattering and can be calculated from SM as follows:

$$m_b^2 = 8\pi(f_\alpha) r_e^2 \lambda^2 SM \left(\frac{\lambda D_{eff}}{2\pi}\right)^{\frac{5}{6}},$$
(7)

where f_α is 1.12m=, r_e is the radius of an electron, λ is wavelength, SM is the scattering measure and D_{eff} is the effective distance which is 0.25* D.

Several studies on the effect of the ISM on narrowband extraterrestrial signals have been conducted. Cordes and Lazio (1991) describe a number of observable scattering phenomena that would affect monochromatic signal detection. Multipath diffractive scattering results in angular and temporal broadening and rapid intensity fluctuations, and refractive scattering causes long-term intensity fluctuations, angular wander, and multiple images. The authors mention spectral broadening, a well-known phenomenon that results from Doppler shifts due to the transverse velocity of the ISM with respect to the observer, which was originally considered by Drake and Helou (1977), but they suggest that it is a less important phenomenon for detecting SETI signals. They conclude that the rapid diffractive intensity scintillations are the most important consideration for SETI signals.

Building on the body of knowledge amassed by pulsar studies, we can use computer modeling to simulate the effects of the IISM on broadband signals. For distances less than 1 kpc we expect the scattering to be dominated by a single region in the line of sight. To this end, we have employed a two-dimensional version of the three-dimensional simulation code model created by Coles and colleagues to run simulations under various conditions (Coles, Filice, Frehlich, and Yadlowsky 1995).

5.0. Investigations Using Modeling

To gain insight into the effect of propagation of a broadband signal through the IISM, the use of a 2D scattering code has been most useful. The 2D scattering code simulates the interaction of a plane or spherical wave with an ionized region in the ISM (see Figure 6.1). The code creates a 2-D mesh upon which to perform the simulation that is divided into (nx, ny) sampling intervals (dx, dy). A random, thin phase screen with a Kolmogorov wavenumber spectrum is generated and a plane or spherical wave interacts with the phase screen and creates an angular spectrum of plane waves with phase variations. This is represented by the FFT of the wave plus phase variations due to the screen. The resulting angular spectrum of plane wave is then propagated to the plane of the observer using the Fresnel kernel. At the plane of the observer, the resulting electric field is calculated by the inverse FFT (Coles, Filice, Frehlich, and Yadlowsky 1995).

The simulation can be run with a number of frequency channels providing a "data-cube" in x, y, and f. The strength of scattering can be adjusted and is limited by memory and compute time to ($m_b^2 < 100$). The mesh size (nx, ny) must be changed to capture the entire spectrum of plane waves generated at the phase screen or the images become blocky. Once the resulting complex electric field has been generated, the data can be evaluated in a number of

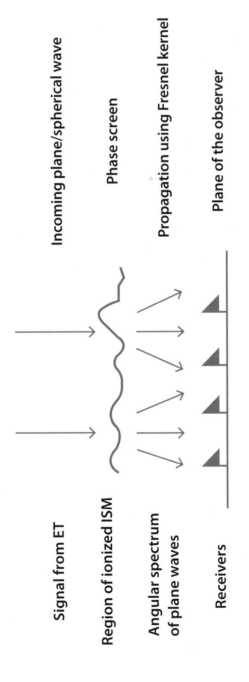

Signal from ET

Region of ionized ISM

Angular spectrum
of plane waves

Receivers

Incoming plane/spherical wave

Phase screen

Propagation using Fresnel kernel

Plane of the observer

Figure 6.1. Diagram of the path of incoming signal through the IISM.

ways to observe the intensity and phase variations of the electric field in the trajectory of the receiver. Dynamic spectra can be created at multiple frequencies by running the simulation for each frequency, extracting the center slice and stacking these slices on top of one another. By doing this one can examine each frequency channel spatially or temporally and estimate the spatial/temporal scales and the frequency scales.

A comparison of weak and strong scattering regimes can be seen in Figure 6.2. The phase changing screen is shown in the top panel and represents a mesh size of 2048 × 2048 samples separated by a distance of .01. The first column of panels, 1a and 1b, represent a strong scattering regime (m_b^2 = 10) and the second column, 2a and 2b, represent weak scattering (m_b^2 = 0.3). The intensity in 1a shows very small diffractive scales (magnitude ~22) and larger, clumpy refractive scales, while the intensity in 2a shows a single scale of the order of r_f. This is the case with weak scattering where the diffractive and refractive scales approach the Fresnel scale.

It is instructive to compare scattering at widely different wavelengths from the same scattering screen. To simulate this scenario, the m_b^2 must be calculated at each wavelength of interest. Simulations were run at 21 cm with an m_b^2 = 2 and at 3 cm with an m_b^2 = .007. Intensity and dynamic spectra were plotted for each case. Figure 6.3 shows the phase screen used for the simulation (Figure 6.3a), the intensity of the entire simulation (Figure 6.3b), and the dynamic spectrum (Figure 6.3c). Each plot has a color scale bar that indicates either phase change in radians for the phase screen or amplitude normalized to the input amplitude for the other 3 plots.

In the case of the 21 cm simulation, the r_f = 4.02 × 10^5 km for a source at 100 pc and a screen at 50 pc. The Fresnel scale is represented in 100 samples and r_{diff} = .56r_f and r_{ref}. The diffractive scale is seen as the bright spots (magnitude ~ 18) on Figure 6.3b and has a spatial scale of 2.25 × 10^5 km. Using v = 10 km/s for a potential ET source, the temporal scale for the diffractive regime is 375 minutes or 6.25 hours. In comparison, the same simulation run at an m_b^2 = 10 gives a diffractive scale of 8.64 × 10^4 km and a time scale of 2.40 hours.The 3 cm simulations were run using the same phase screen (3a) and m_b^2 = .007. All of the other parameters remained the same. In Figure 6.4, the results from this simulation are shown as intensity (Figure 6.4a) and dynamic spectrum (Figure 6.4b). This scattering regime is very weak, as can be seen from the single scales in the intensity plot and the overall correlation across the frequency grid. The relevant scale here is r_f which is 8.57 × 10^4 km and both r_{diff} and r_{ref} approach r_f. The weak scattering associated with the higher frequency 3 cm radiation creates a channel through which a greater percentage of the intensity reaches the plane of the observer. In other words, more of the undeviated waves along the LOS reach the receiver. From these simple simulations, it would be prudent for ET to

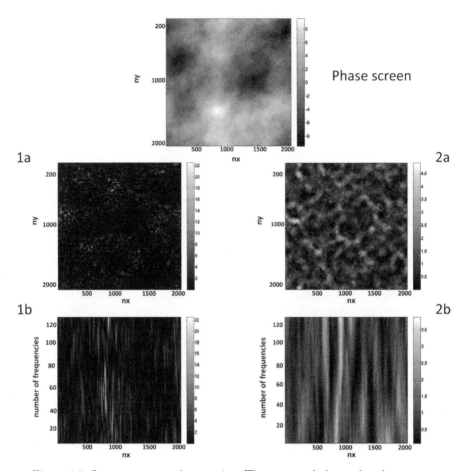

Figure 6.2. Strong versus weak scattering. The top panel shows the phase screen and the scale represents the phase change in radians. Panels 1a and 1b are intensity and dynamic spectrum at $m_b^2 = 10$. Panels 2a and 2b represent intensity and dynamic spectrum of weak scattering at $m_b^2 = 0.3$. Scales for the lower 4 panels represent amplitudes that are normalized to the input amplitude.

send their signals at higher carrier frequencies thereby greatly reducing the modifications introduced by passage through the IISM.

6.0. Conclusions and Future Work

This chapter begins the investigation of spread spectrum engineered signals for interstellar communication and the effect of the IISM on such signals.

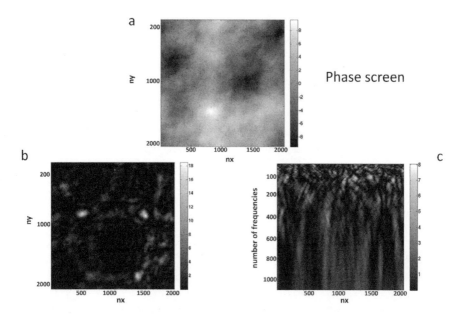

Figure 6.3. Scattering at 21 cm. The phase screen (a), the intensity (b), and dynamic spectrum (c) for the 21 cm line. For this simulation, m_b^2 = 2, nf = 1024, $nx \times ny$ = 2048 $\times$ 2048, and $dx = dy$ = 01. The dlam value is 1.2.

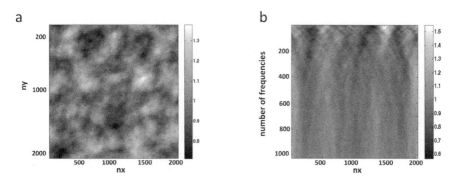

Figure 6.4. Scattering at 3 cm. The phase screen (3a) is the same. For this simulation, m_b^2 = .007, nf = 1024, $nx \times ny$ = 2048 $\times$ 2048, and $dx = dy$ = 01. The dlam value is 1.2. Panel a shows the intensity at each point on the grid. Panel b shows the dynamic spectrum.

Propagation through the IISM may alter these types of signals to an extent that detection sensitivity is reduced and/or detection computational requirements are increased. While the effects of the IISM on broadband signals may be an impediment, they can also offer insight into signal design and parameterization (such as bandwidth, time duration, and carrier frequency) tailored to accommodate these effects, and thus narrow the most favorable search space. To delineate these search dimensions, the IISM can be modeled to gain a better understanding of how a broadband signal would be altered spatially/temporally and in frequency. To this end, cases of strong and weak scattering have been simulated and compared for two wavelengths, 21 cm and 3 cm. The results indicate a strong case for using a higher carrier frequency. By characterizing the bandwidth and time duration over which the dispersive and time-varying medium is coherent, the target space for a search can be narrowed.

The next step in this study will be to look at how the phase varies at the plane of the observer to get a sense of frequency and time coherence. It is valuable to understand where in time and frequency the phase of the incoming signal is stable. One approach is to determine over what scales the phase values of the electric field at the plane of the receiver are statistically dependent (when these values become statistically independent, the phase is no longer coherent). Overall, progress is being made toward a greater understanding of how the IISM affects broadband in the context of interstellar communication.

Works Cited

Coles, W. A., J. P. Filice, R. G. Frehlich, and M. Yadlowsky. 1995. Simulation of wave propagation in three-dimensional random media. *Applied Optics* 34 (12): 2089–2101.

Coles, W. A., B. J. Rickett, J. J. Gao, G. Hobbs, and J. P. W. Verbiest. 2010. Scattering of pulsar radio emission by the interstellar plasma. *The Astrophysical Journal* (submitted).

Cordes, J. M., and T. J. Lazio. 1991. Interstellar scattering effects on the detection of narrow-band signals. *The Astrophysical Journal* 376: 123–34.

Cordes, J. M., and W. T. Sullivan III. 1995. Astrophysical coding: A new approach to SETI signals. I. Signal design and wave propagation. In *Progress in the Search for Extraterrestrial Life*, ed. G. Seth Shostak, Vol. 74.

Drake, F. D., and G. Helou. 1977. The optimum frequencies for interstellar communication as influenced by minimum bandwidth. NAIC Report (unpublished).

Peterson, R. L., R. E. Ziemer, and D. E. Borth. 1995. *Introduction to spread-spectrum communications*. New Jersey: Prentice-Hall.

Ramo, S., J. R. Whinnery, and T. Van Duzer. 1993. *Fields and waves in communication electronics.* New York: John Wiley and Sons.

Rickett, B. J. 1990. Radio propagation through the turbulent interstellar plasma. *Annual Review of Astronomy and Astrophysics* 28: 561–605.

Tu, S. J., and E. Fischbach. 2005. A study on the randomness of ϖ. *International Journal of Modern Physics C* 16(2): 281–94.

The Next Steps in SETI-ITALIA Science and Technology

Stelio Montebugnoli, Cristiano Cosmovici, Jader Monari,
Salvatore Pluchino, Giovanni Naldi, Marco Bartolini,
Andrea Orlati, Emma Salerno, Francesco Schillirò,
Giuseppe Pupillo, Federico Perini, Germano Bianchi,
Mattia Tani, Leonardo Amico

1.0. Introduction

The Italian Medicina Radioastronomy Station is located in the Northern part of Italy, near Bologna, and is composed of two radiotelescopes: the 32 m Very Long Baseline Interferometry (VLBI) dish and the Northern Cross, a large T-shaped parabolic/cylindrical antenna (30,000 m²) as shown in Figures 7.1 and 7.2 respectively. The Medicina dish presents about 800 m² of collecting area. It can work in the radio astronomical bands included in the 1.4–23 GHz range. It is equipped with a very high frequency resolution SERENDIP IV real-time spectrometer (0.6 Hz), developed at the University of California at Berkeley and operating since 1998. The function of this back-end is to operate in piggyback mode (namely, in parallel to ongoing astronomical observations), so that SETI observations can be performed twenty-four hours a day, 365 days a year at extremely low cost. In addition, the same back-end can monitor Radio Frequency Interference (RFI). Even though up to now 1420 MHz has been considered the best frequency for SETI observations, we believe that other frequencies should be investigated as well.

The Northern Cross is a 564 × 640 m T-shaped UHF array characterized by 30,000 m² of collecting area and a 4x4 arcmin beam cross-section. This is one of the larger collecting areas in the northern hemisphere, and it operates with a 2.7 MHz bandwidth (some receivers offer either 5 or 15 MHz bandwidth as well) centered at 408 MHz (λ = 73.5 cm). Due to its large collecting area, this array could be suitable to look for very weak signals from extraterrestrial intelligence.

Some parts of the Northern Cross have been refitted with new technologies and architectures to prepare UHF demonstrators for the Square Kilometre

Figure 7.1. Medicina 32 m VLBI dish.

Figure 7.2. A winter view of the Northern Cross Array.

Array (SKA; Montebugnoli, Bianchi, Bortolotti, Cattani, Maccaferri, Cremonini, Roma, Roda, and Zacchiroli 2006) and a 800 m² superstation for the Low Frequency Array (LOFAR; Zaroubi and Silk 2005), operating from 120 to 240 MHz, to perform some tests of long baseline interferometry at low frequency.

2.0. Science and Technology

At present the 32 m VLBI dish is scheduled with a wide variety of activities. One of these is the ITASEL (Italian Search for Extraterrestrial Life) program (Cosmovici, Montebugnoli, Maccone, Monari, and Flamini 2006), funded by the Italian Space Agency (ASI) and aimed at the search for new exoplanets. Following the detection of a water spectral line (22.35 GHz) due to a maser phenomenon that occurred during the SL9 comet impact on Jupiter in July 1994 (Cosmovici, Montebugnoli, Orfei, Pogrebenko, and Colom 1996), the ITASEL program attempts to detect water maser emissions from the high atmosphere of extrasolar planets. Up to now no evidence of the presence of water has been obtained; however, further investigations could lead to a successful result due to the new high sensitivity observational setup, consisting of a new multifeed low-noise receiver, along with a fast data acquisition and processing system.

The Northern Cross Array, due to its very large collecting area, is characterized by an extremely high sensitivity. This feature is particularly effective for detecting the radio emissions of some carbon isotopes (at approximately 408 MHz). Observations of the C252α (in CAS A) were already performed in 2000 in the framework of a collaboration between IRA and the Russian Academy of Science (Smirnov, Poppi, Cortiglioni, Montebugnoli, and Maccaferri 2001). These observations were particularly critical because of the weakness of the line and the very high power level of the background. This required a high dynamic range in both analog (receiver) and digital (high number of bits of the A/D converter) domains.

In order to be able to face crucial observations for both SETI and the detection of chemical elements for bioastronomical research, a new extremely fast and reconfigurable data acquisition and processing system was designed.

The core of the system, namely SPECTRA-1, is based on fast Field Programmable Gate Arrays (FPGAs) from Xilinx™ that exploit the flexibility of the software along with the velocity of the hardware. A block diagram of the system is shown in Figure 7.3. The input is composed by two A/D converters running at about 100 MS/sec with a high dynamic range (14 bits → 84dB). Data can both be processed in the time domain, if requested, and sent to a Polyphase Filter Bank (PFB). The filtered data can be squared and integrated in order to provide power spectra through the PCI bus (64 bit, 66

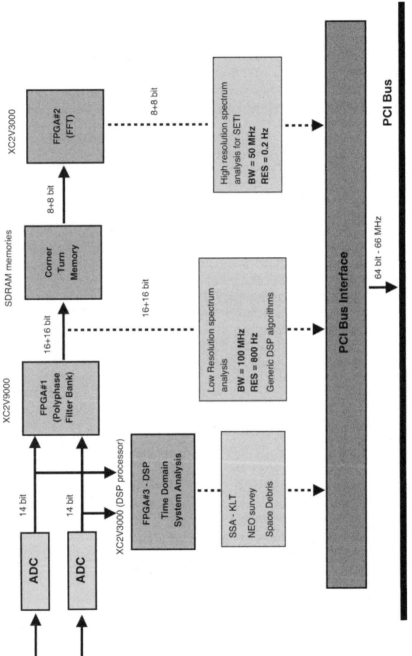

Figure 7.3. Schematic block diagram of the acquisition and processing system SPECTRA-1.

MHz). This way, a programmable spectrometer with a channel to channel rejection of more than 60 dB is obtained.

At the same time the output of the PFB fills up a Corner Turn Memory (CTM) with M spectra of N channels each. This way, implementing the FFT on each column, we obtain a spectrum composed by MxN channels. In the SPECTRA-1 board (Figure 7.4), a 64 million channel spectrometer for SETI observations can be obtained by filling up the CTM with 8K spectra at 8K channels each.

The SPECTRA-1 spectrometer (Figure 7.5) is conceived to operate (for SETI activities) in piggyback mode to the dish antenna activities, and it was planned to operate for more than three hundred days/year (the remainder time is taken by standard maintenance). Considering that such a spectrometer is connected to the 32 m VLBI dish, where a 16 MHz bandwidth is available (independently from the selected receiver), the frequency resolution is given by the 16MHz/64,000,000 = 0.25 Hz, while the time resolution is relatively

Figure 7.4. Photo of the SPECTRA-1 board.

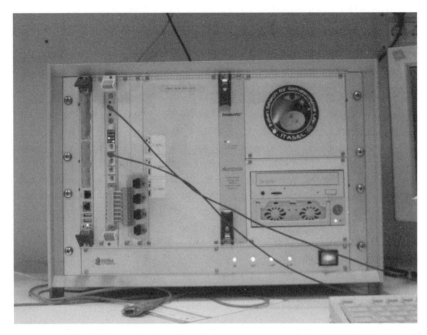

Figure 7.5. The PCI crate equipped by the host computer and the SPECTRA-1 board.

low (about four seconds). Since SPECTRA-1 has been programmed to provide data in the same format of the SERENDIP IV data, the existing post-processing and graphical display software can be used as well.

In order to check the sensitivity of the new spectrometer, a detection test was performed on an extremely weak signal—the carrier coming from the Voyager-1 launched in 1977 and now at 105 AU, or approximately 16,000,000 km away from the Earth. After several hours of integration, the signal was extracted from the noise (see Figure 7.6).

In order to facilitate data interpretation and to introduce alternative methods to search for possible extraterrestrial radio signals, the use of the large UHF Northern Cross transit telescope is considered as well. Sky observations, performed at least within one or two months, could provide for each day a number of submatrices labeled according to the observing start and stop sidereal time in which that portion of sky is transiting inside the beam. The whole set of submatrices will be characterized by an averaged spectrum on each row per each day per each transit interval of time. Keeping constant the transit antenna declination, a coherent signal coming from a definite position of the sky, would produce a "flag on" in the same submatrix

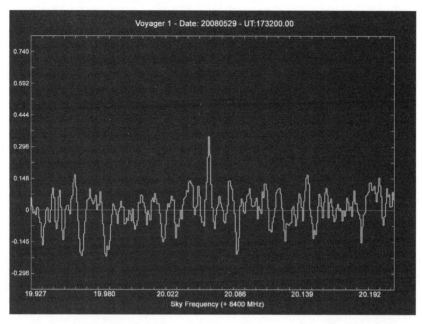

Figure 7.6. Weak signal from the Voyager-1 spacecraft, traveling from a distance of 105 AU from the Earth.

at the same sidereal time, that is, the same point in the sky (see Figure 7.7). If the extraterrestrial transmitter is at the position of the sky labeled 1, the flag does not switch on into the spectrum of the relative submatrix. When it is at position 2 (transit on the local meridian) it switches on and then off when in position 3. The same figure clearly shows the flag switched on every day while Radio Frequency Interference (RFI), highlighted by the white circles, is present in many subsequent spectra independently from the sky transiting through the beam. In addition, this detection could also be considered already "confirmed" since it comes from the same region of the sky and observed regularly for many days.

3.0. Data Processing

So far, the core of SETI data processing has used classical Fast Fourier Transform (FFT). However, the change from time to frequency domain algorithm is valid only if it is applied to linear and stationary signals. Investigations carried out at the Medicina labs proved that the Karhunen-Loève Transform (KLT) under certain conditions could work more efficiently

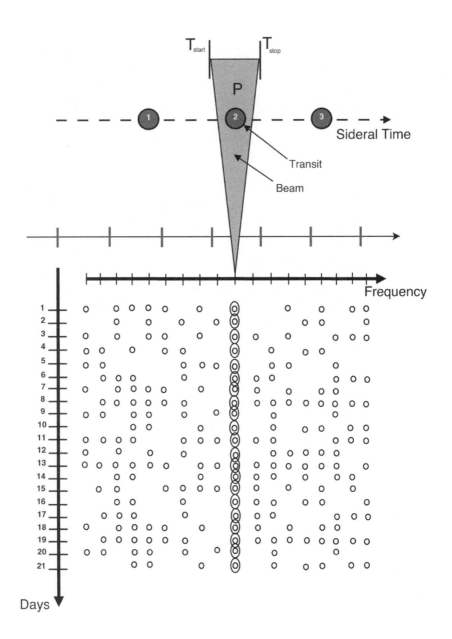

Figure 7.7. Subsubmatrix related to the transit of the "extraterrestrial transmitter" on the local meridian.

than the FFT since it extracts the orthonormal basic functions from the autocorrelation matrix. This way, the functions are always optimized, since they are extracted from the signal itself (Montebugnoli, Monari, Bortolotti, Cattani, Maccaferri, Poloni, Pari Orlati, Righini, Poppi, Roma, Teodorani, Caliendo, Maccone, Cosmovici, and D'Amico 2006). The KLT suffers from some limitations, mainly because it only operates properly with high signal-to-noise (S/N) ratios. In order to increase the "detection efficiency," an eigenspectrum evaluation from functional method is under development. In order to evaluate an eigenspectrum of a portion of signal to decide whether or not it contains a coherent signal, it is possible to evaluate a pseudospectrum that can approximate the expected eigenspectrum T.

$$
C = \begin{bmatrix}
R_{xx}(0) & R_{xx}(1) & \ldots & R_{xx}(m-1) & .. & 0 & R_{xx}(m-1) & .. & R_{xx}(1) & R_{xx}(0) \\
R_{xx}(1) & R_{xx}(0) & R_{xx}(1) & & .. & R_{xx}(m-1) & \ldots & 0 & R_{xx}(m-1) & \ldots & R_{xx}(1) \\
R_{xx}(2) & R_{xx}(1) & R_{xx}(0) & & & & & & & \\
& & & & & & & & & \\
0 & & \ldots & & 0 & & & & R_{xx}(1) & R_{xx}(0)
\end{bmatrix}
$$

It is possible to create a circulant matrix from the autocorrelation vector R of m values, filling this one with a certain number n-m of zeroes. The n eigenvalues of the circulant matrix are then approximated by the function:

$$f(x) = R(0) + 2\,R(1)\cos(x) + 2\,R(2)\cos(2x) + \ldots\ldots\ldots$$
$$+ 2\,R(m)\cos((m-1)x)$$

calculated by $x = j*pi/(n+1)$, for $j = 1,\ldots,n$. The more n is elevated, the more the function $f(x)$ is convergent to the eigenvalue of C, and the more the eigenspectrum C is similar to the original eigenspectrum T.

This is coherent with practice because the correlation function vector is constructed with time acquisition of samples. Therefore, the first elements of the autocorrelation function can be considered filled with a number of zeros. However, by using a low number of filling zeroes, it is possible to interpolate the function $f(x)$ and recover the T eigenspectrum profile, with a quite good approximation and a discrimination for high order eigenvalues, meaning that coherent signals are present.

In the case of nonlinear and nonstationary signals, as for instance, a very fast Doppler variation or a chirped signal, alternative transforms need to be investigated. Obviously, in this case it is not conceivable to integrate in time in order to increase the S/N ratio. Therefore, a single-shot very high efficiency transform needs to be investigated. The Institute of Radioastronomy received from NASA (Goddard Space flight Center, MD) the permission to evaluate, for a limited period of time, the Hilbert Huang Transform computation software within a Software Usage Agreement (SUA GSC14591).

Very preliminary results indicate that the HHT has good possibilities to be used in SETI observations under the above mentioned circumstances. Figure 7.8 shows the suitability of the HHT algorithm to handle nonlinear signals. In the time (X axis), frequency (Y axis) and energy (Z axis, indicated by shading tonality) space the information about the distribution is relatively clear. Further investigations are required to understand how the algorithm is suitable in low S/N ratio conditions. As previously mentioned, this is a crucial situation since nonlinear signals do not allow integration to obtain better detections.

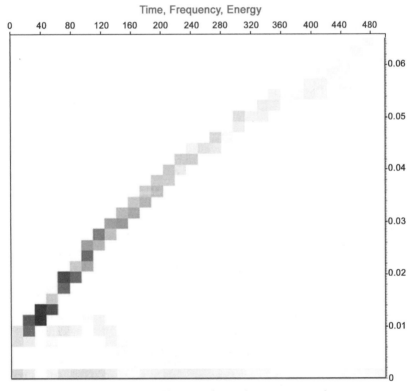

Figure 7.8. HHT output (time, frequency, energy).

4.0. Next Generation Data Processing System

The future development of the SETI-Italia program is based on the use of an extremely powerful programmable data processing platform. This represents the state of the art solution since it allows testing of different algorithms to extract signals from the noise.

A reduced version of such a flexible, powerful, and modular platform is under programming in the framework of the ITASEL and Space Debris programs (ASI) and of the Square Kilometer Array Design Study (SKADS) funded by the EU. The system was developed by the CASPER group at the University of California at Berkeley. It is composed if very fast A/D converters (8bit@ 1GS/sec), IBOBs serializer board, and a BEE2 processing board. Such a board is powered by a cluster of 5 FPGAs rated at 1 TOps/sec (Brodersen, Chang, Wawrzynek, Werthimer, and Wright 2004; Droz 2005). Figure 7.9 shows a block diagram of the cluster (several BEE2s can be parallelized), whereas Figure 7.10 shows the general view of this test bench at the Medicina radiotelescopes.

5.0. Conclusion

An overall review of the present and future SETI-Italia program has been presented. Generally speaking, future developments of this program represent a challenge for scientific and technological knowledge. Much attention must also be given to new observational methodologies, new algorithms for data processing, and extremely fast programmable electronics.

Note

This chapter is an adaptation of Montebugnoli, S., C. Cosmovici, J. Monari, S. Pluchino, L. Zoni, M. Bartolini, A. Orlati, E. Salerno, F. Schillirò, G. Pupillo, F. Perini, G. Bianchi, M. Tani, and L. Amico. 2010. The next steps in SETI-Italia science and technology. *Acta Astronautica*.

Works Cited

Brodersen, B., C. Chang, J. Wawrzynek, D. Werthimer, and M. Wright. 2004. BEE2: A multi-purpose computing platform for radio telescope digital signal processing applications. Paper presented at the Square Kilometre (SKA)/Berkeley Wireless Research Center Meeting.

Droz. P.-Y. 2005. *Physical design and implementation of BEE2: A high-end reconfigurable computer.* MS thesis, Department of Electrical Engineering and Computer Sciences, University of California, Berkeley.

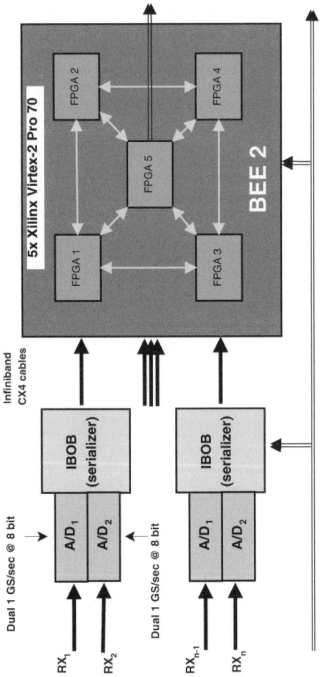

Figure 7.9. Block diagram of a single BEE2 FPGAs cluster.

Figure 7.10. The system is visible in the rack on the right side.

Cosmovici, C. B., S. Montebugnoli, C. Maccone, J. Monari, and E. Flamini. 2006. *The ITASEL Project (Italian Search for Extraterrestrial Life)*. Paper presented at the 57th International Astronautical Congress, Valencia, Spain.

Cosmovici, C. B., S. Montebugnoli, A. Orfei, S. J. Pogrebenko, and P. Colom. 1996. First evidence of planetary water maser emission induced by the comet/Jupiter catastrophic impact. *Planetary and Space Science* 44 (8): 735–39.

Montebugnoli, S., G. Bianchi, C. Bortolotti, A. Cattani, A. Maccaferri, A. Cremonini, M. Roma, J. Roda, and P. Zacchiroli. 2006. Italian SKA test bed based on cylindrical antennas. *Astronomische Nachrichten* 327 (5–6): 624–25.

Montebugnoli, S. J. Monari, C. Bortolotti, A. Cattani, A. Maccaferri, M. Poloni, P. P. Pari, A. Orlati, S. Righini, S. Poppi, M. Roma, M. Teodorani, D. Caliendo, C. Maccone, C. B. Cosmovici, and N. D'Amico. 2006. SETI-Italia 2003 status report and first results of a KL Transform algorithm for ETI signal detection. *Acta Astronautica* 58: 222–29.

Smirnov, G. T., S. Poppi, S. Cortiglioni, S. Montebugnoli, and G. Maccaferri, 2001. Search for radio recombination lines at 408 MHz with the Northern Cross. *Astronomical and Astrophysical Transactions* 20 (2): 203–206.

Zaroubi, S., and J. Silk. 2005. LOFAR as a probe of the sources of cosmological reionization. *Monthly Notices of the Royal Astronomical Society: Letters* 360 (1): L64-L64.

Project SAZANKA: Multisite and Multifrequency Simultaneous SETI Observations in Japan

*Shin-ya Narusawa, Mitsumi Fujishita, Hiroki Akisawa,
Kenta Fujisawa, Yasuhide Fujita, Takahiro Fukuzumi,
Hiromi Funakoshi, Hiroyuki Geshiro, Hideo Hara,
Kenji Hashimoto, Tsutomu Hayamizu, Ryo Iizuka,
Kazumasa Imai, Takeshi Inoue, Masayuki Kagami,
Kazuhisa Kageyama, Takeshi Kamitamari,
Masahiro Koishikawa, Shouta Maeno, Hidehiko Matsuo,
Takashi Miyamoto, Masaki Morimoto, Hiroyuki Naito,
Sumio Nakane, Takeshi Nakashima, Masami Okyudo,
Takaaki Oribe, Takaaki Ozeki, Makoto Sakamoto, Yasuo Sano,
Naoko Sato, Masayuki Tachikawa, Yoshimasa Tai,
Setsuro Takahara, Yoshitaka Takahashi, Mikimasa Takeuchi,
Naoto Tatsumi, Akihiko Tomita, Shinji Toyomasu,
Naoki Toyoshima, Makoto Watanabe, Takeshi Yada,
Ryoji Yamada, Michinari Yamamoto, Hideyo Yokotsuka*

1.0. SETI in Japan

We carried out the world's first multisite and multifrequency simultaneous SETI observations with fourteen radio (22.0 MHz–8.3 GHz) and twenty-seven optical telescopes in Japan. We searched in the direction of the Cassiopeia region (R.A. = 03h07m, Dec. = +58d02m), following Project META (Megachannel Extra Terrestrial Assay), with monitoring taking place on November 11 and 12, 2009, UT 12:00–17:00. No radio signal above noise level was recorded. However, the first test of this network of multisite and multifrequency simultaneous SETI observations was successful. We begin this chapter by briefly introducing SETI observations that have been carried out in Japan.

1.1. IR SETI

On December 15, 1991, J. Jugaku and K. Noguchi conducted the first Japanese SETI observations. They searched for partial Dyson spheres with the 1.3 m Infrared telescope of the Institute of Space and Astronautical Science (Jugaku et al. 1995).

1.2. Radio SETI

One of the authors (M. Fujishita) and his students observed Beta Gem for seventy-five minutes with two antennas of the Solar-Terrestrial environment Laboratory of Nagoya University at 327 MHz from December 16 to 20, 1999 (Fujishita et al. 2006). Moreover, they used the 10 m single dish of Mizusawa Station of National Astronomical Observatory, Japan, from March 1 to 5, 2005. The observed frequency is 8.4 GHz. Targets were the location of the "Wow!" signal, the META region, and candidates of partial Dyson spheres (Fujishita et al. 2006).

1.3. Optical SETI

Another author (S. Narusawa) has continued Optical SETI (OSETI) observations for fifty-six nights with the 2 m telescope "NAYUTA" of the Nishi-Harima Astronomical Observatory(NHAO) since 2005 (Narusawa and Morimoto 2007). The NAYUTA is the largest optical telescope in Japan (Fig. 8.1). This OSETI project uses a spectroscopic method based on Reines

Figure 8.1. The 2 m NAYUTA telescope of the Nishi-Harima Astronomical Observatory.

and Marcy (2002). The detection limit of lasers is 10 peta W. We selected as target stars whose planets would plausibly be in the Habitable Zone.

2.0. Simultaneous Observations in 2005

During the radio observations at Mizusawa in 2005 mentioned above, simultaneous observations were scheduled with the NAYUTA equipped with a CCD Camera (SBIG/ST9). The purpose of these radio-optical simultaneous observations is the same as described in the following section. The view of NAYUTA CCD system (3.5 * 3.5 arcminutes) is smaller than that of the 10 m antenna (FPBW = 13 arcminutes) for the target region. Therefore, the target field of the NAYUTA had to be changed to the view of the 10 m antenna during the simultaneous observations. Because of the bad weather, the NAYUTA could observe for only seventy minutes on the first night. During the simultaneous observations, four small radio fluctuations over estimated noise level were recorded. Unfortunately, when these events were detected, the NAYUTA telescope was just moving. Therefore, an optical event was not recognized during this period. This project demonstrated the feasibility of multisite simultaneous observations for SETI.

3.0. Purpose and Method

3.1. Main Purpose

In the near future, we want to join worldwide SETI observations. Therefore, we organized the multisite and multifrequency simultaneous SETI observations in Japan as a first step toward entering a worldwide network.

3.2. Radio Observations

In order to identify a signal of extraterrestrial origin, we employed multisite radio observations. Multifrequency and multimethod radio observations were adopted to help distinguish between natural phenomena and extraterrestrial intelligence (ETI).

It is often argued that a narrow carrier signal is the most likely signal to be received from ETI, because it requires only one parameter: frequency. However, we have not discovered any clear signals from ETI in these fifty years since Project OZMA, based on this assumption, was carried out in 1960. We suggest that ETI may use other types of signals to inform us of their existence (Fujishita et al. 2006). For example, Ultra-Wide Band (UWB) signals are generally advantageous from the standpoint of signal to noise ratio, as shown by the Global Positioning System. Based on a similar idea, we carried out multisite and multifrequency radio SETI observations.

3.3. Optical Monitoring

We scheduled the multisite optical monitoring observations to occur simultaneously. If radio antennas had detected candidate signals, we would have checked the optical images (CCD, Digital camera, and Video) from the simultaneous observations. These optical observations may help clarify the nature of radio events, whether natural phenomena (e.g., flares of the surface of the star), artificial signals of human origin (e.g., military satellites), or ETI. Two specialists (R. Iizuka and H. Naito) joined this project as persons in charge of artificial satellite identification and new object identification, respectively.

There are more than three hundred optical observatories in Japan and about eighty of them are part of the Japan Public Observatory Society (JAPOS). We invited JAPOS to participate in this project. As a result, more than twenty optical observatories participated (Table 8.1). The Distribution of radio and optical observatories is shown in Fig. 8.2. No such large-scale SETI observations have been carried out elsewhere in the world.

3.4. Operation

The simultaneous SETI observations were organized by a working group (S. Narusawa, M. Fujishita, T. Inoue, and M. Morimoto) of this project as follows:

1. The observational schedule was announced to the astronomical community, and interested observers were able to participate freely.

2. Data are owned by each observer, as agreed upon prior to observations.

3. If radio observers detected one or more candidate signals, then all optical observers would be informed about the time of receipt though the radio coordinator (M. Fujishita) and the Headquarters (S. Narusawa) of the project. After checking optical data (video or CCD images), results would be disseminated to all members through the headquarters. A flow chart of the network of this project is shown in Fig. 8.3. This process was also clearly established prior to observations.

4. Project members used e-mail for communication. M.Tasumi served as administrator of the mailing list.

5. Before observations, we informed news media about the project.

Table 8.1. Optical Monitoring Systems of Project SAZANKA

Observatory (location)	Diameter of telescope [cm]	Instrument (*)
Kihara (Nayoro, Hokkaido)	28	C
Sendai (Sendai, Miyagi)	130	C
Jyoudodaira (Fukushima City)	40	V and
Jyoudodaira (Fukushima City)	25	C
Toyama (Toyama City)	100	V
Anpachi (Anpachi, Gifu)	70	V
Kawabe (Hidakagawa, Wakayama)	100	V
Wakayama University (Wakayama City)	60	C
Misato (Kimino, Wakayama)	(f180mm lens)	V
Nishi-Harima (Sayo, Hyogo)	200	V (Hi-Vision)
Hoshinoko-Yakata (Himeji, Hyogo)	15	C
Kakogawa City Youth Outdoor Center (Kakogowa, Hyogo)	10	V
Nishiwaki Earth Science Museum (Nishiwaki, Hyogo)	81	V
BALLOON YOKA (Yabu, Hyogo)	40	V
Yamamoto (Mimasaka, Okayama)	10	D
Ryuten (Akaiwa, Okayama)	40	V
Saji (Tottori City)	103	C
The Shimane Nature Museum of Mt. Sanbe; Sahimel (Ooda, Simane)	60	C
Kuma Kogen (Kumakogen, Ehime)	60	C
Minami-Aso Luna (Minamiaso, Kumamoto)		82 V
Sakamoto-Hachiryu (Yatsushiro, Kumamoto)	30	D
Yamada (Hikawa, Kumamoto)	20	D and
Yamada (Hikawa, Kumamoto)	10.2	D
Tachikawa (Kumamoto City)	15	D
Kageyama (Kumamoto City)	25	V
Sendai Space Hall (Satsumasendai, Kagoshima)	50 or 28	V
TOKARA Nakanoshima (Toshima, Kagoshima)	60	V

* V:Video; C: CCD camera; D: Digital camera

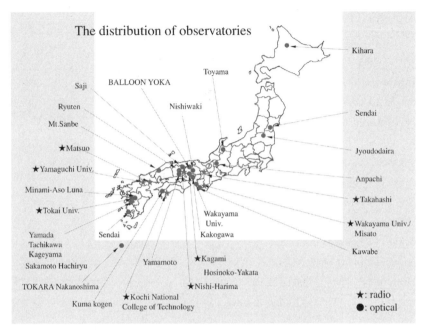

Figure 8.2. The distribution of observatories.

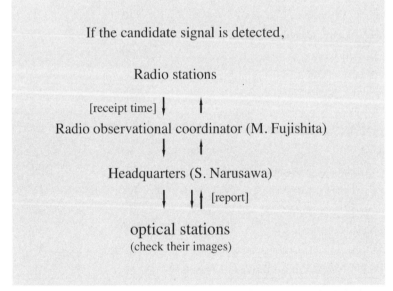

Figure 8.3. A flowchart of the network of Project SAZANKA.

4.0. The Post Detection Guidelines

4.1. The Post Detection Protocol of the International Academy of Astronautics (IAA)

If we had detected a candidate signal, then we would have observed the post detection SETI protocol of the International Academy of Astronautics (IAA). The second clause of this protocol notes that "[t]he discoverer should inform his/her or its relevant national authorities." Therefore, we discussed who or what the "relevant national authorities in Japan" would be at a SETI workshop held at NHAO on November 4, 2007. Before observations we decided these authorities would be:

1. Cabinet Office, Government of Japan, and if necessary:
   i. Ministry of Internal Affairs and Communications
   ii. Foreign Ministry
   iii. Ministry of Education, Culture, Sports, Science, and Technology
   iv. National Public Safety Commission

2. National Astronomical Observatory of Japan

3. Japan National Committee for Astronomy, Science Council of Japan

4.2. The Manual for Communicating with News Media

Almost all Japanese, including the press, are unaware of SETI. Therefore, before we announced the project to news media, we needed to make a manual of how we should communicate with the media, which was distributed to all observatories. The manual also includes procedures in case we found a candidate signal from ETI while being covered by the media.

5.0. Rehearsal Observations

Rehearsal observations of simultaneous SETI were carried out March 28, 2009, UT12:00–17:00, mainly to check the network management. We selected 55 Cnc for a target because one of the five planets of this solar-type star orbits in the Habitable Zone (Fischer et al. 2008). Five antennas of four radio stations (Wakayama University/Misato, Yamaguchi Univrsity, Agawa Jovian Radio Observatory, and NHAO) participated in these observations. Due to the weather conditions, we could observe with only seven optical telescopes (Anpachi, Misato, Kakogowa, Hoshinoko-Yakata, Ryuten, Mt.

Sanbe, and NHAO). During this rehearsal, the OSETI observations were carried out with NAYUTA, and photometric observations were performed with a 15cm telescope of Hoshinoko-Yakata Astronomical Observatory. Network management was successful; however, no radio or optical candidate signals were detected.

6.0. Project SAZANKA

6.1. Observations

Multisite and multifrequency simultaneous SETI observations in Japan were carried out November 11 and 12, 2009, UT 12:00–17:00. We selected these days based on their convenience for participating observatories. In addition to these observations, we used twenty-seven optical systems, including NAYUTA, to monitor the same region at the same time. We named this project "SAZANKA." Sazanka (scientific name *Camellia sasanqua*) is a flower that blossoms during the observing period of these observations.

6.2. Radio Observatories

We used a total of fourteen antennas at eight radio observatories for this project as follows:

1. Yamaguchi University
   frequency: 8302 MHz (BW: 4 MHz)
   polarization: simultaneous observation of right- and
      lefthanded circular polarization
   antenna: 32 m single dish (Fig. 8.4)
   method: spectroscopic observation with FFT
   frequency resolution: 61 Hz
2. Tokai University Space Information Center (TSIC)
   (antenna 1)
   frequency: 8302 MHz (BW: 4 MHz) (agree with
      Yamaguchi University for confirmation)
   polarization: right-handed circular polarization
   antenna: 11 m single dish (Fig. 8.5)
   method: spectroscopic observation with spectrum
      analyzer and total power observation
   frequency resolution: 10 KHz
   (antenna 2)
   antenna: 5 m single dish (Fig. 8.6)
   parameters are same as an 11 m dish

Figure 8.4. The 32 m single dish of Yamaguchi University.

Figure 8.5. The 11 m single dish of Tokai University.

Figure 8.6. The 5 m single dish of Tokai University.

3. Wakayama University/Misato Observatory
   frequency: 1420 MHz (BW: 5 MHz, not including H I
   emission)
   polarization: linear polarization
   antenna: 8 m single dish (Fig. 8.7)
   method: spectroscopic observation with spectrum analyzer
   frequency resolution: 15.625 KHz
4. Takahashi Radio Station
   frequency: 1420 MHz (BW: 0.8 MHz)
   polarization: lefthanded circular polarization
   antenna: 6 m single dish (Fig. 8.8)
   method: spectroscopic observation with FFT

Figure 8.7. The 8 m single dish of Wakayama University/Misato Observatory.

Figure 8.8. The 6 m single dish of Takahashi Observatory.

5. Matsuo Radio Station
   (antenna 1)
      frequency: 1420 MHz (BW: 2.6 KHz)
      polarization: horizontal linear polarization
      antenna: 1 m single dish
      method: spectroscopic observation, capturing one
        spectroscopic image every minute
   (antenna 2)
      frequency: 38.2 MHz (BW: 2.2 KHz)
      polarization: horizontal linear polarization
      antenna: 1/2 l dipole
      method: spectroscopic observation, capturing one
        spectroscopic image every minute

6. Kagami Radio Observatory
   (antenna 1)
      frequency: 1420 MHz (BW: 2MHz)
      polarization: horizontal linear polarization
      antenna: Yagi with fourteen elements
   (antenna 2)
      frequency: 1420 MHz
      polarization: perpendicular linear polarization
      antenna: discone (for measurement of background level)
   (antenna 3)
      frequency: 31.086 MHz (1st night), 32.780 MHz (2nd
        night)
      polarization: perpendicular linear polarization
      antenna: discone
   (antenna 4)
      frequency: 31.086 MHz (1st night), 32.780 MHz (2nd
        night)
      polarization: horizontal linear polarization
      antenna: 1/2 l dipole (2nd night only)

7. Nishi-Harima Astronomical Observatory (NHAO)
   (antenna 1)
      frequency: 38.0 MHz
      polarization: horizontal linear polarization
      antenna: dipole
   (antenna 2)
      frequency: 22.0 MHz
      polarization: horizontal linear polarization
      antenna: 3 elements Yagi

8. Agawa Jovian Radio Observatory of Kochi National College of Technology
   frequency: 32.5 MHz (BW: 5 MHz)
   polarization: righthanded circular polarization
   antenna: log periodic antenna with nine orthographic elements (Fig. 8.9)

6.3. Target Region and Estimation

Horowitz and Sagan (1993) carried out Project META (Megachannel Extra Terrestrial Assay) with the Harvard/Smithsonian 26 m radio telescope. A strong line signal (seven times the threshold power) was detected at 1420 MHz on November 17, 1989, from the direction of the constellation Cassiopeia. An antenna's beam is thirty arcminutes at 1420 MHz. We selected the same field for the target of Project SAZANKA (Fig. 8.10 and Fig. 8.11). We observed the following coordinates as the central position:

R.A. (J2000): 03h07m
Dec. (J2000): +58d02'.

Figure 8.9. The log periodic antenna of Agawa Jovian Radio Observatory of Kochi National College of Technology.

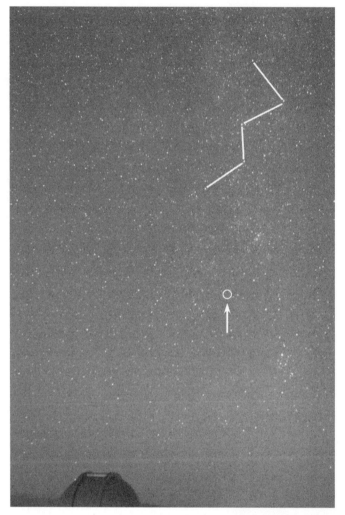

Figure 8.10. The target region (circle). (Photo: Nishi-Harima Astronomical Observatory).

The threshold power of the strongest signal for META observation was 0.8–15 * 10^16 [W] assuming that the signal source was located one thousand light years away (Horowitz and Sagan 1993). Therefore, if our observation is carried out under conditions equivalent to META, and if we find a signal source within one hundred light years, we could detect the effective signal with a 1 m dish.

Figure 8.11. The image of the target region (32.5 * 32.5 arc-minutes). (Photo: Sendai Astronomical Observatory).

6.4. Results and Summary

During the observing run, we were able to obtain effective data through all radio observatories. The analysis of the obtained data were carried out by the respective observatories. There was no signal above 2.82 Jy (the upper limit of flux density) with the 32 m dish of Yamaguchi University. No event was seen simultaneously at the two antennas in TSIC. A total of six signals (4–5s) were detected at 1.4224 GHz at Wakayama University/Misato Observatory on the first day. However, these signals were also detected in the "off" position. No radio signal above the background was detected at Agawa Jovian Radio Observatory of Kochi National College of Technology. Finally, no valid event above the noise level was recorded at any other observatory. In sum, we did not find candidate signals from an extraterrestrial civilization during the two-day project.

Due to the weather condition, we could observe with only eight (Anpachi, Hoshinoko-Yakata, Sakamoto-Hachiryu, Kageyama, Yamada, Minami-Aso Luna, and NHAO) and only seven (Kihara, Toyama, Saji, Mt. Sanbe, Sakamoto-Hachiryu, Kageyama, and Yamada) optical observatories on the first and second nights, respectively. It was not necessary to check optical images formally, because no candidate radio signals were detected. One of optical observers (K. Kageyama) replayed video images obtained independently and found two moving objects. They were identified as Russian rocket debris (Molniya/A-2-e/SL-6 and proton/SL-12) by a person in charge of artificial satellite identification. As mentioned above, these two objects did not have an influence on radio observations.

In summary, simultaneous SETI observations were carried out with fourteen radio and twenty-seven optical telescopes in Japan on November 11 and 12, 2009. No such large-scale SETI observations have been carried out elsewhere in the world. No candidate radio signals from ETI were recorded. However, we believe our operation of the network of multisite and multifrequency simultaneous observations was successful. Now we are willing to join in worldwide SETI observations.

Acknowledgments

The authors thank Mieko Sasaki, Nanto Yusei, Toshihiro Omodaka, Hiroshi Imai, Koitiro Maeda, and the staff of all participating observatories for their help with observations.

Works Cited

Fischer, D. A., G. W. Marcy, R. P. Butler, S. S. Vogt, G. Laughlin, G. W. Henry, D. Abouav, K. M. G. Peek, J. T. Wright, J. A. Johnson, C. McCarthy, and H. Isaacson. 2008. Five planets orbiting 55 Cancri. *The Astrophysical Journal.* 675: 790–801.

Fujishita, M., S. Narusawa, M. Fujishita, and T. Kawase. 2006. SETI activities at Kyushu Tokai University. *Journal of the British Interplanetary Society.* 59: 346–48.

Horowitz, P., and C. Sagan. 1993. Five years of project META: An all-sky narrow-band radio search for extraterrestrial signals. *The Astrophysical Journal.* 415: 218–35.

Jugaku, J., K. Noguchi, and S. Nishimura. 1995. A search for Dyson spheres around late-type stars in the solar neighborhood. In *Progress in the search for extraterrestrial life, Astronomical Society of the Pacific Conference Series,* ed. G. Seth Shostak, 74: 381–85.

Narusawa, S., and M. Morimoto. 2007. Optical SETI observations with the NAYUTA telescope. *Annual Report of the Nishi-Harima Astronomical Observatory* 17:1–3.

Reines, A. E., and G. W. Marcy. 2002. Optical search for Extraterrestrial Intelligence: A spectroscopic search for laser emission from nearby stars. *Publications of the Astronomical Society of the Pacific.* 114: 416–26.

Harvard's Advanced All-sky Optical SETI

Curtis Mead and Paul Horowitz

1.0. Introduction

Since 1961, when the possibility of using lasers for communicating between the stars was first suggested (Schwartz and Townes 1961), laser technology has made rapid advancement. In fact, given current technological capabilities, a pulsed laser powerful enough to be used in a communication link spanning hundreds to thousands of light years could be built by a government entity or collaboration of universities at a pricetag on the order of a couple of billion dollars (Howard et al. 2004). If there are any civilizations in our galaxy at least as advanced as we, might they attempt to contact other similarly advanced civilizations with powerful laser beams? In the past two decades, a handful of surveys have explored this possibility by building optical receivers for pulsed laser beacons (Coldwell 2002; Kingsley 1996; Wright et al. 2001; Lampton 2000). Most of these surveys look at one star at a time for tens of minutes, and are limited in how many stars they can observe by their small fields of view and small number of light-sensing elements. Harvard's All-sky Optical SETI survey aims to search the entire sky of the northern hemisphere for pulsed laser beacons by using a much larger $1.6° \times 0.8°$ field of view and 512 pixel pairs. This chapter describes the recent efforts to retrofit the All-sky camera with a new electronics package to vastly increase the camera's capabilities.

1.1. What is All-sky Optical SETI?

Harvard's All-sky instrument consists of a 1.8 meter primary quasi-newtonian telescope—a light bucket—and a camera able to detect very short flashes of light. It is a transit instrument, meaning the telescope only moves in elevation. Over the course of a night, as the earth turns, the stars sweep across the imaging plane. The camera and telescope have a fairly wide field of view, totaling about 150 percent of the full moon's area, split up into 512 pixel pairs. The light sensing elements are 64 pixel Hamamatsu photomultiplier tubes (PMTs). Sixteen Hamamatsu PMTs are divided onto two identical imaging planes using a beamsplitter with a coincident pulse on matching pixels triggering data collection. Attached to the light-sensing front end are custom circuit boards that digitize the analog PMT outputs at 600 million samples per second (Msps) on all 512 pixel pairs in real-time looking for pulses—a flash light. A distant source of bright flashes of light is the signature of one possible mode of communication over interstellar distances, namely, using a powerful pulsed optical laser as a transmitter. The entire Harvard All-sky system is tucked into a 30 ft by 15 ft roll-off roof building in Harvard, Massachusetts.

2.0. Events

Four years of observations have netted 1454 bright flash events during 3226 observing hours. The recorded data includes the position of the telescope, state of all observatory and camera electronics, time of the event, individual pixel or pixels that were "triggered" (recorded the event), and the digitally sampled photomultiplier tube output waveform for each triggered pixel. The data set of all events is analyzed by software to compile event statistics, and each event is looked at individually by a human to determine if it is consistent with a distant laser pulse. To be a possible ETI detection, the event must have the following characteristics:

1. Triggered pixels must be adjacent and on the same photomultiplier tube.

2. Waveforms of the pixel pairs much be matched to within one analog-to-digital conversion (ADC) level.

3. Neither pixel pair waveform can exhibit reversed polarity, the signature of electrical breakdown or other anomalous noise.

4. If multiple pixels are triggered, the trigger timing must match to within 100 nanoseconds.

Events that don't obey these rules have a variety of sources—random photon pileup from bright sky background (e.g., from the moon or clouds), bright stars, airplane strobe lights, satellites, or cosmic ray induced Cherenkov flashes. It is possible, in some instances, for background sources to mimic distant laser pulses. Section 2.1 gives one example how of this can happen.

2.1. BIC6

The event labeled BIC6 is representative of one type of event that is typically received, defined by two or more nonadjacent pixels crossing the third voltage threshold (~6 photoelectrons) being triggered in less than ten nanoseconds. BIC6 was captured by three nonadjacent pixels on the same PMT. The sampled PMT pixel output data indicates a peak pulse height of ~17 photoelectrons in each pixel in a ten nanosecond window. This is one of the brightest flashes observed to date. The pixel nonadjacency indicates that this flash of light did not come from a distant point source, but is more likely an atmospheric event. The angular extent of the image is at least seven arcminutes. The most logical explanation is that BIC6 is the Cherenkov light from a cosmic ray–induced extensive air shower.

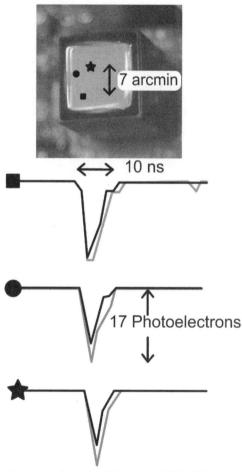

Figure 9.1. Three pixels were "triggered" by event BIC6. The placement of the pixels (square, circle, and star) is overlaid on an image of the front of one PMT. Raw waveform data from the three pixels is shown, dark and light traces representing the two matched pixels in a pair.

2.2. Extensive Air Showers

Cosmic ray–induced extensive air showers are produced when a fast-moving particle, most commonly a proton (Zombeck 2007, 310), enters the Earth's atmosphere and sheds its energy by interacting with atmospheric gases. These high energy collisions sap energy from the primary by producing hadronic secondary particles such as pions, kaons, and neutrons (Rao and Sreekantan 1998). The secondary particles share the energy lost in the collision and go on to engage in their own high energy collisions or decay and turn into muons and gamma rays. Collisions brought on by the high energy secondary particles produce yet more hadrons. This cascade process continues with exponentially more collisions and decays and results in an avalanche of particles and radiation plunging toward the ground along the same direction as the original cosmic ray. Each high speed charged particle above a certain energy threshold in an extensive air shower creates electromagnetic Cherenkov radiation. The radiation manifests itself as a cone of visible light, peaked in the blue, with an opening angle determined by the index of refraction of the medium in which the charged particle is traveling (2.6° for air at sea level.) Cherenkov radiation from extensive air showers can be detected by telescopes on the ground, and forms the basis of VHE (Very High Energy) Gamma Ray Astronomy. The Cherenkov light from a cosmic ray will appear as a short flash of light arriving at the telescope (Weekes 2003). Cosmic rays are constantly buffeting the Earth, and due to interstellar magnetic fields, they arrive isotropically from all directions (Ibid., 25).

2.3. Cosmic Ray Simulations

Because cosmic ray–induced extensive air showers can contribute to unwanted triggers such as BIC6, they are an important background source for pulsed optical SETI surveys. To aid in the understanding of how they trigger the All-sky camera and the trigger efficiency versus cosmic ray energy, simulations were performed using CORSIKA (Heck 1998), a program for simulating extensive air showers.

Showers produced by protons of energy from 1 Tev to 1000 Tev were simulated. The pattern of light on the ground and the image as seen in the sky from a telescope's perspective of one example shower was extracted from the CORSIKA simulation data and is shown in Figure 9.2 and Figure 9.3.

The top of Figure 9.4 is a close-up of the Cherenkov sky image in Figure 9. 3, the white dots representing where on the All-sky imaging plane the Chernkov photons land. In order to cause the All-sky camera to trigger, that is, to record the waveform of the Cherenkov flash, the photon density must be higher than a certain threshold. In the case of the current camera,

Figure 9.2. Simulation of a 10 Tev cosmic ray proton shows a radially symmetric distribution of Cherenkov light on the ground. Image scale is 1km by 1km. The position of the viewer, as seen in Figure 9.3 is at the location indicated by the black square.

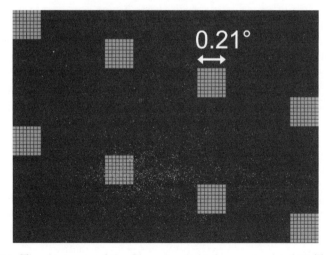

Figure 9.3. The sky image of the Cherenkov light from a simulated 10 Tev cosmic ray proton, overlaid on the imaging plane showing the 8 pixelated PMTs, appears as a faint streak when viewed 100 meters from the shower axis. The location of the viewer for this image is shown as a black square in figure 9.2.

that threshold is three or four photoelectrons per pixel. Thresholding the simulated sky image at the All-sky camera's minimum sensitivity gives an image as seen in the middle section of Figure 9.4. Here, the sky image of the cosmic ray shower's Cherenkov light is overlayed on a picture of one of the All-sky camera's sixteen photomultiplier tubes. Multiple nonadjacent pixels would be triggered if the camera were to see this simulated cosmic ray. Based on the multiple nonadjacent pixels, this event would be characterized by our analysis as a cosmic ray–induced trigger and would not be categorized as a possible ETI detection. What if the simulated cosmic ray were to come from a slightly different angle? A different angle of incidence simply translates the telescope image. As seen in the bottom of Figure 9.4, only one or possibly two adjacent pixels would record the event. Like most cosmic ray Cherenkov flashes, it would appear as a short duration (nanosecond scale) flash of light. Instead of seeing the whole 1° wide image coming from the cosmic ray, it now looks (to the All-sky camera) like a compact location on the sky, exactly the type of characteristics that would occur if the light were to come from a distant pulsed laser transmitter. An upgrade to the current Harvard All-sky camera is being prepared in order to distinguish between the aberrant cosmic ray hit, such as the one just described, and a flash from a distant pulsed laser source (a possible ETI detection). The new camera will allow readback of real-time photon counts and waveform capture for all 512 pixel pairs when a single pixel is triggered. By having all pixel waveform data available when one pixel is triggered, events having a modest angular extent on the sky will be recognized as being cosmic ray induced. With reduced cosmic ray background events, the camera is allowed a reduced threshold and therefore a higher sensitivity.

3.0. Advanced All-sky Optical SETI

The new Advanced All-sky camera will be a retrofit of the current camera's electronic subsystem aimed at attempting to reduce the cosmic ray induced Cherenkov flashes as a background source. The electronic subsystem includes photomultiplier output signal amplifiers, circuits to filter the signals for pulses and enforce pixel pair matching, and electronics for sending the captured data to a separate database computer. These functions are currently handled by eight daughterboards—each containing four Pulsenets (Howard 2006), the custom integrated circuit designed to handle digital sampling and filtering of the signals from All-sky's 512 pixel pairs—and a motherboard which the daughterboards plug into. The motherboard contains circuitry for programming the Pulsenets and communicating control signals to the daughterboards and Pulsenets. A PC104 single board receives commands over ethernet and forwards them on to the motherboard using the ISA-like PC104 bus. The architecture of a single motherboard and eight daughterboards is maintained

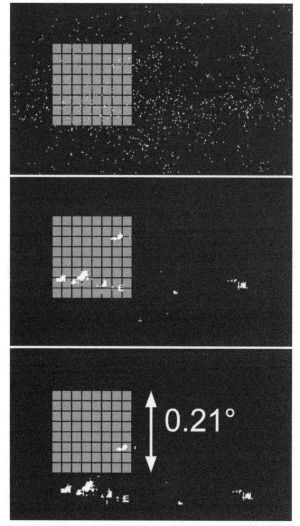

Figure 9.4. Top: Close up of figure 9.3, each white pixel representing 5 visible light photons. For size reference, the sensitive area of an 8x8 pixelated photomultiplier tube is shown in the background. Middle: Areas in top image with a density greater than 50 Cherenkov photons per PMT pixel are highlighted in white. These areas would cause the electronics to "trigger"—capture the waveform of the relevant pixel for a short period of time. Bottom: Same simulated cosmic ray as above but incident at a slightly different angle. Only one or two adjacent PMT pixels (gray squares) would register the Cherenkov light (highlighted in white). Because a distant pulse laser transmitter would also cause one or possibly two adjacent pixels to trigger, this cosmic ray event could not be immediately distinguished from a possible ETI pulsed laser event without the new Advanced All-sky camera.

for the Advanced All-sky upgrade. Pulsenets are replaced by Xilinx Virtex®-5 FPGA's (Field Programmable Gate Array) which communicate directly over an ethernet network, obviating the need for the PC104 computer.

The job of the All-sky camera is to detect bright flashes of light, record as much information as possible about the flash—the intensity of the light over time, where in the sky the light originated, and the exact time it occurred—and send that information to a database computer. To illustrate this process, lets follow a simulated flash through the new Advanced All-sky system.

3.1. Optical Front End

Light entering the All-sky camera comes in through a glass entrance window and meets a beamsplitter, which divides the image plane into two copies. Each image plane has an array of pixelated photomultiplier tubes (PMTs), which converts incident light flashes into current pulses on the corresponding pixel anode. The current pulse travels down coaxial cables and onto the camera's daughterboards.

3.2. Amplification

After entering the daughterboard through a surface-mount connector, the current pulse is amplified by the NEC 2710TB, a 32 dB RF amplifier with 1 GHz of bandwidth. The pulse is then sent to one of the four on-board Virtex-5 FPGA's to be analyzed.

Figure 9.5. The eight daughterboards of the Advanced All-sky camera each contain four Virtex-5 FPGAs, 128 signal amplifiers, a 16-channel DAC for setting flash converter threshold voltage levels, four ethernet jacks, four 128MB DDR chips, and a myriad of voltage regulators.

3.3. Analog to Digital Conversion

The first stage of processing for a pulse entering the Virtex-5 is analog-to-digital conversion (ADC). The analog voltage pulse must be converted to a stream of numbers so that it can be processed by the FPGA. The type of analog-to-digital conversion used in the Advanced All-sky system is called parallel conversion. Eight comparators receive the pulse signal and compare the pulse level with eight reference levels. Separating the levels appropriately, a three-bit (eight level) digital approximation of the analog input signal is obtained. This type of analog to digital converter is not particularly accurate but it is very fast. The Advanced All-sky camera ADCs will operate at 700 million conversions per second (700 Msps) or about 17 percent faster than the old camera. The LX110 Virtex-5 has four hundred differential I/O pairs available; each Virtex-5 must have thirty-two ADC channels so 256 differential pair inputs are used for the parallel conversion. The simulated flash that we are following has gone from a flash of light, to a packet of electrons, to a pulse of current flowing down a coaxial cable, to an amplifier, into the Virtex-5 FPGA, and now it is represented as a stream of three-bit numbers, each separate number corresponding to the intensity of the light flash at a single moment in time.

3.4. Pulse Filtering

The three-bit data stream must be filtered to find the flashes of light because they occur only rarely; most of the time the data stream is just zeros. Logic inside the filtering sections of the Virtex-5 FPGAs waits for a flash that crosses an adjustable threshold. When the numbers representing the intensity of the light flash are high enough, the camera will trigger or start saving the data stream by putting it into an internal memory buffer inside the FPGA. The data stream cannot be saved continuously because the amount of data would be too large; 1024 channels of 700 Msps three-bit data equates to 250 GB of data every second. As of this writing, that would fill up the largest commonly available commercial hard drive in eight seconds and cost $112,000 to store an hour of data. As stated, most of this information is useless and only the interesting timespans—the bright flashes—are stored. Each trigger results in eleven microseconds of digitized output of all pixels to be stored. The pulse filtering logic also enforces the pixel pair matching constraint; the light levels from matched pixels must be within one digital conversion reference level. A "coincidence" is registered if a matched pulse is received in both pixels and the intensity is enough to cross the trigger threshold reference level. With the new Advanced All-sky camera, a coincidence in any one of the 512 pixel pairs can be used to trigger the other 511 pixel pairs in the system or some subset thereof.

FPGAs such as the Virtex-5 are reprogrammable. Therefore, the triggering and pulse filtering logic can be adjusted as needed. As the Harvard All-sky Optical SETI survey evolves, more is learned about its trigger sensitivity and trigger efficiency with respect to cosmic rays and other noise sources. As new information is learned, it can be folded into the camera design as necessary to achieve higher sensitivity or as a way to experiment with new triggering and filtering methods. This feature is new to the Advanced All-sky camera. The previous camera iteration was built around a custom ASIC which, once fabricated, could not be changed. Flexibility to change the design is a major advantage of the Advanced All-sky camera.

3.5. Data Forwarding

As soon as the memory buffer for the triggering event is full, a soft-core processor called Microblaze™ inside the FPGA is notified via an interrupt signal. The Microblaze processor packages the event data into a TCP packet and sends the packet to a database server that is waiting to receive trigger event messages over the local ethernet network. Each Virtex-5 has a 10/100 MB ethernet MAC core peripheral and external PHY that is used by the Microblaze processor to communicate over ethernet. All thirty-two Virtex-5 FPGAs are connected together and to the database server using an ethernet network switch.

3.6. Camera Control

Control of the Advanced All-sky Optical SETI camera is realized through messages sent to the Virtex-5 FPGAs over ethernet. Each FPGA runs a lightweight Web server that can serve HTML pages to a browser and receive POST messaging from the browser. POST messages contain instructions for the Web server and other software running on the Microblaze processor of each Virtex-5. The instructions include the control messages that dictate the camera's behavior on any given observing run. Options include setting the analog-to-digital conversion threshold level spacing, defining which threshold level will cause a pixel to trigger and begin recording waveform data, controlling whether a coincidence will trigger other pixels in the camera array, and forcing a system reset.

4.0. Concluding Comments

Upgrading to the Advanced All-sky Optical SETI camera will provide much additional capability and allow for future expansion and design changes when needed. During observations with the original camera, much was learned about operating this one-of-a-kind survey while the limitations of the camera hardware became increasingly apparent. The Advanced All-sky camera addresses the known limitations and embodies all that was learned in the last four years of the All-sky survey. Its flexible design using FPGAs will ensure that it will remain a relevant optical SETI instrument for at least another four years. Prototype boards are currently being being manufactured and we anticipate that production boards will be installed by mid 2011.

Acknowledgments

The authors are ever grateful to The Planetary Society and the Bosack-Kruger Charitable Foundation for their steadfast support. Much gratitude is extended to Xilinx for their very generous offer to supply the Virtex-5 FPGAs.

References

Cocconi, G., and P. Morrison. 1959. Searching for interstellar communications. *Nature* 184: 844–46.

Coldwell, C. 2002. A Search for interstellar communications at optical wavelengths. PhD diss., Harvard University.

Heck, D., et al. 1998. CORSIKA: A Monte Carlo Code to simulate extensive air showers. *Report FZKA 6019, Forschungszentrum Karlsruhe.* http://www-ik.fzk.de/corsika/physics_description/corsika_phys.html.

Howard, A. W., et al. 2004. Search for nanosecond optical pulse from nearby solar-type stars. *The Astrophysical Journal.* 36 (2): 1270–84.

Howard, A. W. 2006. Astronomical searches for nanosecond pulses. PhD diss., Harvard University.

Kingsley, S. 1996. Prototype optical SETI observatory. *Proceedings of the Search for Extraterrestrial Intelligence (SETI) in the optical spectrum II, Jan31-Feb1, 1996.* Proc. SPIE 2704: 102–16.

Lampton, M. 2000. Optical SETI: The next search frontier. In *Bioastronomy '99—A new era in bioastronomy,* ed. G. Lemarchand and K. Meech. *Astronomical Society of the Pacific Conference Series* 213: 565–70.

Rao, M. V. S., and B. V. Sreekantan. 1998. *Extensive air showers.* Singapore: World Scientific Publishing.

Schwartz, R., and C. Townes. 1961. Interstellar and interplanetary communication by optical masers. *Nature* 190: 205–208.

Weekes, T. C. 2003. *Very high energy gamma ray astronomy.* UK: IOP.

Wright, S. A., Drake, F., Stone, R. P., Treffers, D., and Werthimer, D. 2001. An improved optical SETI detector. *Proceedings of the Search for Extraterrestrial Intelligence (SETI) in the Optical Spectrum III* 4273: 173–77.

Zombeck, M. V. 2007. *Handbook of space astronomy and astrophysics.* New York: Cambridge University Press.

The OZ OSETI Project

Ragbir Bhathal

1.0. Introduction

In the evolutionary sequence of life, fireflies (*Lampyridae*) in the tropical regions of the earth were probably the first living organisms to use light as a means of communicating with the community of fireflies. The chemically produced light that is emitted from the lower abdomen has a wavelength between 510 to 670 nanometers. Fireflies use a pulsed beam of light to draw attention and to signal to other species of their kind. In fact, biologists have found that they use the pulsed light signals for mating purposes. Unfortunately, the light also reveals their location for would-be predators.

It was only in the nineteenth century that human beings thought of using light as a means of communicating with alien beings. The Viennese astronomer, Joseph von Littrow proposed that we should use light to inform extraterrestrial intelligent beings of our presence. His scheme was a very simple one. He proposed that we dig huge trenches about 30 km across in the Sahara Desert, in the shape of triangles, circles, squares, etc., and fill them with kerosene and set it alight. Extraterrestrial beings would make out the geometric figures and realize that earthlings were mathematically minded. Charles Cros, a French physicist, suggested that an array of mirrors be used to reflect sunlight to Mars. However, the schemes never got off the ground (Drake and Sobel 1992). Karl Gauss, the great mathematician and one-time director of the Gottingen Observatory also had an interest in contacting extraterrestrial beings, and proclaimed that the discovery of a signal from ETI would be a discovery greater than the discovery of America by Christopher Columbus (Bhathal 2000).

We had to wait until the second half of the twentieth century, when optical SETI came back on the agenda in a rather curious way. Physicists in the United States and Russia were actively engaged in the new field of quantum electronics and in 1964 the Nobel Prize for physics was awarded to U.S. physicist Charles Townes and the Russian physicists Aleksandr Prokhorov and Nicolay Basov for their research and development of the maser, which led to the development and construction of the laser. In a seminal paper published by Schwartz and Townes (1961) in the British journal *Nature*, they argued for an optical SETI search strategy as an alternative to the radio search strategy. This suggestion fell on deaf ears as radio SETI had become the paradigm technique for any searches for ETI by radio astronomers. In fact, this year marks the fiftieth anniversary of Drake's first attempt to search for ETI in the radio spectrum. By 2000 a number of dedicated university-based optical SETI projects led by Horowitz (Horowitz, Coldwell, Howard, Latham, Stefanik, Wolff, and Zajac 2001; Howard, Horowitz, Wilkinson, Coldwell, Groth, Jarosik, Latham, Stefanik, William, Wolff, and Zajac 2004), Werthimer (Wright, Drake, Stone, Treffers, and Werthimer 2001), and Drake (Stone, Wright, Drake, Munoz, Treffers, and Werthimer 2005) in the United States, and by Bhathal (Bhathal 2008) in Australia have taken on the challenge of searching for ETI in the optical spectrum.

2.0. OZ OSETI Genesis

The OZ OSETI project had its genesis at a lecture given by Townes at the Capri astrobiology conference in 1996. He once again tried to convince the astronomers and astrobiologists present at the meeting that we should be considering doing SETI in the optical spectrum. But the skeptics dismissed the idea. On his return flight home to Sydney, the author did a back-of-the-envelope calculation and was convinced that lasers could be used for interstellar communications and if there were ETI in outer space they would use lasers to communicate with other ETI civilizations and the Earthlings (Davids 2002) in addition to microwaves. So was born the OZ OSETI project at the OZ OSETI observatory at the University of Western Sydney in Campbelltown, a small town about sixty kilometers from the center of Sydney with relatively low levels of light pollution.

3.0. The Observatory

The OZ OSETI project began in 1996 with sporadic tests and observations of a few southern stars with a 0.8 m off campus telescope. But it was soon realized that if one wanted to mount a serious search for ETI it was necessary to have a dedicated observatory that was located at a convenient and accessible location. This led to the establishment of a dedicated observatory at the university's campus at Campbelltown. The observatory was built with a combination of funds from the private sector, the government, and a university research grant. The observatory was opened in Year 2000, the year of the Olympics in Sydney. This year marks the tenth anniversary of the opening of the observatory and the first dedicated observations for ETI in the optical spectrum in Australia and in the southern hemisphere. It is the longest dedicated search for ETI in the optical spectrum in this region.

4.0. Reasons for Pulsed Approach

In his *Nature* paper Townes had suggested searching for a continuous laser beam for ETI. However, this method has shortcomings and has to use fairly complex equipment and heterodyning (photo-mixing). A much simpler search strategy is to search for a nanosecond pulsed laser signal. This was first suggested by Monte Ross (Ross 1965, 2009) as being the most effective and efficient search strategy. It uses a direct detection strategy, which eliminates the need to know the exact wavelength or any magic wavelength or frequency. The nanosecond signal has less background noise to compete with and it makes any detected signal even more obvious as being artificial. The narrow beamwidth and pointing accuracy of a laser is of tremendous advantage for any receiving or sending civilization. The beam can be made just wide enough to cover the habitable zone of the target star. The other advantage is that the telescope can be used as a photon bucket since phase information is not required. This means that one can use less accurately figured optics that are much cheaper than the optics of a comparable size telescope intended for imaging. The nanosecond pulse can outshine its target star by several orders of magnitude. Extinction and smearing of pulsed laser signals due to interstellar grains do not pose a serious problem for this search strategy. The above reasons provided the rationale for undertaking the OZ OSETI project. The OZ OSETI project is based on the assumption that an ETI civilization is capable of sending nanosecond pulses with peak beacon powers of the order of between 10^{15} W and 10^{18} W.

5.0. The Equipment

The equipment used for the search uses 0.4 m and 0.6 m optical telescopes. The detecting equipment uses Hamamatsu R7400U series PMTs with a spectral response range from 300 to 650 nm and a rise time of 0.7 ns. The search is centered on 550 nm. The outputs from the PMTs are fed into a coincidence circuit to eliminate any false signals. Software on a computer interface allows for the loading and saving of measurement data which are analyzed either at the time of the observations or carried out in the laboratory at a later time. The PMTs are sealed off in a compartment that has a container containing silica crystals to absorb any humidity, to prevent any discharges. The instrumental setup is insensitive to terrestrial interference from lightning flashes or other extraneous fast flashes of light. Before an observing run the equipment is tested with a flashing LED to ensure that it is in working order.

6.0. Observations

To date, more than two thousand F, G, and K type southern stars and about thirty southern globular clusters have been observed. The stars and globular clusters lie in the declination range between −20° to −80°.

Data collected in a nights observing session are analyzed with the use of an fft program on MATLAB. The results to date have all been negative. However, in December 2008 we detected a sharp laser look-alike signal from the southern globular cluster 47 Tucanae. We dubbed it the "Is it ET ?" signal. Subsequent follow-up observations over the last six months have not revealed any trace of the signal. Millisecond pulsars have been detected by radio astronomers in 47 Tucanae (Camilo, Lorimer, Freire, Lyne, and Manchester 2000). At one point it was thought that we might have discovered an optical counterpart of a radio pulsar. But this has now been discounted. We have also discounted the fact that the signal was an ETI signal. We have checked our equipment rigorously for any fault but have not found any inherent fault with the equipment. After about six months of follow-up observations we have not seen the signal again. We have now dismissed the signal as a false signal.

7.0. Using SETI for Indigenous Education

The OZ OSETI project also has an educational component in keeping with the university's policy of engaging with the community it serves. The university's Kingswood Campus (about 40 km from the Campbelltown Campus) is located in an area that has a fairly large Aboriginal community, which has children of school-going age. According to the teachers in the area 70 percent of the Aboriginal students drop out of school before the end of Year 10 (the end of the compulsory secondary schooling—Years 7 to 10). The Aboriginal community believes that their children should be encouraged to study science so that they can take up careers in science and engineering.

The Drake equation ($N = R\, f_p\, n_e\, f_l\, f_i\, f_c\, L$) is an excellent way to motivate young students to take an interest in science. With this in mind we developed an astronomy program with a number of activities for the students. We used the physics and engineering laboratories at the university and the observatory to carry out physics/astronomy activities and viewed the night sky through the fully computerized telescopes at the observatory. The aim of the project was to improve the scientific literacy of Aboriginal students by using the above activities. We also used ideas from Aboriginal astronomy (a part of their social-cultural heritage) and scientific astronomy to heighten their curiosity about the natural world. Like their ancestors who lived on the Australian continent for more than forty thousand years we showed them how to carry out naked eye astronomy and build a similar knowledge system about the night sky as their ancestors had done in times gone by. This approach gave the students a sense of pride in what their ancestors had achieved with naked eye astronomy. It was a knowledge system which they could readily identify with.

We ran the program for a group of twenty students from the lower secondary classes (Year 7 and 8). The program was carried out in the second semester of the university year. We envisaged that the program would develop the following skills: mathematical, scientific (developing and testing the ideas), measuring, using and manipulating scientific equipment, observing the night sky with the naked eye and through telescopes, drawing graphs and drawing inferences from the graphs, using art to express ideas, communicating their ideas orally, and appreciating two different astronomical knowledge systems (Aboriginal and modern). The performance indicators we used were as follows: knowledge gained by students of modern scientific astronomy, Aboriginal astronomy, and of the methods and processes of science.

The projects included: constructing a mobile solar system with information they had gathered about the properties of the various planets in the solar system, including a discussion as to why Pluto had been demoted as a planet, properties of reflection and refraction of light from mirrors and glass prisms, including the construction of a simple telescope with two lenses, measuring the distance to two simulated stars in the laboratory, creating craters like those found on the Moon by dropping steel balls of various sizes onto a tray of flour, and the search for life in the universe. When discussing the planets we discussed the planet Venus not only from the point of view of modern-day astronomy but also how it is seen in Aboriginal culture. In Aboriginal culture the Morning Star (Venus) is associated with death and in northern Australia the Aborigines conduct Morning Star ceremonies for the dead. Experiments with light showed the students how scientists test theories in physics. As part of viewing the night sky with the naked eye the students were asked to find the Southern Cross and write a story about it as if they were living about a thousand years ago. Their ancestors saw the Southern Cross either as a shark (the Pointers) chasing a sting ray (the five stars of the Southern Cross) or as the good man Mirrabooka who had been placed in the sky by Biame (a spirit ancestor or Godlike figure) to look after his people. The activities on the search for life in the universe were according to the students the most enjoyable experience. They discussed how and what kind of messages they would send to ETI. After having drawn their messages they were asked to tell the class about it and defend their messages.

The students were surveyed before and after participating in the program. The results were both pleasing and surprising. The students found astronomy to be an interesting activity, especially viewing the night sky with the telescopes. This is very similar to what we found in the general population who visit the observatory on astronomy nights for the public. The students found the experiments they did at the university much more interesting and thought provoking than the ones they do at school. Many students found the experiments they do at school boring and hardly interesting. Only 10 percent of the students agreed with the statement "Science is about ideas and experiments to test whether they are right or wrong" at the start of the program, whereas at the end of the program 90 percent agreed. It was also interesting to find that more students (70%) were considering carrying on to do the Higher School Certificate (a certificate that allows them to gain entry to a tertiary institution of education) after participating in the program than beforehand (20%). Overall, the outcome of the program was positive in changing their attitudes to science and its processes and their attitude to further education.

8.0. Future Directions

We believe that for optical SETI to be successful we need to use larger telescopes. To this end we are planning to acquire a 1 m telescope which will essentially be used as a light bucket and run from a private observatory in country New South Wales. We are also looking into the possibility of using three or four 0.4 m optical telescopes at this observatory to increase our coverage of the search. Rather than limiting ourselves to a wavelength of 550 nm we intend to cover a wider frequency range. We are planning to build a multichannel photometer coupled to a low dispersion spectrograph which will enable us to make observations over various wavelengths. The search strategy will still be based on targeted objects as before. We are also exploring an educational program for targeted school groups which will encompass the broader field of astrobiology, which will allow us to cover the multidisciplinary nature of the search for life in the universe. We have found that Drake's equation is an excellent way to motivate and enhance the scientific and mathematical literacy of students. We see astrobiology with its most intriguing question, "Is there life in the universe," as an excellent way to get more students to study science and pursue scientific and engineering careers. It is the best steppingstone to science in the twenty-first century, just as the Moon landing was to a generation of young children in the late 1960s.

9.0. Conclusion

This year is the tenth anniversary of the OZ OSETI project—the only dedicated OSETI project in the southern hemisphere. To date the results of our search have not picked up any ETI signals in the optical spectrum. It would appear that our search has just begun and that if we do not continue we will never know if ETI are out there.

Acknowledgments

We thank our sponsors for their donations. We also thank Jo Davids, Abishek Tiwary, Anna Tiwary, Leon Darcy, Sharon Tate, Rhonda Gibbons, Paul Toner, Jeff Scott, and Keith Rynott for their assistance.

Works Cited

Bhathal, R. 2000. The case for optical SETI. *Astronomy & Geophysics* 41 (1): 25–26.

———. 2008. Searching for ET. *Journal and Proceedings of the Royal Society of NSW* 141: 33–37.

Camilo, F., D. R. Lorimer, P. Freire, A. G. Lyne, and R. N. Manchester. 2000. Observations of 20 millisecond pulsars in 47 Tucanae at 20 centimeters. *The Astrophysical Journal* 535: 975–90.

Davids, D. 2002. Australian optical SETI. *The Physicist* 39: 79–81.

Drake, F., and D. Sobel. 1992. *Is anyone out there?* New York: Delacorte Press.

Horowitz, P., C. Coldwell, A. Howard, D. Latham, R. Stefanik, J. Wolff, and J. Zajac. 2001. Targeted and all-sky search for nanosecond optical pulses at Harvard-Smithsonian. In *Proceedings SPIE 4273: The Search for Extraterrestrial Intelligence (SETI) in the optical spectrum (III),* ed. Stuart A. Kingsley and Ragbir Bhathal, 4273: 119–27. Bellingham, WA: SPIE.

Howard, A. W., P. Horowitz, D. T. Wilkinson, C. M. Coldwell, E. J. Groth, N. Jarosik, D. W. Latham, R. P. Stefanik, A. J. William Jr., J. Wolff, and J. M. Zajac. 2004. Search for nanosecond optical pulses from nearby solar-type stars. *Astrophysical Journal* 613: 1270–84.

Monte, R. 1965. Search via laser receivers for interstellar communications. *Proceedings of the Institution of Electrical and Electronic Engineers* 53 (11): 1780.

———. 2009. *The search for extraterrestrials: Interpreting alien signals.* Chichester, UK.: Praxis.

Schwartz, R. N., and C. Townes. 1961. Interstellar and interplanetary communication by optical masers. *Nature.* 190: 205–208.

Stone, R. P. S., S. A. Wright, F. Drake, M. Munoz, R. Treffers, and D. Werthimer. 2005. Lick Observatory Optical SETI: Targeted search and new directions. *Astrobiology* 5 (5): 604–11.

Wright, S. A., F. Drake, R. P. Stone, R. Treffers, and D. Werthimer. 2001. An improved optical SETI detector. In *Proceedings SPIE 4273: The Search for Extraterrestrial Intelligence (SETI) in the optical spectrum (III)*, ed. Stuart A. Kingsley and Ragbir Bhathal, 4273: 173–77. Bellingham, WA: SPIE.

The New Telescope/Photometer Optical SETI Project of SETI Institute and the Lick Observatory

*Frank D. Drake, Remington P. S. Stone,
Dan Werthimer, Shelley A. Wright*

1.0. Introduction

We are proposing to construct a telescope array equipped with very fast photometers to search for transient celestial phenomena. It will operate in the visible (350–850 nm) and near-infrared (NIR; 1000–1550 nm), and will have a time resolution as short as a nanosecond. The goal is to conduct dedicated searches for such phenomena with high sensitivity, large sky coverage, and robust rejection of false positives. This project will delve into a portion of observational "phase space" that has been little explored, but that may have undiscovered important phenomena, as predicted in Martin Harwit's (1981) book *Cosmic Discovery: The Search, Scope and Heritage of Astronomy.*

2.0. Motivation

Within the last decade, there has been growing interest in searches for rapid (of the order of a nanosecond) optical pulses emanating from transmitting beacons from extraterrestrial intelligence (e.g, Werthimer et al. 2001; Wright et al. 2001; Howard et al. 2004; Stone et al. 2005). Pulsed lasers can outshine our Sun by at least four orders of magnitude in brightness, and may offer one of the best means for interstellar communication, as originally suggested by Schwartz and Townes (1961). In the past few years, these types of lasers have been developed for a number of applications on Earth and are well within our current technological capabilities. Research groups at Lawrence Livermore National Laboratory are now creating laser pulses with peak power of the order of petawatt for times lasting many picoseconds. Such pulses, when concentrated into a narrow beam by a large telescope such as the current 8–10 m reflectors, create a photon flux in their beams so powerful that they easily outshine all the light of the host star. These high power pulses would be easily detectable with meter-class telescopes even across great distances within the Milky Way. A great advantage of these lasers is that they can be narrowly beamed, thus providing the highest amount of transmitted power flux and information per unit energy. Therefore, using such a laser offers a promising means for interstellar communication, and it is possible that an advanced extraterrestrial civilization would communicate via pulsed lasers.

In order to detect such interstellar signals, they must be distinguishable from other background signals, including terrestrial, atmospheric, and astrophysical phenomena. There are currently no known astrophysical objects that produce nanosecond or shorter optical pulses (Howard et al. 2004). However, this phase space has been poorly explored for short pulses from a variety of compact celestial objects such as neutron stars, novae, supernovae, black holes, and active galactic nuclei (AGN). Suggestions that such searches could bear important fruit come from, for example, the fact that radio pulses observed in the pulsars PSR0950+08 and PSR1133+16 are unresolved in time, and shorter in duration than a microsecond. This implies that very high energy densities exist in the emitting regions, in turn suggesting the possibility of associated optical radiation. Similarly, "giant" radio pulses from the Crab Nebula are unresolved down to nanosecond intervals (Hankins et al. 2003). Recently, new transient surveys (e.g., Palomar Transient Factory; Law et al. 2009) are still discovering new classes of compact objects and supernovae, and finding new optical transients on ever shorter time scales (Kasliwal and Kulkarni 2009). Therefore, searches for pulsed laser signals from extraterrestrial intelligence have the distinct advantage of potential for unveiling exciting new areas of time-domain astrophysics.

Near-infrared (1000–3500 nm) astronomy has also boomed in the last decade with more advanced IR detectors offering higher quantum efficiencies and lower detector noise. At the same time, near-infrared lasers are being developed and explored for a variety of applications, in particular within the telecommunications industry with both lasers and detectors. One of the major advantages for interstellar communication of using longer wavelengths is the decrease in interstellar extinction, which is of particular importance for communicating close to the plane of the Milky Way. For instance, at visual wavelengths (~600 nm) there are thirty magnitudes of extinction looking toward the Galactic Center, whereas at IR wavelengths (~1600 nm) there are only two magnitudes of extinction. In addition, Galactic background from warm dust peaks in the mid- to far-infrared, at wavelengths beyond the near-infrared. The near-IR therefore offers a unique window with both less interstellar extinction and less background from our galaxy and from Earth's atmosphere. A search in the near-IR with fast response instruments would explore an entirely new phase space for rapid transient events from celestial objects and potential communications from extraterrestrial intelligence.

We are proposing to build a fast-photometer instrument and telescope system with large sky coverage, dedicated to searches over extended times for brief optical and near-infrared pulses, as is likely required for success in a search for phenomena that are transient and with unknown repetition intervals. The instrument package will include multiple telescopes and photometers, and will be located in an existing dome at Lick Observatory, on a presently unused telescope mount. In its first phase, this instrument suite will have the light-collecting power of a 1 m class telescope, and will solely be designed to feed rapid response photometers. An auxiliary optical imaging camera will serve for field identification, and will provide a permanent pointing record during driftscans.

In our past research we have used the 40 in. Nickel telescope at Lick Observatory with our team's first instrument using an optical triple photometer as detector (Wright et al. 2001; Stone et al. 2005). This original program started in December 2000 and has been used to search for optical (350–850 nm) nanosecond pulses from individual stars. Each subject star was observed for a total time of ten minutes. In this program we searched for hypothetical nanosecond time scale laser pulses from intelligent activity in the vicinity of 4605 stars of spectral types F to M in the Milky Way. This instrument used an innovative approach with great sensitivity for detecting such brief pulses, while reducing the number of false positives that plagued earlier systems. This original program was limited by being a targeted search. In recent years our access to the Nickel telescope has become limited as competition for time has grown due to the availability of remote observing—hence the need for a new and dedicated instrument.

3.0. Previous Optical Fast-Transient Searches

Early programs to detect nanosecond optical flashes used single detectors, which were greatly troubled by false positive signals. These distracting internal events can be produced in several ways, including radioactive decay in the envelopes of the photomultiplier tubes (PMTs) universally used for detection of brief pulses; sparks from corona discharge; or by cosmic ray hits. For our original detection system, a very successful means to avoid such false positives was developed by one of us (Dan Werthimer).

The method depends on dividing the captured light into three streams, the photons of each stream being delivered to a separate PMT. The stream of electrical pulses delivered from the PMTs are carried along electrical paths of carefully controlled equal length to a coincidence detector, which will be triggered if three pulses arrive simultaneously. Thus, spurious photon events produced within individual PMTs are ignored. A coincidence will only occur if there is strong deviation from Poisson statistics, which is the signature of a simultaneous burst of many synchronized photons from a laser. Such a deviation from Poisson statistics can give strong suggestion of technological origin of the photons.

This method only works if the streams of pulses from the detectors are not too rich in photons. In practice this is controlled by setting appropriate thresholds in the electronics which then pass only pulses of a certain minimum strength. Note that just dividing the raw photon stream into two channels does not give good results—statistically, it is much more likely to find time coincidences between two photons in two streams than coincidences of three photons in three streams. Four streams and a search for quadruple coincidences would be even better. In practice, with our previous detector the richness of the photon streams was such that an accidental triple coincidence should only occur once in several months. This was actually our experience (Stone et al. 2005). Most of these "triples" were subsequently found to be associated with system malfunctions or incorrect settings of thresholds. Of course, when one detects a triple coincidence, it must be taken seriously and observations should then focus on the target where the "triple" occurred. After several years of observing, we were left with only one unexplained triple, and follow-up observations did not reveal any confirming triples from that possible source. It is very interesting to note that this method of detecting a serious deviation from Poisson statistics can result in a valid positive result from the capture of only a few photons. In principle, three photons are enough. In practice, the fact that quantum efficiencies are less than 100 percent, and some photon captures are rejected by the threshold system, mean that more than three photons are required for success. This is discussed in detail below. Still, this high sensitivity to a particular form of signal is striking. It is truly remarkable that a major discovery can be made as a result of detecting only a few photons.

A lesson from these years of experience is that much more observing time is needed to cover more stars. Repetition rates may be very slow for transmitting lasers. Eventually, a robotic and/or remotely operated system is called for. As helpful as this would be, we are not proposing this here due to funding limitations. We seek to enhance our search with a new instrument that will: (1) increase our sky coverage with a wider field of view and hence increase the number of stars/objects observed, (2) extend our search into the near-infrared, (3) employ more sophisticated recording and post-processing techniques such as are similarly used for transient searches in radio data, (4) increase on-sky efficiency by drift scanning, and (5) increase our time on-sky by having a dedicated telescope and instrument package.

This dedicated instrument will enter new and as yet unexplored phase-space.

4.0. The New Optical and IR Fast-Transient Instrument

We intend to build an instrument that incorporates its own light collector and will be affixed to an available telescope mount, which is not currently in use. Exclusive use of the telescope assembly will greatly increase our on-sky time, as will be necessary in a search for perhaps rare transient phenomena. Instead of looking at one star at a time, we are proposing to look at a larger area of the sky to search for transient signals that may be arising from any of the large number of stars in our field of view. Should highly advanced civilizations no longer be tied to nearby stellar sources, as has been suggested, our driftscan strategy will include those locations as well. To extend wavelength coverage into the IR as far as funds permit is very desirable, and we are proposing here to extend into the near-infrared since it is a low-cost enhancement. We intend at some time in the future to expand our capability to search for phenomena into the mid-infrared (10 μm) at a different, higher and dryer observing site, but do not propose that here for cost reasons.

We will search not only for transient phenomena from technological activity, but also from natural objects that might produce very short time scale "flashes" such as pulsars, AGNs, black holes, gamma ray bursts, and other esoteric objects. We will also enhance our instrument with the ability to record detailed time behavior of any flashes detected, and will provide post-processing algorithms for searching within the data.

We are proposing a new instrument that simultaneously minimizes costs, improves sensitivity, and provides even greater resistance to false positives. This new approach uses an ensemble of smaller telescopes, each with its own visual-NIR detector system. Since collecting area costs less per unit area with smaller telescopes, this array concept will allow a large collecting area at minimum cost. This approach sacrifices image quality, but image quality is of no importance for this project. Furthermore, we propose to use "commodity" telescopes to keep costs low. Specifically, we are proposing to use ten Meade 14 in. Schmidt-Cassegrain Optical Telescope Assemblies (OTAs), model LX200-ACF. These ten OTAs will be arranged in five pairs, like a stack of five double-barreled shotguns installed on a single equatorial mount. In most observing modes, each OTA will be observing the same field of view, where the maximum beam size is 10x10 arcminutes. The advantages of this are as follows:

1. Low cost for total collecting area, which will be that of a 1.2-meter conventional telescope (larger than the previously employed Nickel Reflector at Lick Observatory);

2. Totally independent photon streams, eliminating the possibility that a source of pulsed light within any one component will make its way into more than one photon stream and contribute to a false positive;

3. Since the light for each of these 14" OTAs is not divided three ways, as in the old system, and has a smaller collecting area the photon count rate in each detector will be 0.37 times less. This means that we can set thresholds so that nearly three times as many photons can be retained than before and still avoid "photon pile-up." In the simplest picture, this means we will be able to detect signal photons with a minimum flux about half that of our old system. In a way, this is as if we will have gone from a 40 in. telescope to nearly a 60 in. telescope. This simplified picture of improvements in sensitivity is analyzed in more detail below.

Each OTA is to be equipped with a dichroic mirror that will split visible light (350–850 nm) and near-IR (1000–1550 nm) and direct the light into separate high quantum efficiency, short response time (ns) PMTs and avalanche photo diodes (APDs), respectively. We have checked with Meade and they report 80–90 percent throughput of the OTA at optical wavelengths and 50–60 percent at near-IR wavelengths. In total there will be twenty detectors in the telescope array, ten in the visible and ten in the near-IR, that will sample most wavelengths from 350 nm out to at least 1550 nm. The longest wavelength limit will be chosen at the time of instrument construction, and will depend upon selection of the most optimal near-infrared detectors. Although there are PMTs that respond to both visible and IR photons, they are expensive and have significantly lower quantum efficiencies. Furthermore, limiting the spectral coverage in the detectors reduces the photon count rate in each channel and gives the same qualitative increase in sensitivity, as described in point 3 above. Indeed, it is better, even from a cost standpoint, to use two detectors per OTA to get the best sensitivity and also to provide information as to whether a signal arrived in the visible or NIR spectrum.

In the optical we expect to use PMTs from Hamamatsu Photonics, as in the previous system. Our previous experience with these has been excellent— they provide very low dark current, high quantum efficiency, and very fast response times of less than a nanosecond. The typical detector size (~10 mm) will yield a maximum 10x10 arcminute field of view. We will not specify the exact PMT model here, since new and improved ones are often being offered. We will wait until this project is underway to make our final choice.

Near-IR APDs have been under rapid development, primarily due to telecommunications demand. We are currently negotiating with two companies, Laser Communications and Voxtel, which offer APDs with high quantum efficiency (~70–80%), large near-IR bandwidth (ranging from 1000–1700 nm), and fast-response times from 1 to 2.5 Ghz. These APDs have a sufficient window size (>5 mm) for our telescope beam, and include both a cooler and power supply for each system package. We plan to use reimaging optics to compress the larger field of view onto the smaller (compared to the optical PMTs) near-infrared APD detectors.

The proposed ten-detector instrument is a substantial improvement compared to our current three-detector system. As described below, ten detectors provide better sensitivity, a reduced improved false alarm rate, and enable larger flux rate capabilities, so that observers can successfully target bright sources or large fields of view, even with high background count rates.

There will be significant improvements in sensitivity over our previous system. In our previous three-detector system, described in detail by Wright et al. (2001), the light from a single telescope is split three ways using two beamsplitters, and the instrument searches for a coincident photon detection in all three detectors. This "all-three" coincidence requirement results in sensitivity loss due to small number Poisson statistics: photons do not necessarily split three ways in the beamsplitter. For instance, all three photons might go into one detector, or two photons go to one detector and one to another, or the three photons could be split three ways. Only in this last case would our instrument detect this event. In an "all-three coincident" instrument, on average, 5.5 photons are needed to yield a 50 percent probability of transmitting one or more photons to all three detectors.

In the proposed instrument, we will configure the electronics to require at minimum a coincidence from *any* three of the ten detectors. Comparing this to our previous triple-photometer systems, this means that the probability more than doubles that three photons will split over three of the ten detectors, and will provide a factor of roughly 2 improvement in sensitivity (assuming no differences in collecting area, quantum efficiency, etc). In fact, not only do we gain in sensitivity, the system significantly gains in sky coverage compared to our previous instrument. The largest field of view for the telescope-detector system is 10x10 arcminutes compared to our previous targeted search with only a 1x1 arcminute aperture. With an increase in the number of detectors we will be able to take full advantage of the field-of-view offered by these OTAs.

This new instrument will yield a substantial improvement in avoiding false positives, while gaining in sky coverage and bandwidth. The maximum count rate is determined by the acceptable false alarm rate and the number of detectors that are required to be coincident. The equation for the false alarm rate is as follows:

$$F\text{ (False alarm rate)} = Cr^k p^{(k-1)}$$

where k is the number of detectors in which synchronized, detected, photons are required for an acceptable coincidence;

> r is the detected photon rate from each detector;
> p is the coincident pulse width, typically 1 nS;
> C is the number of different ways that H number of photons can hit N detectors with a k coincidence required (e.g., C decreases as k increases with the same number of N detectors).

Observers typically adjust parameters for a constant false alarm rate, typically one false alarm per year of observation time. Note that k does not have to be the total number of detectors in a system (e.g., we might choose to set $k = 3$ even though there are ten telescopes). Or to get a very robust detection we could set the coincidence level higher than $k = 3$ for the ten detectors, even up to requiring that all ten detectors be triggered for $k = 10$. Choices of higher k are extremely unlikely to be false alarms, and thus allow for a powerful diagnostic within our system. Of course, it is possible in software to search for simultaneity of events with various required coincidences.

For a given false alarm rate, the maximum count rate the instrument can handle is a very steep function of the number of detectors required for coincidence (k). For instance, if observers configure the instrument for $k = 4$, the instrument can handle one hundred times higher count rate than $k = 3$. If we configure $k = 5$, the instrument can handle one thousand times higher count rate, with only a small price in sensitivity. Being able to adjust the number of required coincidences given a desired false alarm rate (i.e., once per year) and considering the total flux from any given field will allow observers to target very bright objects and large fields of view with maximum sensitivity. Figure 11.1 shows two plots comparing the false alarm rate per year versus instantaneous field of view (area on-sky in arcmin^2) for N = 3 detectors and $k = 3$ coincidences and between our new N = 10 detector system with $k = 6$ coincidences. These plots immediately illustrate that in order to achieve larger field of view searches need a higher number of detectors to reduce false positives and to gain in sensitivity on-sky. As previously described, our system will allow us to adjust k required coincidences and detector thresholds based on our selected field size and total flux from any given object without increasing our false alarm rate.

Another important benefit of this new system will be the introduction of a very flexible digital backend that will greatly increase the capabilities.

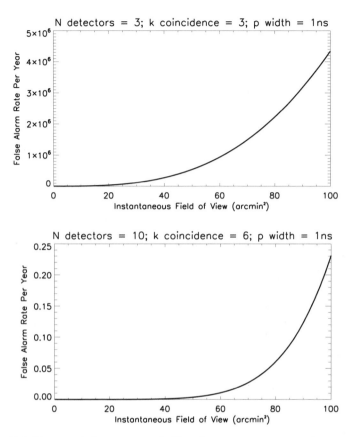

Figure 11.1. Top panel shows expected false alarm rate per year vs. instantaneous field of view for a three-detector (N = 3) instrument that requires a coincidence between all three detectors (k = 3). Bottom panel shows expected false alarm rate per year vs. instantaneous field of view for a ten-detector instrument (N = 10) that requires a coincidence between any six of the ten detectors (k = 6). For both of these plots we use the same pulse width (p = 1 nanosecond), optical bandwidth (350–850 nm), average photon rate from the sky, collecting area (diameter of 14"), instrument throughput, and quantum efficiency. The photon rate (photons s^{-1} nm^{-1} m^{-2} $arcmin^{-2}$) was observed empirically at Lick Observatory using the Nickel telescope CCD camera near the Galactic plane. The sky background will change with respect to the Galactic plane, stellar density, extinction, atmospheric background fluctuations, and observations with respect to moon and Zodiacal light. However, with the average photon rate, the immediate power of using multiple detectors becomes apparent since we are able to gain in significant sky coverage. We will easily be able to observe with the full 10 × 10 arcminute field of view of the OTAs with our ten telescope-instrument design. We note that the false alarm rate assumes the instrument is on-sky continuously for an entire year of observing time. Of course, this would take many years to accomplish given dark time, weather, and other overheads.

153

We plan to use a programmable Field-Programmable Gate Array (FPGA) roach board to serve as a digital backend, which will allow us to improve sensitivity by implementing sophisticated real-time detection algorithms. The FPGA will record both the ten optical and ten near-IR detector waveforms with 1 ns timing resolution. Observers can then examine these recordings to help understand post facto the nature of any detection.

In our existing system at Lick Observatory, the detection algorithm is very simple—all detector signals must be above a programmable threshold to detect an event. In the proposed new instrument, we can implement a variety of detection algorithms. For example, the instrument could trigger when any one of the ten detector signals exceeds a high threshold and any two or three other detector signals exceeds a lower threshold. Or we could cross-correlate detector signals and threshold the product of any number of the ten detectors. We and other authors have simulated several potential algorithms to evaluate the tradeoffs in sensitivity and false alarm rate.

Note also the false alarm rate is a very steep function of the pulse width p. Broader pulses lead to many more chance-timing coincidences. By using interpolation algorithms to locate the centroid of the pulse, we could potentially reduce the uncertainty in photon arrival time by a factor of 10, from $p = 1000$ to 100 picoseconds. This requires detectors that have low timing jitter. Timing jitter in PMTs is dominated by variable delay in the electron cascade from photocathode to anode. We will select PMTs that have extremely low delay variability. Microchannel plate (MCP) detectors have better performance in photon input to anode output delay variability, but MCP detectors are expensive, usually have low quantum efficiency, and are beyond the scope of this project.

5.0. Site Location of New Telescope and Instrument

We have a long association with Lick Observatory, the site of our previous search for celestial transients. Michael Bolte, Director of UCO Observatories, which includes Lick Observatory, has generously allowed us to install the new telescope-photometer in a presently unused but well-sized dome at Lick. This is the dome housing the "Carnegie Astrograph," a telescope consisting of two long focal-length refractors that were used for about fifty years for proper motion studies. The Astrograph dome is immediately adjacent to the new Automated Planet Finder telescope, and very close to the Shane 3 m telescope, both of which are heavily involved in searching for other planetary systems. Thus, our telescope will be in good company, figuratively. Such use of the dome is contingent upon securing funding for other aspects of the project.

It will be necessary to remove the two astrograph telescope tubes. The remaining mount of the Astrograph is very large and exceptionally robust since it was built to support the two massive refractors. It will carry our telescope array with ease. It is a classical equatorial mount, with consequent ease of pointing and tracking. The large dome and sturdy mount will allow us to add additional telescope-photometer clusters with relative ease in the future should funding become available.

We hope that, after full operation is achieved, we can find additional funding to implement remote and/or automated observing. Lick already has considerable experience with this, having installed the very successful Katzman Automatic Imaging Telescope (KAIT), which searches for supernova events robotically, and having successfully implemented remote observing capability on the 1 m Nickel and 3 m Shane telescopes.

6.0. Observing Strategy

We intend to employ two observational methods: drift scan and targeted search. Astronomers and biologists alike have in the recent past been misguided and limited by "reasonable" assumptions about what kinds of possible planetary systems nature might allow, and under what conditions life can exist on Earth-like planets. By drift scanning we will avoid similar limiting preconceptions about viable habitats for extraterrestrial intelligence (Stone et al. 2005).

The primary observing program will consist of a search for nanosecond time scale pulses from a large number of stars of the Milky Way. This will be done by observing with a drift scans with a typical aperture size of 10x10 arcminutes.

In our normal drift scan search mode, we will use a sub-sidereal tracking rate that will allow a ten-minute dwell per sky element. When possible, we will give preference to scans within five degrees of the Galactic equator in order to maximize stellar densities within the field of view. A small auxiliary telescope will allow initial field acquisition, as well as periodic saved reference frames which will assure that we can determine the precise field in view upon acquisition of interesting simultaneous pulses. As with our prior experiment, criteria for such pulses of interest will be predetermined. When those criteria are met, the following actions will automatically occur: the observer will be immediately alerted, pulse shapes as seen by each detector will be recorded for later analysis, an additional time-stamped location reference image will be triggered and saved, and the telescope will revert to sidereal rate so that we may continue to monitor that patch of sky for some significant additional period.

We will use the targeted search mode to monitor objects of special interest. These may be objects that seem relatively likely as possible astrophysical sources of short duration high energy pulses, or they may be objects such as known planetary systems, especially (as seems imminent) should we find nearby Earth-analogues.

Note that we will use no filters in our photometer. This is based on the fact that lasers now in existence on Earth can make a nanosecond flash, which if collimated into a narrow beam by a 10 m telescope, will outshine a nearby star at all wavelengths combined by a large margin. This fortuitous circumstance circumvents the usual need in search for extraterrestrial intelligence (SETI), particularly radio SETI, to choose a limited range of wavelengths for the search. As in all exploratory projects, we expect the paradigm of the observing program to evolve with time as a result of experience.

7.0. Management Plan

7.1. Project Time Line

All of the technology needed to construct this telescope exists and is readily available from commercial sources as commodity items. We will design and construct an interface to hold the OTA units in place on the existing telescope mount. This is easily within the capability of the Lick shops, and we would expect to have the unit made there. Installation of the optical components and electronics will be done by our group with student participation, probably with support if needed from our associates at Lick and UC Berkeley. Considering that the components of the telescope are readily available, with no very demanding installation protocols, we anticipate that the telescope can be constructed and in full operation in less than one year.

The observational program of the telescope will be carried out by the authors, with assistance from both graduate and undergraduate students. This was our very successful style of operation with our previous optical SETI project. As before, we will obtain assistance from interested and talented students from nearby colleges who wish to participate in the project and learn from it. Some serve as interns, some come from the SETI Institute's "Research Experience for Undergraduates" (REU) program, and some wish to conduct senior or master's theses. One author (Shelley Wright) became an expert at OSETI while building and using our first photometer as her senior thesis project at UC Santa Cruz. Potential sources of students, which we have used before successfully, are UC Berkeley, UC Santa Cruz, and San Jose State University. Students from nearby Santa Clara University and Stanford University may also wish to participate.

7.2. Guest Scientists

We will be looking for what are likely to be very rare events produced from a few of very many candidate objects. Opportunity to observe for extended periods is not available on suitable general-purpose telescopes; this situation has been the prime driver for this project. The primary observing program for our telescope will call for much dedicated observing time. Nevertheless, we expect that there will be other very attractive projects that call for a telescope such as we are proposing. An example of such a project in which some of us actively participated was the 2005 NASA Deep Impact mission, for which UCLA researcher David Lynch utilized our predecessor instrument for high-time resolution observations of the intentional collision of an impactor with comet 9P/Tempel. We will greet such proposed innovative projects with enthusiasm, and make instrument time and scientific support available to external scientists at an appropriate level.

8.0. Student Involvement

This project easily lends itself to student involvement. Some construction of the telescope and instrumentation, as well as testing and calibration of the instrument will not demand high levels of technical expertise. In our previous OSETI project we utilized six undergraduates extensively in the building and observing programs, all of whom are coauthors on resulting papers. This was very successful, and the students benefited greatly from the experience. Indeed, we need students to participate as much as possible to keep the ongoing observational program running well. As stated previously, we can attract talented students from local universities to participate as interns or observers. They also can be useful in preparing software to do advanced analyses of the data we will obtain. Each summer we have access to a number of REU students from the very successful SETI Institute REU program. We anticipate using at least two students each summer from this program. This program gives high priority to diversity, as do the university intern programs we might draw from, and this will automatically help us to attain diversity in the students working with us. Our location on the edge of the very ethnically rich Silicon Valley means the student populations from which we draw are diverse as well.

9.0. Education and Public Outreach

A major motivation for the participation of Lick Observatory lies in the remarkable public outreach potential of this project. It at once has enormous appeal to the technologically sophisticated populace of nearby Silicon Valley, as well as great intrinsic excitement and motivational interest for the young.

All astronomy projects having to do with new and perhaps strange phenomena are of wide public interest. This is particularly true of any project aimed at learning of life in the universe, especially intelligent life, and the recent growth of astrobiology. In addition, the aim to search for transient signals from exciting objects such as black holes will be popular with the general populace. This project will serve as an example of unusual things that can be accomplished with high technology, and how we can make instruments that can do amazing things, such as record a signal in a mere billionth of a second. It will open eyes to the surprising things that might be happening and observable within the universe. Nor will this be lost on the media, and we expect that there will be a great deal of media attention, probably more than is usually commensurate with the overall cost of a project, as was the case with our previous one.

As with historical radio SETI, we expect that this project will find its way into the science curricula at high schools and universities. Many important features of our universe are revealed through considerations of optical and NIR SETI. How far can a reasonable signal be detected, and what are the limitations on this? Why can't we send interstellar signals consisting of nanosecond pulses at radio wavelengths, whereas this is entirely feasible at optical and infrared frequencies? How do the remarkably powerful lasers work that are being invented on Earth? Why do infrared wavelengths work better than optical wavelengths for very long distance communication in the Milky Way? How would a black hole or other compact celestial objects emit fast pulses of light? Questions like these, which might seem in isolation to be boring to many students, become very provocative when they are associated with the possible detection of and communication with other civilizations. They work to attract young people into science careers.

This instrument will offer abundant educational activities for students. For example, each summer about sixteen students participate in the SETI Institute's REU program. They work with staff scientists on the scientists' projects. Some of these students will work with the telescope, and all will have some contact with it. Similarly, during the academic year there will be opportunities for students from the San Francisco Bay Area to experience the telescope in action, and even to do some observing with it. Both undergraduate and graduate students may participate in the construction and testing of new equipment for the telescope. An example of one such project is the construction of a light "pulser," a device that can generate nanosecond flashes of light. The device would be used to verify that the telescope-photometer is working correctly, and to calibrate the telescope sensitivity. This is of a difficulty that is suitable for a student project.

Acknowledgments

We are grateful to Boyd Multerer for the generous gift of the initial funding for this project.

Works Cited

Hankins, T. H., J. S. Kern, J. C. Weatherall, and J. A. Ellek. 2003. Nanosecond radio bursts from strong plasma turbulence in the Crab pulsar. *Nature* 422 (6923): 141–43.

Harwit, M. 1981. *Cosmic discovery: The search, scope, and heritage of astronomy.* New York: Basic Books.

Howard, A., and P. Horowitz. 2001. Is there "RFI" in pulsed optical SETI? *Proceedings of SPIE—The International Society for Optical Engineering* 4273: 153–60.

Kasliwal, M. M., and S. Kulkarni. 2009. Palomar Transient Factory: Transients in the local universe. *Bulletin of the American Astronomical Society* 41: 419.

Law, N. M., S. R. Kulkarni, R. G. Dekany, E. O. Ofek, R. M. Quimby, P. E. Nugent, J. Surace, C. C. Grillmair, J. S. Bloom, M. M. Kasliwal, L. Bildsten, T. Brown, S. B. Cenko, D. Ciardi, E. Croner, S. G. Djorgovski, J. Van Eyken, A. V. Filippenko, D. B. Fox, A. Gal-Yam, D. Hale, N. Hamam, G. Helou, J. Henning, D. A. Howell, J. Jacobsen, R. Laher, S. Mattingly, D. McKenna, A. Pickles, D. Poznanski, G. Rahmer, A. Rau, W. Rosing, M. Shara, R. Smith, D. Starr, M. Sullivan, V. Velur, R. Walters, and J. Zolkower. 2009. The Palomar Transient Factory: System overview, performance, and first results. *Publications of the Astronomical Society of the Pacific* 121 (886): 1395–1408.

Schwartz, R. N., and C. H. Townes. 1961. Interstellar and interplanetary communication by optical masers. *Nature* 190 (4772): 205–208.

Stone, R. P. S., S. A. Wright, F. Drake, M. Muñoz, R. Treffers, and D. Werthimer. 2005. Lick Observatory optical SETI: Targeted search and new directions. *Astrobiology* 5 (5): 604–11.

Werthimer, D., D. Anderson, S. Bowyer, J. Cobb, E. Heien, E. Korpela, M. Lampton, M. Lebofsky, G. Marcy, M. McGarry, and D. Treffers. 2001. The Berkeley radio and optical SETI program: SETI@home, SERENDIP, and SEVENDIP. *Proceedings of SPIE—The International Society for Optical Engineering* 4273: 104–109.

Wright, S. A., F. Drake, R. P. S. Stone, R. Treffers, and D. Werthimer. 2001. An improved optical SETI detector. *Proceedings of SPIE—The International Society for Optical Engineering* 4273: 173–77.

Large-Scale Use of Solar Power May Be Visible across Interstellar Distances

Louis K. Scheffer

1.0. Introduction

Energy from the sun provides a very large and renewable source of energy (Hoffert et al. 2002). Great efforts are being made to make this technology cheaper and more efficient (Butler 2007). These structures may cover a substantial fraction of a planet's surface, and hence intercept a large amount of solar radiation. To maximize output for a given cost, the arrays are oriented roughly normal to the incoming solar radiation. Since the Sun subtends a small angle as seen from a planet, the position, and hence reflections, of all these solar panels are highly correlated. This results in very bright reflections, though obviously only in some directions, as total energy is conserved. Exoplanet characterization missions, such as *Darwin* and *Terrestial Planet Finder*, will examine the reflections and emissions of extra-solar planets, and can potentially see these reflections.

The strongest reflections from photovoltaic cells will be in the infrared (IR), between the visible and thermal portions of the spectrum. The albedo of solar cells must be low for visible wavelengths, since the cell absorbs visible light to generate power. However, solar arrays are very reflective in the infrared ($\lambda > 1.2\mu$), at least for current Earth technology. This is because infrared photons do not have enough energy to be harvested, since they have less energy per photon than the bandgap of the material used. These photons go through the cell, reflect off the substrate, and return through the cell to generate the reflection (Zhao and Green 1991; Basore 1990). This effect is similar to the "red edge" (Horler et al. 1983) seen when observing photosynthetic surfaces, and for exactly the same reason—the shorter wavelengths are being harvested for their energy. On the long wavelength end, the contrast will be reduced as the wavelengths extend into the thermal IR, as the planet's glow exceeds the reflecting incident light (this happens at roughly $\lambda > 7\mu$ for Earth-like planets around a star like the Sun) (Woolf 2000). Both ends of this "sweet spot" will depend on the system under examination—on the short end, presumably extraterrestrial engineers would tune their solar cells to their star's peak wavelengths as we do, and the thermal IR cutoff will be determined by the temperatures of the star and planet. Fortunately for our hopes of observing this phenomenon, most techniques that aim to analyze the light from planets also work in the infrared, since the contrast between the planet and the star is greater than in the visible spectrum, and many spectral features useful in the search for biology fall within this region of the spectrum.

2.0. Power Falling on Solar Panels

The world's total energy usage, as of 2010, is about 15 TW. Suppose a civilization generates this power by photovoltaic panels. Assuming a 10 percent efficiency for collection, storage, and transmission, this means that about 150 TW would need to fall on the solar collectors. This is entirely feasible since about 172 PW falls on the Earth's surface, so it would require covering roughly 0.1 percent, or one part in a thousand, of the Earth's surface (Hoffert et al. 2002). Raising the entire world population to the standards of the developed countries might require about an order of magnitude more area, still quite practical—the Earth's deserts cover much more than this amount of land, have good insolation, and are not used for agriculture.

There are already ideas, in various stages of planning, to provide a substantial fraction of the Earth's energy from solar. India, for example, has a plan to install roughly 200 GW of solar power by 2050; this implies about 1 TW when extrapolated to the whole Earth. At 100 watts/m^2 this implies an area of 2×10^9 m^2, or about 0.1% of India's area of 3×10^{12} m^2. Another even larger plan, by the consortium Desertec (Breyer and Knies 2009), would use installations in the Sahara to create all electricity needed for Europe and the Middle East.

3.0. Motivational Example

In this section we show through back-of-the-envelope calculations that the reflection from a large but plausible solar installation, of one particular style, could outshine the light reflected from the rest of the Earth. The rest of the chapter then looks at other configurations and their reflections.

As an example of a large but plausible installation, consider a photovoltaic array in one of the world's deserts, sized to produce an average power of 300 GW, or 1.6 percent of the Earth's energy requirements. There is roughly a 6:1 ratio of peak to average power for solar photo-voltaic, to account for night, cloudy weather, changing angles of incidence, and so on, so the peak output of such an array would be about 1.8 TW. At 10 percent efficiency, such an array would need to cover 0.01 percent of the Earth's surface, or a circle about 140 km in diameter. This size of array falls within the range

Figure 12.1. Solar photovoltaic array at Nellis Air Force Base, in Nevada, USA. Since all the panels have a similar orientation, the reflection will be concentrated in one particular direction. This particular array uses single axis trackers turning about a roughly polar axis. Credit: U.S. Air Force photo by Senior Airman Larry E. Reid Jr.

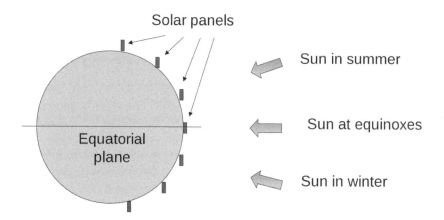

Figure 12.2. Why solar panel orientations are correlated between installations. For maximum power production, each user tries to set their panel normal to the incident radiation. Since the incoming rays are almost parallel, this results in all reflections adding up.

of existing renewable energy mandates, would require little storage as the energy could be used as generated, and could be sited in a number of deserts.

How bright will the reflection be? A very simple estimate of the brightness of the reflection comes from the fraction of the celestial sphere into which the reflection is concentrated. The inverse of this will be the approximate gain. Much more sophisticated models are of course possible, but probably not worthwhile given the uncertainty of the other assumptions.

In the case of a single flat mirror, or an array of accurately aligned panels, then the size of the reflection is dominated by the angular size of the sun, or about 0.5 degrees. This leads to a gain of about 160,000 over isotropic. For an array of the size considered here, the Earth's curvature must be taken into account. If each panel is optimally aligned to the sun, then from North to South on the array, these panels will remain aligned due to the configuration exhibited in Figure 12.2. However, from east to west, each panel will be aligned to its local noon, and so there will be a roughly 1.5 degree spread in the orientations. Added to the 0.5 degree size of the sun, this means the reflection will be roughly 2 degrees east-west by 0.5 degrees north-south, for a total size of about 1 square degree and a total gain of about 40,000.

As the Earth rotates at about 0.25 degrees per minute, from a distant point the reflection is visible for about eight minutes, and during this time it outshines the planet by roughly a factor of 4 (0.01% of area times gain of 40,000).

The area swept out by the reflection during a given day is about 0.5 degrees high by about 270 degrees long, or about 135 square degrees, or about 1/240 of the celestial sphere. However, this arc changes as the Earth goes around the sun, and over the course of a year perhaps 40 percent of the celestial sphere will get a reflection at some time of the year (this depends on the planet's axial tilt, and the 40 percent figure assumes the Earth's tilt is typical).

Reflections such as these could be seen by proposed (but not yet built) telescopes such as Darwin (Fridlund 1999) and Terrestrial Planet Finder (Woolf 2000). These missions, intended to fly by about 2020, aim to examine terrestrial planets around nearby (few tens of light-years) stars by suppressing the light of the central star (by clever interferometry, or apodization, or other techniques) until the planet can be seen directly. Given that these missions could see the unenhanced planet at all, they almost surely could see a 4x increase in brightness. By 2050, presumably techniques at least this good will be in operation. Thus, overall, we see that a year 2050 class solar installation could be noticed by a 2050 class observatory, both based on existing technology. This is the earliest case, at least so far, where a civilization can see another civilization at the same level of development, out to a distance of a few tens of light years.

This ability is not limited to the Darwin and TPF missions. Any technique that works by examining the reflected light from a planet, and is capable of seeing terrestrial size planets, could see these reflections.

4.0. Solar Collector Orientation

The sun moves across the sky from east to west each day, and the elevation of the noonday sun (for Earth) varies roughly ±23 degrees from its elevation at equinox. Photovoltaic systems are most efficient when facing normal to the incident sunlight. The decision of what tracking, if any, to employ is a compromise between simplicity and performance, and many solutions are possible.

As a result, there are many different strategies in use for orienting solar panels. The simplest is to mount the solar panel on the most appropriate existing surface. This is often used when retrofitting existing structures, or when appearance is a primary concern. The next simplest mounting is a fixed orientation, optimized for power production (either when power is needed most, or for year-round total production). Slightly more complex are mounts where the elevation is manually changed on a seasonal basis. The next step up, starting to involve moving parts, are single-axis trackers that follow the sun from east to west, possibly combined with seasonal manual elevation. Dual-axis trackers, the most complex, follow the sun in both axes. Each type of mount creates its own reflection pattern, but all provide considerable gains above isotropic reflection in some directions.

4.1. Most Suitable Existing Surface

One approach, common when retrofitting solar installations, is simply to fix the panel to the most appropriate south-facing (in the northern hemisphere, north-facing in the southern hemisphere) surface. If we assume the most appropriate surface is a roof, the elevation angle will be perhaps within 0.15 radians of optimal. Then the reflection is directed into a belt around the Earth about 0.3 radians high. The total area of this belt is about 2 steradians, compared to the roughly 4π radians available. Thus even this strategy results in a gain of about 6. This could potentially be visible if the area is big, perhaps 0.5 percent of the surface area of the planet or more.

4.2. Fixed Arrays

The most common strategy for a fixed array is to point the panel due south (north in the southern hemisphere), then set the angle from horizontal equal to the latitude of the site. This is the strategy shown in Figure 12.2. Since each installation will face local south (north in the southern hemisphere) this will create the radiation pattern of a cylinder whose axis coincides with the planet's axis. The sun's reflection will be channeled into a narrow band aligned with the equator of the planet. Assuming, as with Earth, that the equatorial plane is not aligned with the orbital plane, the ring of reflections will tilt with the seasons. In the case of Earth, it would tilt ±23 degrees, sweeping out about 40 percent of the sky. The signature would be a twice yearly brightening of the planet's reflection. (Once per year near the ends of the sweep.) On top of this brightening would be a daily modulation depending on the location of the arrays on the planet.

How accurately will the panels be aligned? Each user's alignment may differ, depending on whether the goal is to maximize year-round production, summer production, or winter production. Also, there is no good engineering reason for very accurate alignment, since solar production only falls as the cosine of the misalignment.

In a centralized array, assuming a common technology, one might expect all panels to have orientations within 0.05 radians of each other, good enough to keep alignment losses to well under a percent. Then the reflection is concentrated into 0.0025π steradians of the 4π steradians available. This results in a gain of more than 1600 in brightness, compared with an equal area of nondirectional reflector such as a textured surface. Even with this lesser gain, a somewhat larger array with a solar cell area of 1/1000 of the planet would still generate reflections as bright as the planet itself, if the observer is within the reflection.

Another case might be many arrays, distributed over a continent or portion of a planet. Each of these settings only differs by a few degrees, likely by more than the centralized array since many different mountings might be used. If the band is about 0.1 radian wide (about 6 degrees), then the reflection are contained in band about 0.6 steradians in area, for a gain of about 20, so if solar arrays cover 1 percent of the surface, then a 20 percent bump in brightness would be expected. Since the band is fairly wide in this case, the increased reflection would last about a month.

4.3. Manual Elevation Tracking

In theory, if every user adjusted their panels at the same time, using the same algorithm (a typical one for Earth is to set the panels four times per year, increase the tilt by 15 degrees in the winter, and decrease it by the same amount in the summer, with the tilt equal to the latitude otherwise), then you simply get the same behavior as in the fixed panel case, but with a smaller yearly range of motion. Observers within the smaller band would see perhaps a slightly greater gain, since the skew might be less since a

166

smaller range of compromises is required. Also, the users might see several bumps within the year.

4.4. Single Axis Tracking

The next step up in performance and complexity is single axis tracking, where the arrays track the sun in one dimension. This is almost always the east-west dimension, since the sun traverses a much larger range east-west than north-south, so the potential efficiency gain is much higher. The elevation may be fixed, or reset manually a few times per year as with fixed panels.

. Since power only falls off as cos (Θ), where Θ is the amount of misalignment, exact alignment is not critical—the power only falls off by 0.5 percent for up to 0.1 radian (6 degree) misalignment. Typical systems keep the panels pointed within a few degrees of maximum power.

For installations of this type, the reflection is concentrated on the sun in east-west, but still varies along the year north-south. They generate a much stronger reflection, since the entire planet disk is now reflecting in the same direction. The area covered by the reflection might be 0.1 radian north-south (unchanged from the fixed panel case) in by about 0.1 radians east-west. Therefore, the gain is about 1600. The observer would see this as a bump in planet brightness near the time of secondary eclipse of the planet by its sun. Depending on the tilt of the planet's axis, this can potentially be observed in systems that are far from edge-on and perhaps resolved from the host star.

4.5. Dual Axis Tracking

Dual axis tracking gives the largest efficiency in solar power collection, at the cost of the largest complexity, by keeping the array normal to the incident sublight at all times. Many mechanisms have been developed to keep this alignment, which can result in a 30–40 percent gain in efficiency (Gay et al. 1982; King et al. 2002). Quite different methods are used—clockwork drives and more sophisticated pre-computed ephemerides, differential feedback driven by two or more photosensors, and differential expansion of heated gases or bimetallic strips.

As in single axis trackers, exact alignment is not required and may even be counterproductive, as it may take more power to maintain this alignment than is gained by the higher accuracy. Therefore, a wide range of tracking accuracies is possible. For example, a system that readjusts the panels each hour will have an error that ranges from −7.5 to +7.5 degrees, with an average of 3.25 degrees of misalignment. A simple feedback system (Beltran et al. 2007) has a measured performance of about 3 degrees. So again, as in the single axis case, 0.05 radians might be a typical accuracy.

The observer here will see a gain in brightness similar to, or perhaps slightly greater than, the bump in the single axis case, depending on the accuracy of the elevation tracking. However, unlike the single axis case where the bump may occur in a range around the secondary eclipse, the two axis case will always be nearly aligned with this eclipse.

4.6. Solar Tower and Solar Trough

Two other methods produce electrical power from sunlight by converting it to heat first. These will generate a different type of reflection. There will be no band gap, since the collector simply tries to absorb at all frequencies where it is profitable to do so. The reflection would be in the thermal IR, as determined by the temperature of the collector.

In the solar tower approach, a large field of heliostat mirrors reflect the sun onto a central collector. These systems have excellent alignment characteristics as it is needed for their operation. Any reflections will be limited by the size of the sun seen from Earth, and will create a spot about twice the diameter of the sun, as seen at opposition. This reflection will offer very high gain (about 160,000), but the reflection is always very close to the sun and probably cannot be resolved from it.

Another technology, called the solar trough, is also used for collecting thermal energy. This is a roughly parabolic trough that directs the sunlight to a tube that runs down the axis of the trough. Usually the trough tracks the sun in one axis, normally east-west. This will create a long thin enhanced reflection, centered on the sun in the east-west direction and extending perhaps 30 degrees in elevation from the sun (limited by defocus as you get farther from face-on to the trough).

4.7. Summary

These different patterns of reflection are summarized in Table 12.1. In addition to the yearly effects, there would daily modulation, depending on type of reflector and distribution of facilities on the surface of the planet. For Earth, for example, large-scale solar installations would most likely be in the Sahara Desert of Africa, the Mohave Desert of North America, and the outback of Australia. Depending on which of these cast visible reflections, centralized installations in these spots would lead to between one and three high amplitude (but short) peaks per Earth day. There could also be relatively fast variations from weather, on the order of perhaps thirty minutes. This would indicate that the source is fairly small compared to the size of the planet. Alternatively, arrays might be distributed in a way similar to cities or other infrastructure. In this case, a smoother variation would be obtained,

Table 12.1.

Mounting	Wide × Tall, degrees	Gain	Percent sky covered	Resolved
South facing	180 × 30	8	25%	Yes
Optimum elevation	180 × 4	60	3–30% (depends on tilt)	Yes
Single axis (polar)	4 × 4, moves	3000	3%	Yes
Dual axis	4 × 4, sun centered	3000	3%	No
Thermal trough	4 × 45, sun centered	300	40%	Yes
Solar tower	1 × 1, sun centered	160,000	0.4%	No

This table summarizes different possible solar technologies, the size and gain (over isotropic) of their reflections, the percent of the sky over which reflections might be visible, and whether the peak reflection can happen while the planet can be optically resolved from the host star.

but if the Earth is typical there would still be significant variation, such as might be caused on Earth as North America rotates out of view, to be replaced by the Pacific Ocean.

5.0. Can This Be Visible?

A natural question is whether this is likely to be observed. One specific case was covered in the motivation section, but each case has different considerations. There are two aspects to noticing such a reflection. First, the methods must be sensitive enough to see the reflection against the glare of the star. Second, the observation must be conducted at the right time to see the reflection. Since the power of the reflection is conserved, if the reflection in one direction is increased by a given factor, the area of the sky covered by the reflection must be decreased by at least the same amount. So if the gain for an Earth-like planet would be about 1000, only one in a thousand planets will have the reflection aimed our way at a given time. Thus, timing of observations is critical. The question of whether the star can be resolved from the planet is central to both the questions of visibility and those of timing.

Whether or not the planet is resolved from the star is critical since this implies which of two basic methods can be used for measuring the light from a planet. If the planet and star cannot be resolved, the light from the planet must be analyzed by subtracting the light of the star from the total. This is possible for hot Jupiters, as shown by the Kepler mission, but will be much more difficult for smaller planets in larger orbits. An Earth-size planet is about ten thousand times dimmer than the reflections spotted by Kepler (100x for planet area, 100x for less solar irradiation due to larger distance from the sun). On the other hand, for the subset of such planets that actually transit their parent star, the odds that we are looking at the right time to see a reflection are excellent, since the primary and secondary transit are natural times to study such systems (Deming et al. 2005; Charbonneau et al. 2005; Deming et al. 2006; Richardson et al. 2007).

Alternatively, if the planet can be resolved from the star, the light from the planet can be analyzed alone, or at least with orders of magnitude less background subtraction required. (TPF and Darwin are aiming to decrease the background by factors of at least 10^5.) In this case the reflection will be most likely to be spotted if it occurs at near maximum elongation, which is when these systems will be most easily studied.

Unfortunately, the solar technologies with the brightest reflections happen when the planet is close to the star, as observed by us. Therefore, these measurements would need to be made by subtracting the light of the star.

Since this technique has a low signal to noise ratio even for giant planets close to their star, there is little possibility of using Earth-level observing technology to see Earth-level solar energy technology in this case. Even much larger telescopes may not allow this possibility, since it is fundamentally limited by the noise level of the much brighter star.

As an example of an unresolved observation, consider observing a secondary eclipse of a planet that has a large installation of two axis tracking panels. In this case there are at least three effects—a slow ramp in intensity as the planet approaches the star, an enhanced drop in intensity during the eclipse, and the spectral changes determined by the combination of reflection from the planet and the solar cells. For an Earth-type planet moving in its orbit about one degree per day, a 0.05 radian accuracy means the intensity will start to rise about three days before the transit. With plausible values (0.1% coverage, gain of 1000) the brightness will roughly double. Then, both the planet and the solar reflection will drop to zero as the planet goes behind its star. Both of these would be superimposed on the yearly, roughly sinusoidal, variation expected from the reflected light of the planet (Deming and Seager 2003).

Other techniques, such as fixed orientation arrays, generate less-strong reflections, but they can occur when the planet can be resolved from their host star. Although they have less gain, these can potentially be observed even near maximum elongation, and in these cases Earth-level technology could see Earth-level reflections.

As an example of a resolved observation, consider fixed axis panels, viewed from a point opposite winter solstice and 35 degrees above the ecliptic. In such a case, maximum elongation is reached at vernal and autumnal equinox, and near these times the reflection will be aimed at the observer. At winter solstice, the reflection is aimed above the observer (at a roughly 46 degree angle to the ecliptic), and at summer solstice the reflection is not visible since the observer is seeing the dark side of the planet. With optimistic values, 1 percent coverage and a gain of 60, there would be a roughly 40 percent bump in the reflections for a few weeks around the spring and fall equinoxes, as shown in Figure 12.3. In addition, there would be daily modulation caused by the likely uneven placement of solar arrays on the planet's surface, as shown in Figure 12.4. Since daily modulation of the planet's light curve is also caused by the configuration of continents, ice caps, weather, axial tilt, and other factors, it is already a planned subject of study (Ford et al. 2001), increasing the chances that solar power effects would be seen.

One confounding factor in these predictions might be further advances in photovoltaic technologies. An ideal solar panel would look flat black from all angles, and use all incident radiation to create power. Such an ideal panel would not be noticeable at any great distance.

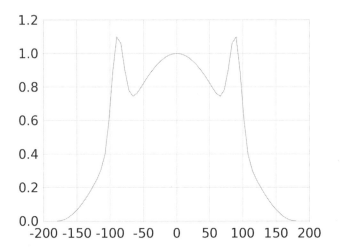

Figure 12.3. Sample brightness curve of a terrestrial planet over the course of a year. Two periods of increased reflectivity are superimposed on the roughly sinusoidal variation from the planet's phase.

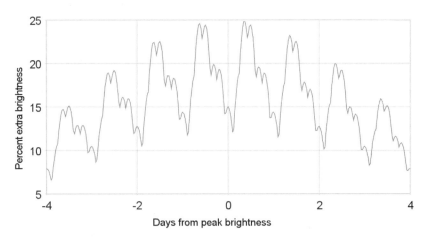

Figure 12.4. Closeup of the peaks from Figure 12.3. In this case the daily structure corresponds to arrays distributed across the planet's continents. The exact form of such variation is unknown, and the data here were created only to illustrate the potential daily variation.

6.0. Implications for the Drake Equation and the Fermi Paradox

The Drake equation is a method used to estimate how many civilizations in the galaxy we might communicate with at any given time. It assumes that communication is deliberate, so that two of the factors included are whether a civilization has the technology to communicate across interstellar distances, and then whether they are willing and able to do so. A civilization that can harness gigawatts of solar power is most probably capable of interstellar communications by radio or optical means should they wish to do so, so the first factor would be largely unchanged. However, the second factor, those willing to communicate, might be very different if the Earth is typical. Although Earth technology is quite capable of strong directed transmissions, there is no consensus about whether we should send and/or what the message should be. On the other hand, there is a growing consensus that renewable energy is an excellent idea, and rapidly improving technology for harvesting it. Therefore, on the current course, the Earth is unlikely to send large-scale explicit messages, but may well construct technologies that are visible across interstellar distances. If we wildly speculate that only 1 percent of civilizations will explicitly communicate, but 50 percent will develop solar power, then our estimate of the number of civilizations we might observe could increase by a factor of fifty. It is also possible that civilizations might not want to advertise their presence, and hence use solar technologies that do not generate detectable reflections. However, if Earth politics is typical and such a design costs more than a basic design, then spending the additional money to avoid reflections seems very unlikely.

7.0. Conclusions

The large-scale use of solar energy may be visible across interstellar distances, and astronomers observing extrasolar planets should be alert to possible signs. This is one the first cases where expected Earth-level technology development will be visible to Earth-level astronomical techniques, in about 2050. This is not dependent on the other civilization doing any more than pursuing their own interest, as we are, and so may result in a higher number of potentially visible civilizations.

Works Cited

Basore, P. A. 1990. Numerical modeling of textured silicon solar cells using pc-1d. *Electron Devices, IEEE Transactions on* 37 (2): 337–43, Feb. 1990. ISSN 0018-9383. doi:10.1109/16.46362.

Beltran, J. A., J. L. S. Gonzalez Rubio, and C. D. Garcia-Beltran. 2007. Design, manufacturing, and performance test of a solar tracker made by a embedded control. *Electronics, Robotics and Automotive Mechanics Conference, 2007. CERMA 2007* (Sept.): 129–34. doi:10.1109/ CERMA.2007.4367673.

Breyer, C., and G. Knies. 2009. Global energy supply potential of concentrating solar power. *Proc. SolarPACES*: 15–18.

Butler, D. 2007. Solar Power: California's latest gold rush. *Nature* 450: 768–69.

Charbonneau, D., L. E. Allen, S. T. Megeath, G. Torres, R. Alonso, T. M. Brown, R. L. Gilliland, D. W. Latham, G. Mandushev, F. T. O'Donovan et al. 2005. Detection of Thermal Emission from an Extrasolar Planet. *The Astrophysical Journal* 626 (1): 523–29.

Deming, D., and S. Seager. 2003. Detecting extrasolar planets in reflected light using COROT and Kepler. In *Scientific Frontiers in Research on Extrasolar Planets*.

Deming, D., S. Seager, L. J. Richardson, and J. Harrington. 2005. Infrared radiation from an extrasolar planet. *Nature* 434 (7034): 740–43.

Deming, D., J. Harrington, S. Seager, and L. J. Richardson. 2006. Strong infrared emission from the extrasolar planet HD 189733b. *The Astrophysical Journal* 644 (1): 560–64.

Ford, E. B., S. Seager, and E. L. Turner. 2001. Characterization of extrasolar terrestrial planets from diurnal photometric variability. *Nature* 412 (6850): 885–87.

Fridlund, C. V. M. 1999. Darwin—The infrared space interferometer. *Proceedings of Darwin & Astronomy—The Infrared Space Interferometer*, 17–19.

Gay, C. F., J. H. Wilson, and J. W. Yerkes. 1982. Performance advantages of two-axis tracking for large flat-plate photovoltaic energy systems. *Conf. Rec. IEEE Photovoltaic Spec. Conf*, 16.

Hoffert, M. I., K. Caldeira, G. Benford, D. R. Criswell, C. Green, H. Herzog, A. K. Jain, H. S. Kheshgi, K. S. Lackner, J. S. Lewis et al. 2002. Advanced technology paths to global climate stability: Energy for a greenhouse planet. *Science* 298 (5595): 981.

Horler, D. N. H., M. Dockray, and J. Barber. 1983. The red edge of plant leaf reflectance. *International Journal of Remote Sensing* 4 (2): 273–88.

King, D. L., W. E. Boyson, and J. A. Kratochvil. 2002. Analysis of factors influencing the annual energy production of photovoltaic systems. *Photovoltaic Specialists Conference, 2002. Conference Record of the Twenty-Ninth IEEE* (May) 1356–61. ISSN 1060-8371.

Richardson, L. J., D. Deming, K. Horning, S. Seager, and J. Harrington. 2007. A spectrum of an extrasolar planet. *Nature(London)* 445 (7130): 892–95.

Woolf, N. 2000. Terrestrial planet finder. *Bioastronomy 99* 213: 143.

Zhao, J., and M. A. Green. 1991. Optimized antireflection coatings for high-efficiency silicon solar cells. *Electron Devices, IEEE Transactions on* 38 (8) (August): 1925–34. ISSN 0018-9383. doi: 10.1109/16.119035.

Interstellar Radio Links Enabled by Gravitational Lenses of the Sun and Stars

Claudio Maccone

1.0. Introduction

The gravitational focusing effect of the Sun is one of the most amazing discoveries produced by the general theory of relativity. The first paper in this field was published by Albert Einstein in 1936 (Einstein 1936), but his work was virtually forgotten until 1964, when Sydney Liebes of Stanford University (Liebes 1964) gave the mathematical theory of gravitational focusing by a galaxy located between the Earth and a very distant cosmological object, such as a quasar.

In 1978 the first "twin quasar" image, caused by the gravitational field of an intermediate galaxy, was spotted by the British astronomer Dennis Walsh and his colleagues. Subsequent discoveries of several more examples of gravitational lenses eliminated all doubts about gravitational focusing predicted by general relativity.

Von Eshleman of Stanford University then went on to apply the theory to the case of the Sun in 1979 (Eshleman 1979). His paper for the first time suggested the possibility of sending a spacecraft to 550 AU from the Sun to exploit the enormous magnifications provided by the gravitational lens of the Sun, particularly at microwave frequencies, such as the hydrogen line at 1420 MHz (21 cm wavelength). This is the frequency that all SETI radioastronomers regard as "magic" for interstellar communications, and thus the tremendous potential of the gravitational lens of the Sun for getting in touch with alien civilizations became obvious.

The first experimental SETI radioastronomer in history, Frank Drake (*Project Ozma*, 1960), presented a paper on the advantages of using the gravitational lens of the Sun for SETI at the *Second International Bioastronomy Conference* held in Hungary in 1987 (Drake 1987), as did Nathan "Chip" Cohen of Boston University (Cohen 1987). Nontechnical descriptions of the topic were also given by them in their popular books (Drake and Sobel 1992, and Cohen 1988).

However, the possibility of planning and funding a space mission to 550 AU to exploit the gravitational lens of the Sun immediately proved a difficult task. Space scientists and engineers first turned their attention to this goal at the June 18, 1992, *Conference on Space Missions and Astrodynamics* organized in Turin, Italy, led by this author. The relevant proceedings were published in 1994 in the *Journal of the British Interplanetary Society* (Maccone 1994). Meanwhile, on May 20, 1993 this author also submitted a formal proposal to the European Space Agency (ESA) to fund the space mission design (Maccone 1993). The optimal direction of space to launch the FOCAL spacecraft was also discussed by Jean Heidmann of Paris Meudon Observatory and the author (Heidmann and Maccone 1994), but it seemed clear that a demanding space mission such as this one should not be devoted entirely to SETI. Things like the computation of the parallaxes of many distant stars in the Galaxy, the detection of gravitational waves by virtue of the very long baseline between the spacecraft and the Earth, plus a host of other experiments would complement the SETI utilization of this space mission to 550 AU and beyond. The mission was dubbed "SETISAIL" in earlier papers (Maccone 1995), and "FOCAL" in the proposal submitted to ESA in 1993.

In the third edition of his book "The Sun as a Gravitational Lens: Proposed Space Missions" (Maccone 2002), the author summarized all knowledge available as of 2002 about the FOCAL space mission to 550 AU and beyond to 1000 AU. On October 3, 1999, this book was awarded the Engineering Science Book Award by the International Academy of Astronautics (IAA).

Finally, in March 2009, the new, four hundred page, comprehensive book by the author, entitled *Deep Space Flight and Communications—Exploiting the Sun as a Gravitational Lens* (Maccone 2009), was published. This book embodies all the previous material published about the FOCAL space mission and updates it.

2.0. Why 550 AU Is the Minimal Distance That "FOCAL" Must Reach

The geometry of the Sun's gravitational lens is easily described: incoming electromagnetic waves (arriving, for instance, from the center of the Galaxy) pass *outside* the Sun and pass within a certain distance r of its center. Then the basic result following from the Schwarzschild solution shows that the corresponding *deflection angle* $\alpha(r)$ at the distance r from the Sun's center is given by

$$\alpha(r) = \frac{4GM_{Sun}}{c^2 r}, \tag{1}$$

Figure 13.1 shows the basic geometry of the Sun's gravitational lens with the various parameters in the game.

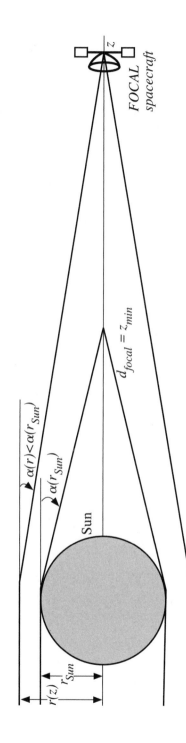

Figure 13.1. Geometry of the Sun gravitational lens with the minimal focal length of 550 AU (= 3.17 light days = 13.75 times beyond Pluto's orbit) and the FOCAL spacecraft position beyond the minimal focal length.

The light rays, that is, electromagnetic waves, cannot pass through the Sun's interior (whereas gravitational waves and neutrinos can), so the largest deflection angle α occurs for those rays just grazing the Sun's surface, that is, for $r = r_{Sun}$. This yields the inequality

$$\alpha(r_{Sun}) > \alpha(r) \tag{2}$$

with

$$\alpha(r_{Sun}) = \frac{4GM_{Sun}}{c^2 r_{Sun}}, \tag{3}$$

From the illustration it should be clear that the minimal focal distance d_{focal} is related to the tangent of the maximum deflection angle by the formula

$$\tan(\alpha(r_{Sun})) = \frac{r_{Sun}}{d_{focal}}. \tag{4}$$

Moreover, since the angle $\alpha(r_{Sun})$ is very small (its actual value is about 1.75 arcseconds), the above expression may be rewritten by replacing the tangent by the small angle itself:

$$\alpha(r_{Sun}) \approx \frac{r_{Sun}}{d_{focal}}. \tag{5}$$

Eliminating the angle $\alpha(r_{Sun})$ between equations (3) and (5), and then solving for the minimal focal distance d_{focal}, one gets

$$d_{focal} \approx \frac{r_{Sun}}{\alpha(r_{Sun})} = \frac{r_{Sun}}{\dfrac{4G\,M_{Sun}}{c^2 r_{Sun}}} = \frac{c^2}{4G} \cdot \frac{r_{Sun}^2}{M_{Sun}}. \tag{6}$$

This basic result may also be rewritten in terms the *Schwarzschild radius*

$$r_{Schwarzschild} = \frac{2G\,M}{c^2}. \tag{7}$$

yielding

$$d_{focal} \approx \frac{r_{Sun}}{\alpha(r_{Sun})} = \frac{r_{Sun}}{\dfrac{4GM_{Sun}}{c^2 r_{Sun}}} = \frac{r_{Sun}^2}{2r_{Schwarzschild}}. \tag{8}$$

Numerically, one finds

$$d_{focal} \approx 542\text{AU} \approx 550\text{AU} \approx 3.171 \text{ light days.} \tag{9}$$

This is the fundamental formula yielding the minimal focal distance of the gravitational lens of the Sun, namely, the minimal distance from the Sun's center that the FOCAL spacecraft must reach in order to get magnified radio pictures of whatever lies on the other side of the Sun with respect to the spacecraft position.

Furthermore, a simple, but very important consequence of the above discussion is that *all points on the straight line beyond this minimal focal distance are foci too*, because the light rays passing by the Sun farther than the minimum distance have smaller deflection angles and thus come together at an even greater distance from the Sun.

And the very important astronautical consequence of this fact for the FOCAL mission is that *it is not necessary to stop the spacecraft at 550 AU. It can go on to almost any distance beyond and focus as well or better.* In fact, the farther it goes beyond 550 AU the less distorted the collected radio waves by the Sun Corona fluctuations. The important problem of Corona fluctuations and related distortions is currently being studied by Von Eshleman and colleagues at Stanford University (Maccone 2009).

We would like to add here one more result that is very important because it holds well not just for the Sun, but for all stars in general. This we will do without demonstration; that can be found in the work of Orta, Savi, and Tascone (1994). Consider a spherical star with radius r_{star} and mass M_{star}, which will be called the "focusing star." Suppose also that a light source (i.e., another star or an advanced extraterrestrial civilization) is located at the distance D_{source} from it. Then ask: How far is the minimal focal distance d_{focal} on the opposite side of the source with respect to the focusing star center? The answer is given by the formula

$$d_{focal} = \frac{r_{star}^2}{\dfrac{4GM_{star}}{c^2} - \dfrac{r_{star}^2}{D_{source}}} \,. \tag{10}$$

This is the key to gravitational focusing for a pair of stars, and may well be the key to SETI in finding extraterrestrial civilizations. It could also be considered for the magnification of a certain source by any star that is perfectly aligned with that source and the Earth: the latter would then be in the same situation as the FOCAL spacecraft except, of course, it is located much farther out than 550 AU with respect to the focusing, intermediate star. Finally, notice that equation (10) reduces to equation (6) in the limity

$D_{source} \to \infty$, that is, (6) is the special case of (10) for light rays approaching the focusing star from an infinite distance.

3.0. The Huge (Antenna) Gain of the Sun's Gravitational Lens

Having thus determined the minimal distance of 550 AU that the FOCAL spacecraft must reach, one now wonders what's the good of going so far out of the solar system, that is, how much focussing of light rays is caused by the gravitational field of the Sun. The answer to such a question is provided by the technical notion of "antenna gain," which stems from antenna theory.

A standard formula in antenna theory relates the antenna gain, $G_{antenna}$, to the antenna effective area $A_{effective}$, and to the wavelength λ or the frequency v by virtue of the equation (refer, for instance, to Kraus 1966, in particular page 6-117, equation [6-241]):

$$G_{antenna} = \frac{4\pi \, A_{effective}}{\lambda^2} . \qquad (11)$$

Now, assume the antenna is circular with radius $r_{antenna}$, and assume also a 50 percent efficiency. Then, the antenna effective area is obviously given by

$$A_{effective} = \frac{A_{physical}}{2} = \frac{\pi \, r_{antenna}^2}{2} . \qquad (12)$$

Substituting this back into (11) yields the antenna gain as a function of the antenna radius and of the observed frequency:

$$G_{antenna} = \frac{4\pi \, A_{effective}}{\lambda^2} = \frac{2\pi A_{physical}}{\lambda^2} = \frac{2\pi^2 r_{antenna}^2}{\lambda^2} = \frac{2\pi^2 r_{antenna}^2}{c^2} \cdot v^2 . \qquad (13)$$

The important point here is that *the antenna gain increases with the square of the frequency,* thus favoring observations on frequencies as high as possible.

Is anything similar happening for the Sun's gravitational lens also? *Yes* is the answer, and the "gain" (one maintains this terminology for convenience) of the gravitational lens of the Sun can be proved to be

$$G_{Sun} = 4\pi^2 \frac{r_{Schwarzschild}}{\lambda} \qquad (14)$$

or, invoking the expression (7) of the Schwarzschild radius

$$G_{sun} = \frac{8\pi^2 GM_{Sun}}{c^2} \cdot \frac{1}{\lambda} = \frac{8\pi^2 GM_{Sun}}{c^3} \cdot \nu. \tag{15}$$

The mathematical proof of equation (14) is difficult to achieve. The author, unsatisfied with the treatment of this key topic given by Einstein (1936), Kraus (1966), and Eshleman (1979), turned to three engineers of the engineering school in his home town: Renato Orta, Patrizia Savi, and Riccardo Tascone. To his surprise, in a few weeks they provided a full proof of not just the Sun gain formula (14), but also of the focal distance for rays originated from a source at finite distance, equation (10). Their proof is fully described by Orta, Savi, and Tascone (1994), and is based on the aperture method used to study the propagation of electromagnetic waves, rather than on ray optics.

Using the words of these three authors' own Abstract, they have "computed the radiation pattern of the [spacecraft's] Antenna+Sun system, which has an extremely high directivity. It has been observed that the focal region of the lens for an incoming plane wave is a half line parallel to the propagation direction starting at a point [550 AU] whose position is related to the blocking effect of the Sun disk (Figure 1). Moreover, a characteristic of this thin lens is that its gain, defined as the magnification factor of the antenna gain, is constant along this half line. In particular, for a wavelength of 21 cm, this lens gain reaches the value of 57.5 dB. Also a measure of the transversal extent of the focal region has been obtained. The performance of this radiation system has been determined by adopting a thin lens model which introduces a phase factor depending on the logarithm of the impact parameter of the incident rays. Then the antenna is considered to be in transmission mode and the radiated field is computed by asymptotic evaluation of the radiation integral in the Fresnel approximantion."

One is now able to compute the Total Gain of the Antenna+Sun system, which is simply obtained by multiplying equations, the two equations yielding the spacecraft gain proportional to ν^3 and the Sun gain proportional to ν:

$$G_{Total} = G_{Sun} \cdot G_{antenna} = \frac{16\pi^2 M_{Sun} r^2_{antenna}}{c^5} \cdot \nu^3 \tag{16}$$

Since the total gain increases with the *cube* of the observed frequency, it favors electromagnetic radiation in the microwave region of the spectrum. Table 13.1 shows the numerical data provided by the last equation for five

Table 13.1.

Line	Neutral Hydrogen	OH radical		H_2O
Frequency ν	1420 MHz	327 MHz	1.6 GHz	5 GHz	22 GHz
Wavelength λ	21 cm	92 cm	18 cm	6 cm	1.35 cm
S/C Antenna Beamwidth	1.231 deg	5.348 deg	1.092 deg	0.350 deg	0.080 deg
Sun Gain	57.4 dB	51.0 dB	57.9 dB	62.9 dB	69.3 dB
12-meter Antenna S/C Gain	42.0 dB	29.3 dB	43.1 dB	53.0 dB	65.8 dB
Combined Sun + S/C Gain	99.5 dB	80.3 dB	101.0 dB	115.9 dB	135.1 dB

Table showing the gain of the Sun's lens alone, the gain of a 12 m spacecraft (S/C) antenna and the combined gain of the Sun + S/C Antenna system the at five selected frequencies important in radioastronomy.

selected frequencies: the hydrogen line at 1420 MHz and the four frequencies that the Quasat radio astronomy satellite planned to observe, had it been built jointly by ESA and NASA as planned before 1988, but Quasat was abandoned by 1990 due to lack of funding. The definition of dB is, of course that N measured in dB equals:

$$N \ dB = 101 Log_{10}N = 10 lnN/ln10.$$

4.0. The 2009 Book by the Author about the "FOCAL" Space Mission

In March 2009, the new, comprehensive, four hundred page book by the author, entitled *Deep Space Flight and Communications: Exploiting the Sun as a Gravitational Lens*, was published. This book embodies all the previous material published about the FOCAL space mission and updated it in view of submitting a formal proposal to NASA about FOCAL. The front and back covers of this book are reproduced in Figure 13.2.

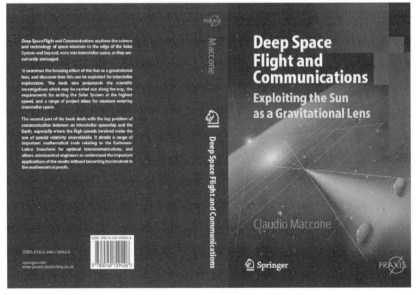

Figure 13.2. Front and back covers of the author's new book entitled *Deep Space Flight and Communications: Exploiting the Sun as a Gravitational Lens* published by Springer-Praxis (Maccone 2009).

5.0. The Radio Link

The goal of this chapter is to prove that only by exploiting the Sun as a gravitational lens we will be able to have reliable telecommunication links across large interstellar distances. In other words, a direct link between different star systems, even if held by virtue of the largest radio telescopes on Earth, will not be feasible across distances of the order of thousands of light years or more. We want to show that only a FOCAL mission in the direction from Earth opposite to that target star system will ensure a reliable telecommunication link across thousands of light years. Namely, we prove that Bit Error Rate (or BER, see site http://www.en.wikipedia.org/wiki/Bit_error_rate) will be unacceptable already at the distance of Alpha Centauri unless we resort to supporting a FOCAL space mission in the opposite direction from the Sun.

In order to face these problems mathematically, we must first understand the radio link among any two stars, and we think that no neater treatment of this subject exists than the book *Radio Astronomy* by the late professor John D. Kraus of Ohio State University (Kraus 1966), which we follow hereafter.

Consider a radio transmitter that radiates a Power P_t isotropically and uniformly over a bandwidth B_t. Then, at a distance r it produces a flux density given by

$$\frac{P_t}{B_t \, 4\pi \, r^2} \tag{17}$$

A receiving antenna of effective aperture A_{er} at a distance r can collect a power given by (17) multiplied by both the effective aperture of the receiving antenna and its bandwidth, namely the received power P_r is given by

$$P_t = \frac{P_t}{B_t \, 4\pi \, r^2} A_e B_t . \tag{18}$$

It is assumed that the receiving bandwidth B_r is smaller or, at best (in the "matched bandwidths" case) equal to the transmitting bandwidth B_t, that is $B_r \le B_t$.

So far, we have been talking about an isotropic radiator. But let us now assume that the transmitting antenna has a directivity D that is an antenna gain in the sense of (11):

$$D = \frac{4\pi \, A_{et}}{\lambda^2} . \tag{19}$$

The received power P_r is then increased by just such a factor due to the directivity of the transmitting antenna, and so (18) must now be replaced by a new equation where the righthand side is multiplied by such a increased factor, that is

$$P_r = \frac{4\pi A_{et}}{\lambda^2} \cdot \frac{P_t}{B_t 4\pi r^2} A_{er} B_r . \tag{20}$$

Rearranging a little, this becomes

$$P_r = \frac{P_t A_{rt} A_{er}}{r^2 \lambda^2} \cdot \frac{B_r}{B_t} . \tag{21}$$

This is the *received signal power* expression. For the matched bandwidths case, that is, for $B_r = B_t$, this is called the Friis transmission formula, since it was first published back in 1946 by the American radio engineer Harald T. Friis (1893–1976) of the Bell Labs. In space missions, we of course know exactly both B_t and B_r and so we construct the spacecraft in such a way the two *bands match exactly*, namely, $B_r = B_t$. So, for the case of telecommunications with a spacecraft (but not necessarily for the SETI case) we may well assume the matched bandwidths and have (21) reducing to

$$P_r = \frac{P_t A_{et} A_{er}}{r^2 \lambda^2} . \tag{22}$$

Let us now rewrite (22) in such a way that we may take into account the gains (i.e., directionalities) of both the transmitting and receiving antennae, that is, in agreement with (11)

$$\begin{cases} G_t = \dfrac{4\pi A_{et}}{\lambda^2} \\[2mm] G_r = \dfrac{4\pi A_{er}}{\lambda^2} \end{cases} \quad \text{that is} \quad \left. \begin{array}{l} A_{et} = \dfrac{G_t \lambda^2}{4\pi} \\[2mm] A_{er} = \dfrac{G_r \lambda^2}{4\pi} \end{array} \right\} \tag{23}$$

Replacing the last two expressions into (22), we find that (22) is turned into

$$P_r = \frac{P_t G_t G_r}{(4\pi)^2 r^2} \cdot \lambda^2 . $$

187

This may finally be rewritten in the more traditional form

$$P_r = \frac{P_t\, G_t\, G_r}{L(r,\lambda)}. \tag{24}$$

if one defines

$$L(r,\lambda) = (4\pi)^2 \cdot \frac{r^2}{\lambda^2} \tag{25}$$

that is the Path Loss (or path attenuation), to wit, the reduction in power density (attenuation) of the electromagnetic waves as they propagates through space. Path loss is a major component in the analysis and design of the link budget of a telecommunication system; see the site http://www.en.wikipedia. org/wiki/Path_loss.

6.0. Bit Error Rate for an "Ordinary" Direct Link with a Probe at the Alpha Centauri Distance

In this section we first define the Bit Error Rate (BER). Then, by virtue of a numerical example, we show that, even at the distance of the nearest star (Alpha Cen at 4.37 AU) the telecommunications would be impossible by the ordinary powers available today for interplanetary space flight. But in the next section we shall show that the telecommunications would become feasible if we could take advantage of the magnification provided by the Sun's gravity lens, that is, if we would send out to 550 AU a FOCAL relay spacecraft for each target star system that we wish to communicate with. And this is the first key new result presented in this chapter.

So, let us start by defining the Bit Error Rate or BER. In telecommunication theory an *error ratio* is the ratio of the number of bits, elements, characters, or blocks incorrectly received to the total number of bits, elements, characters, or blocks sent during a specified time interval. Among these error ratios, the most commonly encountered ratio is the *bit error ratio* (BER)—also called *bit error rate*—that is, the number of erroneous bits received divided by the total number of bits transmitted. At the bit error rate Wikipedia site http://www.en.wikipedia.org/wiki/Bit_error_rate it is shown that the likelihood of a bit misinterpretation

$$p_e = p(0 \mid 1)p_1 + p(1 \mid 0)p_0. \tag{26}$$

(believing that we have received a 0 while it was a 1 or the other way round) is basically given by the "complementary error function" or erfc(x) as follows

$$BER(d, \nu, P_t) = \frac{1}{2} \, erfc \left(\sqrt{\frac{E_b(d, \nu, P_t)}{N_0}} \right). \qquad (27)$$

In this equation one has:

1. d = distance between the transmitting station on Earth and the receiving antenna in space. For instance, this could be the antenna of a precursor interstellar space probe that was sent out to some light years away.

2. ν = frequency of the electromagnetic waves used in the telecommunication link. The higher this frequency, the better it is, since the photons are then more energetic ($E = h\,\nu$). In today's practice, however, the highest ν for spacecraft links (such as the link of the Cassini probe, now at Saturn) are the ones in the Ka band, that is $\nu_{Ka} \approx 32$ GHz.

3. P_t is the power in watts transmitted by the Earth antenna, typically a NASA Deep Space Network antenna 70 meters in diameter.

4. The complementary error function $erf(x)$ is defined by the integral

$$erfc(x) = \frac{2}{\sqrt{\pi}} \int_x^\infty e^{-t^2} dt \qquad (28)$$

(for more mathematics, see the relevant Wikipedia site http://www.en.wikipedia.org/wiki/Complementary_error_function).

5. $E_b(d, \nu, P_t)$ is the received energy per bit, that is, the ratio

$$E_b(d, \nu, P_t) = \frac{P_r(d, \nu, P_t)}{\text{Bit\_rate}} \qquad (29)$$

6. Finally, N_0 is given by the Boltzmann's constant k multiplied by the noise temperature of space far away from the Sun and from any other star. This "empty space noise temperature" might be assumed to equal, say, 100 K.

This is the analytical structure of the MathCad code that this author wrote to yield the BER. Let us now consider the input values that he used in practice:

1. Suppose that a human space probe has reached the Alpha Cen system at 4.37 light year distance from the Sun: then, d = 4.37 light years.

2. Suppose also that the transmitting antenna from the Earth is a typical NASA Deep Space Network (DSN) antenna having a diameter of 70 meters (like those at Goldstone, Madrid, and Canberra), and assume that its efficiency is about 50 percent.

3. Suppose that the receiving antenna aboard the spacecraft is 12 meters in diameter (it might be an inflatable space antenna, as we supposed in Maccone [2009] for the FOCAL spacecraft), and assume a 50 percent efficiency.

4. Suppose that the link frequency is the Ka band (i.e., 32 GHz), as for the Cassini highest frequency.

5. Suppose that the bit rate is 32 kbps = 32000 bit / second. This is the bit rate of ESA's Rosetta interplanetary spacecraft now on its way to a comet.

6. And finally (this is the most important input assumption), suppose that the transmitting power P_t is moderate: just 40 watts.

Then:

1. The gain of the transmitting NASA DSN antenna (at Ka frequency) is about 84 dB.

2. The gain of the spacecraft antenna is about 69 dB.

3. The path loss at the distance of Alpha Cen is 395 dB (a very high indeed path loss with respect to today's interplanetary missions, of course).

4. The power received by the spacecraft at that distance is 2.90 $\times$ 10^{-23} watt.

5. The received energy per bit (lowered by the noise temperature of the space in between the Sun and Alpha Cen) is 1.3 $\times$ 10^{-37} joule.

6. And finally the BER is 0.49, that is, there is a 50 percent probability of errors in the telecommunications between the Earth and the probe at Alpha Cen. if we use such a small transmitting power.

In other words, if these are the telecommunication links between the Earth and our probe at Alpha Cen, then this precursor interstellar mission is *worthless*.

The key point in this example is that, for all calculations, (24) and (25) were used without taking the gain of the Sun's gravity lens into account, because this was a direct link and not a FOCAL mission.

7.0. Bit Error Rate at the Alpha Centauri Distance Enhanced by the Magnification Provided by the Sun's Gravity Lens (FOCAL)

The disappointing BER results of the previous section are totally reversed, however, if we suppose that a FOCAL space mission has been previously sent out to 550 AU in the direction opposite to Alpha Cen so that we now have the *magnification* of the Sun's gravity lens playing in the game.

Mathematically, this means that we must introduce a third multiplicative gain at the numerator of (24): the Sun's gravity lens gain, given by (14) where the Schwarzschild radius of the Sun is given by (7).

This new gain is huge at the Ka band frequency:

$$G_{Sun}(v_{Ka}) = 12444837 \sim 70 \text{ dB} \tag{30}$$

and so the received power (24) at Alpha Cen, with the usual Earth-transmitted power of just 40 watts becomes

$$P_r = 2.9 \times 10^{-23} \text{ watts} \tag{31}$$

and the relevant BER becomes absolutely acceptable:

$$\text{BER} = 0.000000526387845 \tag{32}$$

This should convince anybody that the FOCAL space mission is indispensable to keep the link at interstellar distances equal to or higher than Alpha Cen.

8.0. The "Radio Bridge" between the Sun and Alpha Cen A by Using the Two Gravitational Lenses of Both Just Matched to the Other

In this section we provide one more new result: we define the radio bridge between the Sun and Alpha Cen A by using BOTH gravitational lenses. In other words, suppose that in future we will be able to send a probe to

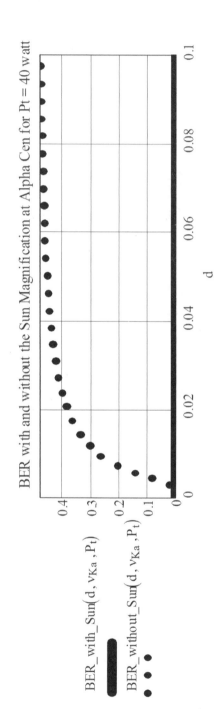

BER with and without the Sun Magnification at Alpha Cen for Pt = 40 watt

BER_with_Sun(d, v_{Ka}, P_t)

BER_without_Sun(d, v_{Ka}, P_t)

DISTANCE of interstellar probe from the Sun (light years)

Figure 13.3. The Bit Error Rate (BER) (upper, dot-dot curve) tends immediately to the 50 percent value (BER = 0.5) even at moderate distances from the Sun (0 to 0.1 light years) for a 40 w transmission from a DSN antenna that is a *direct* transmission, i.e., without using the Sun's Magnifying Lens. On the contrary (lower solid curve) the BER keeps staying at zero value (perfect communications) if the FOCAL space mission is made, so the Sun's magnifying action is made to work.

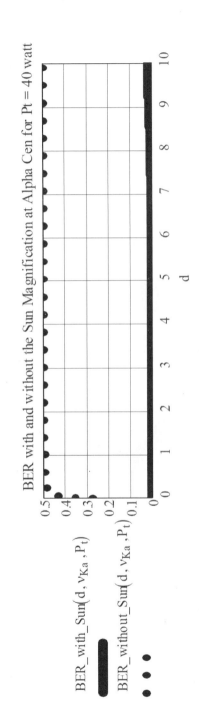

BER with and without the Sun Magnification at Alpha Cen for Pt = 40 watt

BER_with_Sun(d, v_{Ka}, P_t)

BER_without_Sun(d, v_{Ka}, P_t)

DISTANCE of interstellar probe from the Sun (light years)

Figure 13.4. Same as in Figure 13.3, but for probe distances up to ten light-years. We see that at about nine light-years away the BER curve starts being not exactly flat any more, and starts increasing slowly.

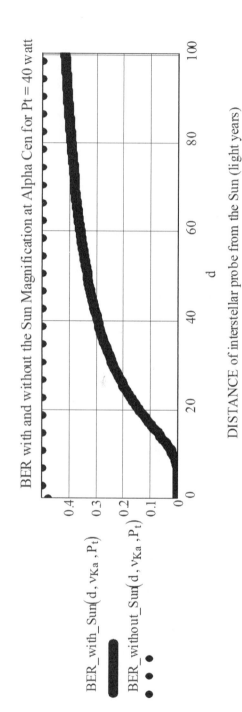

BER with and without the Sun Magnification at Alpha Cen for P_t = 40 watt

BER_with_Sun$\left(d, \nu_{Ka}, P_t\right)$

BER_without_Sun$\left(d, \nu_{Ka}, P_t\right)$

DISTANCE of interstellar probe from the Sun (light years)

Figure 13.5. Same as in Figure 13.4, but for probe distances up to one hundred light-years. We see that, from nine light-years onward, the Sun-BER increases, reaching the dangerous level of 40 percent (Sun-BER = 0.4) at about one hundred light-years. Namely, at one hundred light-years even the Sun's Lens cannot cope with this very low transmitted power of 40 w.

Alpha Cen A and suppose that we succeed in placing this probe just on the other side of Alpha Cen A with respect to the Sun and at the minimal focal distance typical of Alpha Cen A. This distance is NOT 550 AU, obviously, because both the radius and the mass of Alpha Cen A are different (actually slightly higher) than the values of the Sun:

$$\begin{cases} r_{Alpha\_Cen\_A} = 1.227 \ r_{Sun} \\ M_{Alpha\_Cen\_A} = 1.100 \ M_{Sun} \ . \end{cases} \tag{33}$$

Replacing these values into (6) (obviously rewritten for Alpha Cen A), the relevant minimal focal distance is found

$$d_{focal\_Alpha\_Cen\_A} \approx \frac{c^2 r^2_{Alpha\_Cen\_A}}{4GM_{Alpha\_Cen\_A}} \approx 749 \ \text{AU}. \tag{34}$$

The Schwarzschild radius for Alpha Cen A is given by

$$r_{Schwarzschild} = \frac{2GM_{Alpha\_Cen\_A}}{c^2} = 3.248 \ \text{km}. \tag{35}$$

and so the gain, provided by (14), turns out to equal

$$G_{Alpha\_Cen\_A}(\nu_{Ka}) = 4\pi^2 \frac{r_{Schwarzschild\_Alpha\_Cen\_A}}{\lambda_{Ka}} = 13689321. \tag{36}$$

That is,

$$G_{Alpha\_Cen\_A}(\nu_{Ka}) \approx 71 \ \text{dB}. \tag{37}$$

Incidentally, we chose Alpha Cen A, and not B or C, because it has the highest mass, and so the highest gain, in the whole Alpha Cen triple system. The future telecommunications between the Sun and the Alpha Cen system are thus optimized by selecting Alpha Cen A as the star on the other side of which to place a FOCAL spacecraft at the minimal distance of 750 AU. That FOCAL spacecraft would then easily relay its data anywhere within the Alpha Cen system.

Having found the Alpha Cen A gain (37) we are now able to write the new equation corresponding to (24) for the Sun–Alpha Cen bridge. In fact, we must now put at the numerator of (24) three gains:

1. The Sun gain at 32 GHz,

2. The Alpha Cen A gain at 32 GHz, and

3. The 12 meter FOCAL antenna gain at 32 GHz raised to the square because there are two such 12 meter antennas: one at 550 AU from the Sun and one at 749 AU from Alpha Cen A, and they must be perfectly aligned with the axis passing thru both the Sun and Alpha Cen A.

Thus, the received power given by (24) now reads

$$P_r = \frac{P_t G_{Sun} \, G_{Alpha\_Cen\_A} (G_{12\_meter\_antennta\_at\_Ka})^2}{L(r, \lambda)} \tag{38}$$

where, obviously, r equals 4.37 light years and λ corresponds to a 32 GHz frequency.

Let us now go back to the BER and replace (38) instead of (24) in the long chain of calculations described in section 6. Since the received power P_r has now changed, clearly both (29) and (27) yield different numerical results. But now:

1. The link frequency has been fixed at 32 GHz (Ka band), and so no longer is an independent variable in the game.

2. Also the distance d has been fixed (it is the distance of Alpha Cen A) and so it no longer an independent variable in the game.

3. It follows that in (24) and (25) the only variable to be free to vary is now the transmitted power, P_t.

Let us rephrase the last sentence in different terms. Practically, we are now studying the BER as a function of the transmitted power P_t only and, physically, this mean that:

1. We start by inputting very low transmission powers in watts, and find out that the BER is an awful 50 percent, that is, the telecommunications between the Sun and Alpha Cen are totally disrupted. This is of course because the energy per bit is so much lower than the empty space noise temperature.

2. We then increase the transmitted power, at a certain point the BER starts getting smaller than 50 percent. And so it gets

smaller and smaller until the transmitted power is so high that the BER gets down to zero and the telecommunications are just perfect.

3. But the surprise is that for the Sun–Alpha Cen direct radio bridge exploiting both the two gravitational lenses, this minimum transmitted power is incredibly small. Actually, it just equals less than 10^{-4} watts, that is, one-tenth of a milliwatt is enough to have perfect communication between the Sun and Alpha Cen through two 12 meter FOCAL spacecraft antennas. How is that possible?

4. Well, that is the capacity given by gravitational lenses to both explore the universe and make a link with other stars. As a reminder, in 2009 the discovery of the first extrasolar planet in the Andromeda galaxy (M31) was announced because of the gravitational lens caused by something in between.

9.0. The "Radio Bridge" between the Sun and Barnard's Star Using the Two Gravitational Lenses of Both

The next closest star to the Sun beyond the triple Alpha Cen system is Barnard's star (see, for instance, the Web site http://www.en.wikipedia. org/wiki/Barnard's_Star). Let us now repeat for the gravitational lens of Barnard's star the same calculations that we did in the previous section for Alpha Cen A. Then one has:

$$\begin{cases} d_{Barnard} = 5.98 \text{ light years} \\ r_{Barnard} = 0.17 \ r_{Sun} \\ M_{Barnard} = 0.16 \ M_{Sun}. \end{cases} \tag{39}$$

Barnard's star is thus just a small red star, that is actually "passing by" the Sun right now and is not known to have planets around it, As a consequence of the numbers listed in (39), one infers that

$$\begin{cases} d_{focal\_Barnard} = 98 \text{ AU} \\ r_{Schwarzschild\_Barnard} = 0.47 \text{ km} \\ G_{Barnard}(\nu_{Ka}) = 1991174. \end{cases} \tag{40}$$

Especially the gain is important to us:

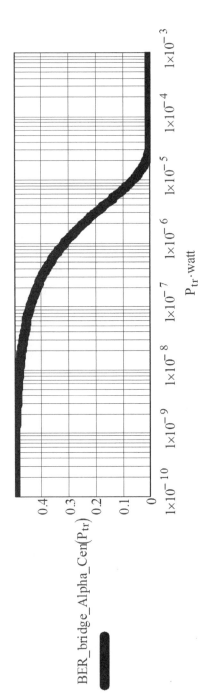

$P_{tr} \cdot$ watt

$BER\_bridge\_Alpha\_Cen(P_{tr})$

Figure 13.6. Bit Error Rate (BER) for the double-gravitational-lens system giving the radio bridge between the Sun and Alpha Cen A. In other words, there are two gravitational lenses in the game here: the Sun one and the Alpha Cen A one, and two 12 m FOCAL spacecrafts are supposed to have been put along the two-star axis on opposite sides at or beyond the minimal focal distances of 550 AU and 749 AU, respectively. This radio bridge has an *overall gain so high* that a mere 10^{-4} w transmitting power is sufficient to let the BER get down to zero, i.e., to have perfect telecommunications. Notice also that the scale of the horizontal axis is logarithmic, and the trace is solid. This will help us to distinguish this curve from the similar curve for the Barnard's Star.

$$G_{Barnard}(v_{Ka}) = 63 \text{ dB.} \tag{41}$$

We replace this into the Bernard's star equivalent of (38), again supposing that two 12 meter FOCAL spacecraft antennas are placed along the Sun-Barnard straight line at or beyond 550 AU and 100 AU, respectively. The result is the new graph of the BER as a function of the transmitted power only, as in Figure 13.7.

10.0. The "Radio Bridge" between the Sun and Sirius A Using the Two Gravitational Lenses of Both

The next star we want to consider is Sirius A. This is because Sirius A is a big, massive bluish star and so it is completely different from both Alpha Cen A (which is a Sun-like star) and Barnard's star, which is a small red star. Data may again be taken from the Wikipedia site http://www.en.wikipedia.org/wiki/Sirius, and one gets:

$$\begin{cases} d_{Sirius\_A} = 8.6 \text{ light years} \\ r_{Sirius\_A} = 1.711 \ r_{Sun} \\ M_{Sirius\_A} = 2.02 \ M_{Sun}. \end{cases} \tag{42}$$

From these data one gets:

$$\begin{cases} d_{focal\_Sirius\_A} = 8.6 \text{ light years} \\ r_{Schwarzschild\_Sirius\_A} = 1.711 \ r_{Sun} \\ G_{Sirius\_A}(v_{Ka}) = 251385723. \end{cases} \tag{43}$$

The important thing is, of course, the gain:

$$G_{Sirius\_A}(v_{Ka}) = 74 \text{ dB.} \tag{44}$$

Then, one replaces this into the Sirius A equivalent of (38), again supposing that two 12 meter FOCAL spacecraft antennas are placed along the Sun-Sirius A straight line at or beyond 550 AU and 793 AU, respectively. The result is the new graph of the BER as a function of the transmitted power only, as in Figure 13.8.

11.0. The "Radio Bridge" between the Sun and Another Sun-like Star Located at the Galactic Bulge Using the Two Gravitational Lenses of Both

Tempted by the suggestion to increase the distance of the second star more and more, and then see what our calculations yield, we now imagine that the

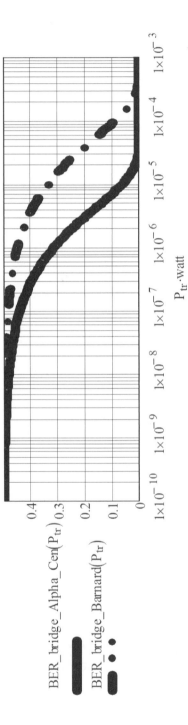

Figure 13.7. Bit Error Rate (BER) for the double-gravitational-lens of the radio bridge between the Sun and Alpha Cen A (solid curve) plus the same curve for the radio bridge between the Sun and Barnard's star (dash-dot curve): for it, 10^{-3} w are needed to keep the BER down to zero, because the gain of Barnard's star is so small when compared to Alpha Centauri A's.

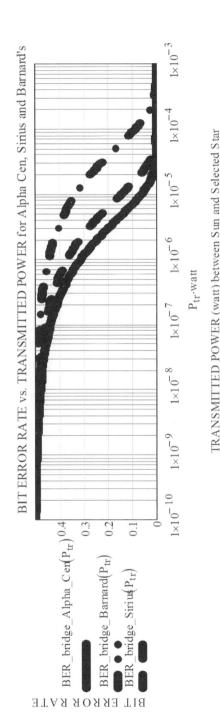

Figure 13.8. Bit Error Rate (BER) for the double-gravitational-lens of the radio bridge between the Sun and Alpha Cen A (solid curve), plus the same curve for the radio bridge between the Sun and Barnard's star (dash-dot curve), plus the same curve of the radio bridge between the Sun and Sirius A (dash-dash curve). From this last curve we see that only 10^{-4} w are needed to keep the BER down to zero, because the gain of Sirius A is so big when compared the gain of the Barnard's star that it gets closer to Alpha Cen A's gain even if Sirius A is so much further out than Barnard's star. In other words, the star's gain and the size combined matter even more than its distance.

second star is Sun-like (i.e., that it has exactly the same radius and mass as the Sun) but is located . . . inside the Galactic Bulge. Namely 26,000 light years away, according to the Wikipedia "Milky Way Galaxy" site http://www.en.wikipedia.org/wiki/Milky_Way. So, the equivalent of (24) and (38) now becomes

$$P_r = \frac{P_t (G_{Sun\_at\_Ka})^2 (G_{12\_meter\_antennta\_at\_Ka})^2}{L(r, \lambda)} \tag{45}$$

and the plot of the BER versus transmitted power is shown in Figure 13.9 as the new, dot-dot curve at the far right of the previous three curves of Alpha Cen A (solid), Barnard's star (dash-dot), and Sirius A (dash-dash). The new dot-dot curve showing the BER of a Sun-like star at the Galactic Bulge is naturally much to the right of the previous three stellar curves inasmuch as the Bulge distance of 26,000 light years is so much higher than the distances of the three mentioned nearby stars (less then ten light years away anyway). The horizontal axis scale is much higher now, since the dot-dot BER curve gets to zero only for transmitted power of about 1000 watts.

12.0. The "Radio Bridge" between the Sun and Another Sun-like Star Located inside the Andromeda Galaxy (M31) Using the Two Gravitational Lenses of Both

We conclude this chapter by calculating the radio bridge between the Sun and another star in the Andromeda galaxy. The distance is now 2.5 million light years, but the bridge would still work if the transmitted power was of higher than about 10^7 watts = 10 Megawatt. This is shown by the new solid curve on the far right in Figure 13.10. This idea may not be as surprising as it might initially appear, given that recently (June 2009) the first extrasolar planet in the Andromeda galaxy was announced to have been discovered just by gravitational lensing. See, for instance, the Web site: http://www.redorbit.com/.../possible_planet_found_outside_our_galaxy/index.html.

13.0. Drake's Alternative Proposal of Power Gain of a Star as a Gravitational Lens

On February 2, 2010, this author received an e-mail message from Frank Drake in which he proposed a new definition of Power Gain of any star as a gravitational lens. In words, Drake proposed the Power Gain of any star to be proportional to the *square* of the ratio of the star's Schwarzschild radius

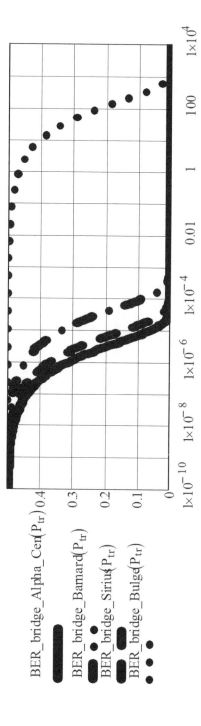

Figure 13.9. Bit Error Rate (BER) for the double-gravitational-lens of the radio bridge between the Sun and Alpha Cen A (solid curve) plus the same curve for the radio bridge between the Sun and Barnard's star (dash-dot curve) plus the same curve of the radio bridge between the Sun and Sirius A (dash-dash curve). In addition, to the far right we now have the dot-dot curve showing the BER for a radio bridge between the Sun and another Sun (identical in mass and size) located inside the Galactic Bulge at a distance of 26,000 light-years. The radio bridge between these two Suns works and their two gravitational lenses works perfectly (i.e., BER = 0) if the transmitted power is higher than about 1000 w.

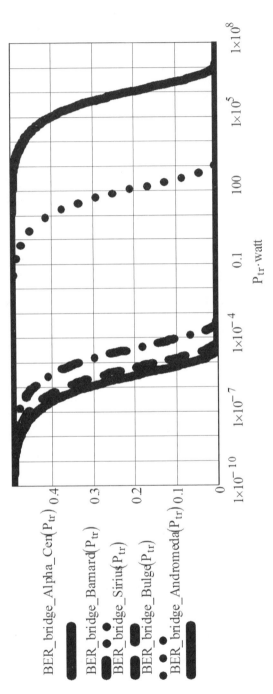

Figure 13.10. The same four Bit Error Rate (BER) curves as shown in Figure 13.9 plus the new solid curve appearing here on the far right: this is the BER curve of the radio bridge between the Sun and another Sun just the same but located somewhere in the Andromeda Galaxy (M 31). Notice that this radio bridge would work fine (i.e., with BER = 0) if the transmitting power was at least 10^7 w = 10 Mw.

to the transmission's wavelength, that is,

$$G_{Sun} \propto \left(\frac{r_{Schwarzschild}}{\lambda} \right). \tag{46}$$

Drake declared this to be the *"true" Power Gain*, whereas (14) was just to be regarded by him as the *voltage gain*, in full analogy to what engineers say when they make the distinction between power and voltage for an electrical current. Drake did not say anything about the proportionality factor (46). Following the receipt of his message, this author recomputed all the star gains contained in this paper in order to match them with the previously computed values, respectively. In doing so, however, we was forced to assume a certain proportionality factor in (46), that, in analogy to (14), he assumed to be equal to $4\pi^2$.

To see the impact of using using Drake's new definition of Power Gain, consider the Bit Error Rate (BER) for the double gravitational lens system giving the radio bridge between the Sun and Alpha Cen A. Specifically, let us consider *two different definitions* of a star's *power gain*:

1. The definition given by equation (14), that we shall now call "Kraus Gain" since it was apparently given for the first time by John D. Kraus in his book *Radio Astronomy* in 1966. Kraus, however, stated the formula without proving it in his book.

2. The definition given by Frank Drake in his reported message of February 2, 2010, to this author, which we shall now call "Drake Gain." Notice, however that Drake did *not* say what the multiplicative factor is (4 pi square?) in front of his proposed ratio given by the Schwarzschild radius over the wavelength *both squared*. To make the computation feasible, this author simply *assumed* that the factor is 4 pi square.

To compare these definitions of gain, Figure 13.11 shows two curves: The Kraus Gain (solid *thick* curve), and The Drake Gain (solid *thin* curve, labeled by a D_ in front of the BER, i.e. D_BER).

We immediately see that the Drake Gain yields *far better values* than the Kraus Gain, as predicted by Drake in his e-mail message to the author. The improvement is of the order of *twenty orders of magnitude in the minimal transmitted power*. The results of this author's other new calculations are shown in Figures 13.12–13.16.

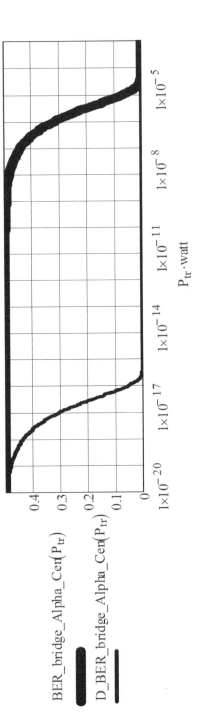

Figure 13.11. Bit Error Rate (BER) for the double-gravitational-lens system giving the radio bridge between the Sun and Alpha Cen A, using both Kraus Gain (solid *thick* curve) and the Drake Gain (solid *thin* curve).

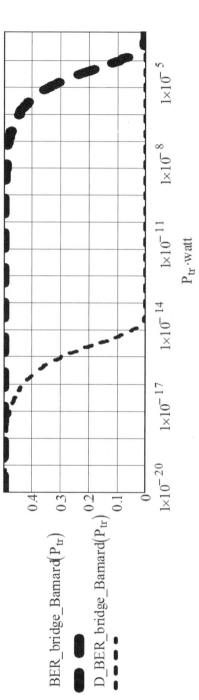

Figure 13.12. The radio bridge between the Sun and Barnard's star by assuming the two different definitions of gain: Drake Gain (*thin dash-dash curve*) and Kraus Gain (*thick dash-dash curve*). Again, the power requested to keep the BER down to zero by using Drake's proposed formula is some *twenty* orders magnitude better than the power requested by the Kraus Gain.

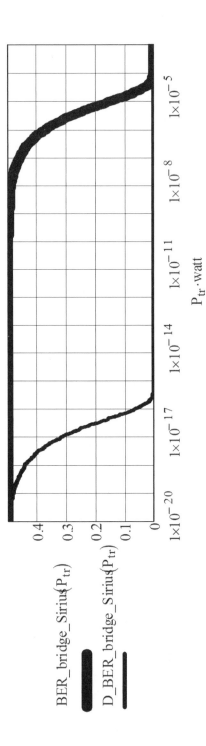

Figure 13.13. The radio bridge between the Sun and Sirius A by assuming the two different definitions of gain: Drake Gain (*thin* solid curve) and Kraus Gain (*thick* solid curve). The same large difference as seen in the previous two figures is evident.

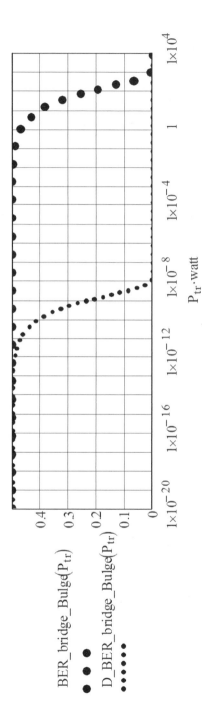

Figure 13.14. The radio bridge between the Sun and a Sun-like star located at the distance of the Galactic Bulge, by assuming the two different definitions gain: Drake Gain (*thin* dot-dot curve) and Kraus Gain (*thick* dot-dot curve). The same large difference as seen in the previous three figures is evident.

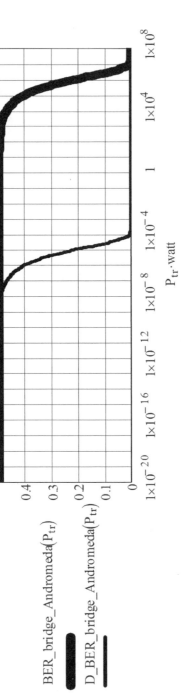

Figure 13.15. The radio bridge between the Sun and a Sun-like star located at the distance of the Andromeda galaxy (M 31), by assuming the two different definitions of gain: Drake Gain *(thin* solid curve) and Kraus Gain *(thick* solid curve). The same large difference as seen in the previous four figures is evident.

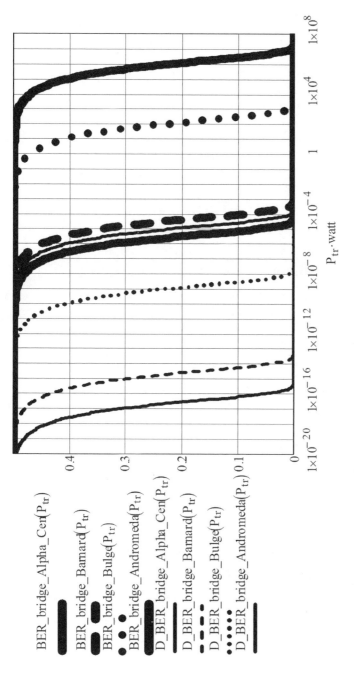

Figure 13.16. Finally, to allow a direct comparison, we plot all the curves of Figures 13.11–13.15 into a single figure. The twenty order of magnitude improvement of the Drake Gain vs. the Kraus Gain is confirmed.

14.0. Conclusion

In these few pages we could just sketch the FOCAL space mission to 550 AU and beyond to 1000 AU. A number of issues still have to be investigated in: (1) the science related to the mission, (2) the propulsion tradeoffs to get there in the least possible time, and (3) the optimization of the telecommunication link. Nevertheless, it plainly appears that the Sun focus at 550 AU is the next most important target that humankind must reach in order to be prepared for the successive and more difficult task of achieving the interstellar flight. In particular, we proved that the FOCAL mission only will allow us to *extend* our telecommunications with spacecraft located in space at least at the distance of Alpha Centauri or higher. This we did by resorting to the notion of Bit Error Rate, that would be zero or nearly zero only if we build up *radio bridges* between the Solar System and the destination target, whether that would be a spacecraft or another star system.

Acknowledgments

The author is indebted to many colleagues for conversations and suggestions, but, in particular, he would like to thank Paul Gilster for maintaining his Centauri Dreams Web site: http://www.centauri-dreams.org.

Note

Portions of this chapter originally appeared in Maccone, C. 2009. *Deep space flight and communications: Exploiting the Sun as a gravitational lens.* New York: Springer.

Works Cited

Cohen, N. 1987. The pro's and con's of gravitational lenses in CETI. In *Proceedings of the Bioastronomy International Conference,* Balatonfüred, Hungary, June 22–27, ed. G. Marx, 395.

———. 1988. *Gravity's lens.* New York: Wiley Science Editions.

Derosa, L., and C. Maccone. 2007. Propulsion tradeoffs for a mission to Alpha Centauri. *Acta Astronautica* 60: 711–18.

Drake, F. 1987. Stars as gravitational lenses. *Proceedings of the Bioastronomy International Conference,* Balatonfüred, Hungary, June 22–27, ed. G. Marx, 391–94.

———. 2010. Personal communication, February 2.

———, and D. Sobel. 1992. *Is anyone out there?* New York: Delacorte Press.

Einstein, A. 1936. Lens-like action of a star by the deviation of light in the gravitational field. *Science* 84: 506–507.

Eshleman, V. 1979. Gravitational lens of the Sun: Its potential for observations and communications over interstellar distances. *Science* 205: 1133–35.

Heidmann, J., and C. Maccone. 1994. AstroSail and FOCAL: Two extrasolar system missions to the Sun's gravitational focuses. *Acta Astronautica* 35: 409–10.

Kraus, J. D. 1966. *Radio astronomy*, 2nd ed. Powell, OH: Cygnus-Quasar Books.

Liebes, S., Jr. 1964. Gravitational lenses. *Physical Review* 133: B835-B844.

Maccone, C, 1993. *FOCAL, A New Space Mission to 550 AU to Exploit the Gravitational Lens of the Sun.* A Proposal for an M3 Space Mission submitted to the European Space Agency (ESA) on May 20, 1993, on behalf of an international team of scientists and engineers. Later (October 1993) reconsidered by ESA within the "Horizon 2000 Plus" space missions plan.

———, 1994. Space missions outside the solar system to exploit the gravitational lens of the Sun. *Journal of the British Interplanetary Society* 47: 45–52.

———. 1995. The SETISAIL Project. In *Progress in the search for extra-terrestrial life: Proceedings of the 1993 Bioastronomy Symposium*, ed. G. Seth Shostak, 407–17. San Francisco: Astronomical Society of the Pacific.

———. 2002. *The Sun as a gravitational lens: Proposed space missions*, 3rd edition. Colorado Springs: IPI Press.

———. 2009. *Deep space flight and communications: Exploiting the Sun as a gravitational lens.* New York: Springer.

Orta, R., P. Savi, and R. Tascone. 1994. Analysis of gravitational lens antennas. *Journal of the British Interplanetary Society* 47: 53–56.

Cost Analysis of Space Exploration for an Extraterrestrial Civilization

Yvan Dutil and Stéphane Dumas

1.0. Introduction

Are some civilizations bound to their home planet because the local gravitational field is too strong? Can space exploration be so difficult that a civilization might just give up before even trying? Are we favored compared to our galactic neighbors? Analysis of the relative cost of space exploration between planets, in light of the physical characteristics of a civilization's planet of origin, might provide some insight into the problem of the Fermi paradox or in a more general way the possibility of physical contact between civilizations. This is not a trivial issue, since there may be some advantage to using space probes as an effective way to communicate between civilizations (Rose and Wright 2004). Such contacts do not need to be done by sentient beings themselves, as advanced automated space probes could achieve the same goal (Bracewell 1960; von Neumann and Burks 1966; Boyce 1979).

Nevertheless, even classical SETI can benefit from space exploration capabilities. For example, it would be helpful to place a SETI observatory on the lunar far side to avoid local radio interference (Heidmann 1994). Ironically, spacefaring capabilities would make this type of SETI project even more attractive, as interference shielding is more complex when interfering sources are themselves in space. Alternatively, an extraterrestrial civilization might want to take advantage of the gravitational focusing of the Sun to increase the sensitivity of its SETI project, which would require significant spacefaring capabilities, as noted in the immediately preceding chapter in this volume (Maccone 2011).

We are constructing our analysis around a simple and sound scenario. Before exploring interstellar space, a civilization must first succeed in achieving two earlier steps of exploration: getting above the atmosphere in the lowest possible orbit and moving between planets within their own stellar system. Once these steps have been mastered, there will be little variation in interstellar mission costs. In consequence, if the cost of launching a satellite in low orbit and reaching another nearby planet is prohibitive, an extraterrestrial civilization might simply give up space exploration and direct its resources elsewhere.

To shed some light on this potential problem, we analyze the relative energy cost of launching a satellite in low Earth orbit and from there reaching the nearest planet, in various potential life-bearing planetary configurations.

2.0. Budgeting Space Exploration

From the engineering point of view, the cost of space exploration is essentially determined by one parameter: the velocity change. Velocity changes themselves are determined by the local gravity field and the maneuvers needed to move from one orbit to another. Three basic maneuvers are needed to perform interplanetary exploration: (1) getting above the atmosphere, (2) circularizing the orbit, and (3) reaching another planet's orbit. It should be noted that leaving the home planet's gravity field requires only $\sqrt{2} = 1.4142$ times the velocity change needed to reach low earth orbit. The same ratio also applies to the problem of escaping the stellar gravity field. In consequence, only those three basic maneuvers need to be analyzed.

If the velocity change is an important factor, it is because it drives the total weight of the launcher. As more velocity is needed, more and more fuel must be carried in addition to the useful payload. The relation between the velocity change (ΔV) and the mass fraction (original mass over the spacecraft mass m_o/m_s) is known as the rocket equation. This equation was derived for the first time by the Russian space exploration pioneer Konstantin Tsiolkovsky in 1903 (Tsiolkovsky 1903). It has this form:

$$\frac{m_o}{m_s} = \exp\left(\frac{\Delta V}{V_e}\right)$$

where V_e is the ejection speed of rocket exhaust.

Since for a chemical rocket, ejection speed is controlled by characteristics of the reactants involved, this parameter can be assumed to be relatively constant for every civilization. Indeed, one of the most effective practical reactions is between hydrogen and oxygen—two very abundant elements forming the water molecule. Alternate propulsion methods, such as ionic and nuclear motors, have higher ejection speeds but are impractical for first stage launch, since they do not provide enough thrust to fight the local gravity. Once in space they are more useful, but take a long amount of time to be effective. Use of such an advanced space propulsion system will concentrate the expense of cost on the launch to low orbit. To avoid unneeded complexity, we will assume that all space maneuvers use chemical propulsion. As we will see later, this assumption bears little impact on our conclusions.

Another factor driving the cost is the size of the smallest possible spacecraft. Technology drives the minimal size of artificial satellites. It is reasonable to assume that the dry weight of a satellite for a given function is relatively constant for any planet, because aerospace technologies have performances very close to those allowed by physics. However, for inhabited spacecraft, the minimal size is driven by the size of the pilot. Unfortunately, we cannot derive the pilot size from first principles based on planetary and stellar parameters. This is why, and for the stake of simplicity, we will assume that the reference spacecraft mass is independent of the local condition.

Before going into orbit, a spacecraft must first get above the atmosphere. Hence, the minimum orbital altitude is controlled by the atmospheric drag. Atmospheric density and the shape and velocity of the satellite are the main factors affecting the drag. As we will see later, orbital velocity does not change much across planets. We also assume that satellite size and shape are constant. In consequence, only the atmospheric density really matters.

Atmospheric density drops exponentially with the altitude. On Earth, for an increase of 7.6 km, the scale height, the atmospheric density drops by a factor of 2.72 on average. The scale height itself is proportional to the atmospheric temperature and inversely proportional to the local gravitational acceleration and average mass of atmospheric molecules. As we expect from habitability considerations, atmospheric composition and temperature will not be vastly different from the Earth's, and scale height will mostly depend on local gravity.

The cost of climbing above the atmosphere is proportional to the energy needed, which is the product of the height by the surface gravitational acceleration. Since scale height is inversely proportional to the surface acceleration, the product of the two is roughly constant for any inhabited planet. Surface pressure also affects the minimal orbital altitude, but its impact is minimal. For example, going up by one scale height will reduce the atmospheric density by a factor of 2.72. Since the lowest orbital attitude is at least twenty scale heights, a small change in altitude can absorb a wide range of surface atmospheric pressure, 0.3 to 3 times Earth's, with minimal impact on the launching cost. In addition, for typical launch system, fighting gravity is only 20 to 25 percent of the velocity change needed to reach the Earth orbit; the cost of going above the atmosphere is essentially independent of the planet of origin. Impact of atmospheric drag is even lower, as it is roughly equal to 10 percent of the gravity for our launcher technology, and optimization of the launch profile minimizes its impact on mission performances.

Clearing the atmosphere is a relatively easy step compared to reaching orbital velocity. To reach orbit, we must reach a velocity that will generate enough centrifugal acceleration to balance the local gravitational acceleration. The following equation describes the relation between the two phenomena:

$$g = \frac{v^2}{R + h} \approx \frac{v^2}{R} \quad \text{when } h <\!\!< R.$$

It is reasonable to assume that the minimal orbital altitude (h) is much smaller than the planetary radius (R) as it is in the case for Earth. For a given planet, the surface gravitational acceleration (g) is proportional to the planet's mass and inversely proportional to the square of its radius: $g \propto \frac{M}{R^2}$.

We can use this formula to estimate the ΔV needed to put a satellite in orbit. Using the previous result, we derive:

$$\Delta V \propto \sqrt{\frac{M}{R}}$$

At this point in our analysis, we need to establish a relationship between the planet's mass and radius. Assuming a constant density, the planetary radius is then proportional to the cubic root of the volume and, as a consequence, the mass ($R \propto \sqrt[3]{M}$). This approximation is roughly correct between 0.3 and 10 Earth masses (Seager et al. 2007; Swift et al. 2010), which covers the range expected for an inhabited exoplanet. In consequence:

$$\Delta V \propto \sqrt{\frac{M}{R}} \propto \sqrt{\frac{M}{M^{1/3}}} \, M^{1/3}$$

The relation indicates that the velocity change needed to reach low planetary orbit is only weakly sensitive to the planet's mass, as velocity change only increases with the cubic root of the planet's mass.

Integrating this result in the cost relation derived previously, we have:

$$\text{cost} \propto \exp(\Delta V) \propto \exp(M^{1/3})$$

In consequence, the relative launch cost compared to the Earth for a 0.3 earth mass ($0.3 M_{\oplus}$) planet would be 0.72 and 1.56 for a three earth mass ($3 M_{\oplus}$) planet. Keep in mind that this is a very simplified analysis of a space mission cost. Nevertheless, it allows us to the reach a simple conclusion: *cost for launching a satellite to low orbit is essentially independent of the planet of origin.* In consequence, it would be neither much easier nor more difficult for an extraterrestrial civilization to put its first satellite in orbit, than it was for us. And keep in mind, that while the launch cost increases with the size of the planet, planet resources also increase even faster (surface $\propto M^{2/3}$). Relative cost would then drop on a larger planet.

Once it has reached the low planetary orbit, the next step for a spacefaring civilization is to move between planets of its own stellar system. Planetary escape velocity is only ~2 times larger than initial launch cost. However, others aspects of the interplanetary exploration scale differently. Indeed, for travel between planets, velocity changes are proportional to the planetary orbital velocity, which scales as:

$$\Delta V \propto \sqrt{\frac{M_*}{d}}$$

where M_* is the mass of the star and d is the planet's orbital distance. We can relate those quantities by imposing the requirement that the planet lies within the habitable zone. To keep the surface temperature roughly constant, a planet circling around a brighter star must be farther from the star, while a planet circling around a fainter star must be nearer to it. This implies that $d \propto \sqrt{L_*}$. We also know the relationship between the mass of a star (M_*) and its luminosity (L_*) (Zeilik, Gregory, and Smith 1992). For a star with $M_* > 0.43 M_\odot$ it follows the relationship $L \propto M^4$, and for a star with $M_* < 0.43 M_\odot$ it follows the relation $L \propto 0.23 M^{2.3}$.

Combining these relations with the formula for orbital velocity, we can derive the following results:

For $M > 0.43 M_*$ $v_0 \propto \sqrt{\frac{M}{\sqrt{M_4}}} \propto M^{-1/2}$ and for $M > 0.43 M_*$ $v_0 \propto \frac{M_*}{0.48 M_*^{1.15}} = 1.444 M_*^{-0.075}$.

Possible stellar mass range is constrained by the minimal lifetime of the star (larger than one billion years) needed to reach the stage of civilization following the formation of the planet and the need to have enough mass to generate nuclear reactions needed to heat the planet. These constraints limit the possible mass range between $0.08 M_\odot$ and 1.8 M. Again, to avoid unneeded complexity, we do not take into account issues such as synchronous rotation, uv radiation, and atmospheric erosion that could restrict this mass range further.

Using the derived velocities from this mass range and introducing them into the rocket equation, we can calculate the cost of interplanetary orbital transfer. For the heaviest star, it is only 77.5 percent of ours and 2.1 times costlier for the lightest star. Again, relative cost is restricted to a relatively narrow range. In addition, it should be pointed out that the extent of this range is largely driven by the smallest stars, the red dwarfs. For example, for a star of 0.43 M the cost increases only by 77 percent relative to the Earth's. Nevertheless, with red dwarfs being the most abundant stars, we cannot simply dismiss them from our analysis.

3.0. Adding Cost

The relative total mission cost is the sum of spacecraft cost, low orbit launch cost, and interplanetary flight cost. Today on Earth, spacecraft cost is roughly five times the low orbit launch cost (Wertz and Larson 1999). As previously, we argue that since aerospace technologies have performances very close to the limits imposed by basic physics, this ratio is relatively constant for any civilization.

From the required ΔV, we can estimate the relative cost between low earth orbit and interplanetary flight. As a benchmark, we use the Earth-Mars orbital ratio, which is likely to be the same for the successive planet in any hierarchical planetary system. We also include the deorbit burn ΔV for the target planet, which is set to be equal to the escape velocity of the planet of origin. Adding those factors, we have derived for the total ΔV:

$$\Delta v_{tot} = 1.82 \, v_{LEO} + 0.2 v_{Earth}$$

The total cost induced by velocity changes is calculated as usual with the help of the rocket equation using the appropriate scaling function described earlier: f_{Leo} for low orbit launch and f_{int} for interplanetary travel. For consistency, we also introduce a scaling cost factor between 1.82 ΔV_{LEO} (14.4 km/s) and 0.2 ΔV_{Earth} (5.95 km/s). The scaling factor is equal to 0.089. The total cost equation is then equal to:

$$cost_{total} = cost_{spacecraft} + f_{LEO}cost_{LEO} + f_{inter}cost_{inter},$$

which reduces to:

$$cost_{total} = 5(f_{LEO}cost_{LEO}) + f_{LEO}cost_{LEO} + 0.089 \, f_{inter}cost_{LEO}$$

and finally to:

$$cost_{total} = (6f_{LEO} + 0.089 \, f_{inter})cost_{LEO}$$

In consequence, the relative cost of space exploration, in the best case, for a star of 1.8 $M_\odot$ and a planet of 0.3 $M_\odot$, would be 28 percent cheaper. In the worst case, $M_* = 0.08 M_\odot$ and a planet of three $M_\odot$, would be only 56 percent more expensive.

It should be noted than the scaling cost factor between the low orbit velocity change and the interplanetary velocity change is a key factor in this discussion. Another orbital scenario could shift the sensitivity from planetary parameter to stellar parameter. For example, if we calculate the cost of sending a probe directly to interstellar space from a planetary surface, the ΔV needed is given by:

$$\Delta \nu_{tot} = 0.41 \nu_{LEO} + 0.41 \nu_{Earth}$$

The scaling cost factor between the $0.41 \Delta V_{LEO}$ (3.28 km/s) and 0.41 ΔV_{Earth} (12.33 km/s) is equal to 42.9. Using the same approach as previously, we derived:

$$\text{cost}_{total} = (6 f_{LEO} + 42.9 \, f_{inter}) \text{cost}_{LEO}$$

In the worst case (big planet and small star), the relative space mission cost would be twice as expensive as on Earth. However, if we restrict the planet of origin to stars with mass larger than 0.43 $M_\odot$, worst case would be only 71 percent more expensive. In the best case (small planet and big star) the cost drops by 23 percent. As we can see, cost change is minimal for a large parameter space.

This analysis also yields an interesting observation. Since our Sun is a relatively big star and hence is relatively rare, our interplanetary exploration cost is among the cheapest possible! This might have provided a head start to our space program, compared to extraterrestrial civilizations.

4.0. Conclusion

Even if this analysis is rather crude, it can provide some insight into the relative cost of space exploration for extraterrestrial civilizations. First, for realistic orbital and interplanetary missions, the relative cost range is rather small (<2) even in a worst case analysis. Nevertheless, this analysis intentionally excluded other factors, such as radiation level in space, that might play an important role. It is also possible that a civilization inhabits a moon of a larger planet. This configuration will certainly increase the difficulty significantly, but, in our opinion, not to the point of making space exploration impossible.

As a general conclusion, the planet and star of origin of an extraterrestrial civilization are likely to have little importance in the development of its indigenous space exploration. In consequence, if there is a physical restriction to interplanetary space travel, it is not caused by the local gravitational potential well but by other factors not examined here.

In addition, failure to initiate a successful space program due to its prohibitive cost cannot be invoked as an explanation for the Fermi paradox.

Communication by physical artifacts cannot be dismissed on this basis either. The same can be said about SETI enabling technology that needs to be implemented in space, which also cannot be dismissed for this reason.

Works Cited

Boyce, C. 1979. *Extraterrestrial encounter: A personal perspective.* New York: Chartwell Books.

Bracewell, R. N. 1960. Communications from superior galactic communities. *Nature* 186: 670–71.

Heidmann, J. 1994. Saha Crater: A candidate for a SETI lunar base. *Acta Astronautica* 32: 471–72.

Maccone, C. 2011. Interstellar radio links enabled by gravitational lenses of the Sun and stars. In *Communication with extraterrestrial intelligence (CETI)*, ed. D. A. Vakoch. Albany, NY: State University of New York Press.

Rose, C., and G. Wright. 2004. Inscribed matter as an energy efficient means of communication with an extraterrestrial civilization. *Nature* 431: 47–49.

Seager, S., M. Kuchner, C. A. Hier-Majumder, and B. Militzer. 2007. Mass-radius relationship for solid exoplanets. *Astrophysical Journal* 669 (2): 1279–97.

Swift, D. C., J. Eggert, D. G. Hicks, S. Hamel, K. Caspersen, E. Schwegler, G. W. Collins, and G. J. Ackland. 2010. Mass-radius relationships for exoplanets: http://arxiv.org/pdf/1001.4851.

Tsiolkovsky, K. E. 1903. The exploration of cosmic space by means of reaction devices (Исследование мировых пространств реактивными приборами). *The Science Review* (5).

von Neumann, J., and A. W. Burks. 1966. *Theory of self-reproducing automata.* Urbana: Univ. of Illinois Press.

Wertz, J. R., and W. J. Larson. 1999. *Space mission analysis and design*, 3rd ed. El Segundo, CA: Microcosm Press.

Zeilik, M., S. A. Gregory, and E. V. Smith. 1992. *Introductory astronomy and astrophysics*, 3rd ed. Fort Worth: Saunders College Publishing.

Understanding the Search Space for SETI

William Edmondson

Many a hearth upon our dark globe sighs after many a vanish'd face,
Many a planet by many a sun may roll with a dust of a vanish'd race . . .
Raving politics, never at rest—as this poor earth's pale history runs,—
What is it all but a trouble of ants in the gleam of a million million
of suns?

—Alfred Lord Tennyson, *Vastness*

1.0. Introduction

The SETI enterprise is best cast as the (1) *scientific* (2) *search* for
(3) *incontrovertible* (4) *evidence* for the (5) *existence* (6) *elsewhere* of (7) *intelligent*
(8) *life*. In what follows we explore part of the multidimensional space set
out in this statement. Our goal is to refine our understanding of what it is
to conduct a search for extraterrestrial intelligence, and in this we are not
alone. Steven J. Dick's book *Life on Other Worlds* (Dick 1998) covers the topic
in exemplary detail, and Paul Davies's new book *The Eerie Silence* (Davies
2010) offers a more recent perspective. Some of the points made below
echo points made by Davies (or indeed pre-echo; they were written before
encountering Davies's work), but in fact our intellectual trajectories though
the search space are different despite our apparent current collocation. It is
as if we are standing close by, but looking out into search space in different
directions. Importantly, it is not possible to do more in a single book chapter
than explore in depth just one or two of the dimensions listed above, but
even within this restriction it is possible to see where my conception of the
search space differs interestingly from that offered by Davies.

2.0. Search Space—I

One perspective on the notion of a search space for SETI is just to explore
the "dimensions" enumerated above. It is not difficult to locate current

activities in relation to these characteristics (cf. papers in journals such as *Astrobiology*, *International Journal of Astrobiology*). Where the notion of a *space* or *dimension* is not self-evident, as for example with (1) or (6), it can be made clear with a little work: (1) can be quantified in the sense of socially sanctioned definitions (the Hubble space-based telescope is "more" scientific than, for example, the use of the radio telescope at Arecibo by an amateur astronomer, and even more than the occasional use of a home-built instrument in one's back yard). Of interest here is the role of technology in shaping the search space by changing the scientific discussion. Optical SETI illustrates this well—the advent of laser technology changed the search space by changing the nature of the science deployed in the enterprise.

Note that *search*—dimension (2) above—is focused on the search strategy, which in part can be considered to be the deployment of the science. As such it is also not readily quantifiable but it nonetheless serves as a dimension in our discussion by virtue of enabling discrimination between different SETI activities. Some of what follows can be understood to be an elaboration of the *search* dimension. The reader should be aware that the word *search* is used both for the general enterprise, and thus as the first term in the acronym SETI and in the title of this chapter, and *also* as just one dimension or factor in the multidimensional/multifactorial account of the activity—search is both the total enterprise and a mere component, a common enough ambiguity in usage but a potential difficulty for those unfamiliar with SETI. My intention is to remind the reader that within the overall activity identified as SETI there is much that is not about searching as such, but which is very relevant to the overall enterprise. Mostly these other factors turn out to be the "science internal" or "domain specific" locations for many assumptions about what it is to do SETI. We will see an illustration of this below, in relation to the *intelligence* dimension.

The issue here is not that new science and technology emerge which can be pressed into the service of SETI (cf. lasers) but that within a deployed science there may be domain internal developments which also change the search space. When we consider factor (6) we have no problem recognizing that the notion of *elsewhere* could be scaled by physical distance. However, work on extremophiles, including the search for *weird life* here on Earth, forces elaboration of both (6) and (8). For example, as scientific knowledge has developed over the last century we have come to recognize that our understanding of Mars was incorrect (e.g., there are no canals but there is probably subsurface water) and we have transformed our expectations regarding life on Europa. The search space, so to speak, is in flux because of the changes within the deployed sciences, but the metaphor is valuable nonetheless.

Occasionally, domains or topics assume an importance that can subsequently be seen as misdirected or tangential. Here the change in the search space is for yet a third reason—the relevance of the deployed science can change. This is akin to realigning and/or rescaling some of the dimensions sketched out above—changing the search space. An example here could be that we come to recognize "messaging" as an incomplete and possibly inadequate approach to SETI. The metaphor of search space is perhaps labored by now, although nice in context. It has the value of reminding people of the multidisciplinary nature of the work, and of uncertainty about how best to proceed. As we will see below, one particular elaboration—that concerning factor (6) *elsewhere*—can be refined to the point where simple distance becomes irrelevant, and implications for other dimensions become important.

The main value of this rather simplistic approach is that it emphasizes the multifactorial nature of the scientific challenge. Another recently published perspective on the complexity of the multidisciplinary effort that underpins SETI can be found in a recent issue of *Astrobiology* (Fridlund and Lammer 2010) wherein ten articles cover the scientific and technological challenges in terms of evolution of life, formation of planetary systems, exoplanet search/characterization technologies and so forth (see for example Schneider, Léger, Fridlund, White, Eiroa, Henning, Herbst, Lammer, Liseau, Paresce, Penny, Quirrenbach, Röttgering, Selsis, Beichman, Danchi, Kaltenegger, Lunine, Stam, and Tinetti 2010). The publication of the journal issue as the *Astrobiology Habitability Primer* (Fridlund and Lammer 2010) illustrates our theme but also reinforces the scientific and cultural importance of SETI. We will return to this point in the concluding comments.

3.0. Search Space—II

The reasons for writing this chapter are not that we should perhaps constrain SETI and/or our futurology to the eight topics noted above, or to the themes in the *Astrobiology Habitability Primer* (Fridlund and Lammer 2010). Cases could be made for doing both these things, especially in the context of the three forces for change in the search space identified above (deploying new science; developments in deployed science; reassessing/revaluing deployed science). Rather, my point is that there are substantial search efforts being made on the basis of a range of variably credible assumptions relating to factor (7) in the list above—*intelligence*—efforts that need to be evaluated and discussed. These efforts can in their turn be set out in multifactorial form—but here the result is an elaboration of the *search* dimension: *active versus passive, direct versus indirect, electromagnetic spectrum*. However, because there are issues arising from general assumptions concerning the nature of intelligence and motivations for searching and signaling, it makes sense to review those issues first and then move on to consider the factors listed above.

4.0. Intelligence

In SETI efforts past and present one can find a range of conjectures about the presumed intelligence for which one is searching. These include the idea that ETI will be "advanced" in some nonspecific way (with equally vaguely defined notions of "advanced" technologies), sometimes linked to the idea that their civilization will be older. Related to this is the idea that ETI's intelligence may be radically "different" or even "postbiological" (Dick 2008; Davies 2010). Such discussions resemble science fiction narratives involving humongous intelligences or disembodied brains (cf. Stapledon 1930)—the link between science fiction and the science of SETI is stronger than many would like, and is certainly unavoidable (cf. Dick 1998, ch. 4). Additional aspects of such discussions centre on the sensory world of a possible ETI—for example, will an ETI necessarily have eyes and ears that respond to physical stimuli in the way our organs do, and over frequency ranges that match ours (cf. Hoffman, in press).

In fact, it is possible to address such speculations in considerable detail. It can be argued that intelligence is necessarily embodied, with sensory apparatus and articulators for interacting with the physical world in which the embodied intelligence resides (see Edmondson 2010a). The argument extends to the assertion that a psychophysiological universal exists—the sequential imperative—and this accounts for the functionality of any brain, anywhere. The sequential imperative is simply the requirement that organisms must map intentions and perceptions (brain states) into and out of the necessarily sequential structure of behavior and encounters with the environment. Further, the characteristics of terrestrial biochemistry look to be universal, which suggests that sensory apparatus will work the same way as ours in a general sense, in a physical world rather like ours (silicon biochemistry remains the stuff of science fiction; underwater intelligences will have trouble becoming technologically competent, especially in respect of using electricity). And the problems of embodiment with an exoskeleton suggest that the science fiction fantasy of super-sized insects could never be realized (Berkeley 2010).

Intelligences such as we might now conjecture to exist will have notions of physics and psychology and linguistics, will have technological competences capable of working with a variety of materials, will have varied energy resources (but we do not need to imagine that they will have control over the energy of their star or galaxy), and so forth. And they will have such views of us. Therefore, we do not need to transmit (or assume the transmission of) complex messages: codes, linguistic material, musical content (cf. Vakoch 2010), interstellar Sudoku puzzles, or whatever—there really is no point (cf. Reed 2000). Indeed, one can even propose on plausible grounds that intelligences evolve to the point of stalling the evolution of intelligence (see Edmondson 2010a, 2004) thus calling into question ideas of any kind of "postbiological" intelligence and the notion of "advanced" civilizations of "superior beings"—we will formulate below a working definition of "advanced civilization with advanced technology."

5.0. Three+1 factors

We consider now three conventional factors in SETI work, contextualized by the discussion above: *active versus passive; direct versus indirect; spectrum*. We will then introduce a fourth factor, which, it is suggested, is more useful for thinking about the future of SETI.

5.1. Active versus Passive SETI

Should SETI be conducted on the basis of *active* broadcast of the fact of our existence, or on the basis of a more *passive* approach such as the attempted detection of such broadcasts? This distinction is frequently made but is problematic (cf. Shostak 2009). *Active* mode requires planning to transmit, for some extended period of time, signals with significant power toward selected stars as targets. *Passive* mode requires the use of general purpose astronomical instruments for observing stars generally, or perhaps just a few selected stars. Aside from technological considerations the underlying issue is the same: What does it mean to be a target? We need to identify targets to which to point telescopes. We need to know why we might be a target, and from which stars our status as a target looks good. The answer suggested elsewhere (Edmondson 2010b) is that the *active/passive* distinction is not interesting (and in any case it remains very doubtful that sending messages is a good way to look for intelligent life).

5.2. Direct versus Indirect SETI

Direct SETI is based on the assumption that the search is for a communicative signal, transmitted with intent. Both *active* and *passive* SETI are forms of *direct* SETI (if one accepts that *Active SETI* transmissions from Earth are part of another ETI's SETI). *Indirect* SETI, on the other hand, assumes no intent on the part of the source of any signal. *Indirect* SETI is simply the search for evidence produced by an ETI of its existence. It can be considered in two forms—electromagnetic signals; biomarkers (of which more later).

5.3. Electromagnetic Spectrum

This is familiar territory for many SETI researchers. The primary assumptions are that transmissions from an ETI will be at what we call radio frequencies—typically at or near the 1.42 GHz HI emission line. Such frequencies will be chosen because they are within the transmission window for an atmosphere surrounding a planet such as ours (see, e.g., Dick 1998, 207; Oliver 1979; cf. Zuckerman 1985). Lower transmission frequencies are discussed when considering *indirect* SETI—looking for an ETI's version of our "babble bubble" of transmitted electromagnetic energy currently expanding away from Earth in the form of short wave radio broadcasts, TV transmissions, interplanetary radar, etc. But this intensive leakage from our planet has already peaked as now receivers are so much more sensitive, antennae more directional, and alternative media more widely used (copper and optical fiber cables).

Optical transmissions (OSETI) have been sought using modest equipment (see Howard, Horowitz, Mead, Sreetharan, Gallicchio, Howard, Coldwell, Zajac, and Sliski 2007) on the basis of calculations that show that extremely high power lasers with nanosecond duration pulses can produce at interstellar distances observable pulses of light briefly outshining an accompanying star.

5.4. Asymmetric SETI

The additional factor that is more important than the three factors just discussed is the *asymmetry* of the situation in SETI as conventionally envisaged. It matters that the searching is being done by an Intelligence (ourselves) that does not have an answer to the Big Question (BQ): Are We Alone? Because it is usually assumed that we are probably not alone, and that we simply don't know this as fact, the main presumption in SETI work is that at least one other Intelligence has the answer—they know they are not alone.

In consequence, when addressing our lack of success with SETI we must consider several possibilities. But first, we should probably discount the idea that our technology is not yet sophisticated enough for "listening"—the resources available are constantly improving and if technology is the only issue then we will have cracked it before long. Indeed, we can plausibly argue that we are entering a phase in current conceptions of SETI where technology is not as limiting as the conceptions themselves. This leaves us with the following ideas: (a) ETI exists and to help us answer the Big Question ETI is altruistically trying to communicate with us, but we don't know *where* to look for them (yet); (b) ETI exists and has not yet detected sufficiently advanced life forms on Earth to warrant the effort required to transmit to us (cf. Davies 2010; of course we would still need to address (a) above when ETI fires up its transmitter—but we can't know about that eventuality until long after it has happened); (c) they know we exist but they do not care to help us answer the BQ; (d) such help is too costly and they assume we will answer the BQ in other ways, as indeed they did. Note, this perspective dismisses as fanciful the idea that an ETI might be doing SETI by means of transmitting messages—we would not do this so why should any other Intelligence?

The value of considering the asymmetry in SETI is that it reveals many of the assumptions often ignored in more conventional approaches, and in doing so points the way to a different style of approach to SETI itself. By considering the reasoning of ETI—anthropomorphism licensed by thinking about the way ETI must necessarily cognize—we can argue that the primary issue for SETI is the notion of targets: specification and characterization of *plausible target stars* to which it makes sense to attend. We can assume, on the basis of licensed anthropomorphism, that our notions of plausibility for a target star will be shared by an ETI. The secondary question is *how* we attend to those stars, but currently this issue has assumed primacy.

6.0. Target Stars

An attempt to identify stars as targets for SETI was made by Edmondson and Stevens (2003)—they listed Habstars in alignment with Earth and pulsars. This attempt is characterized as addressing the problem identified in the (a) category in the previous section—ETI knows we exist and assumes we know how to work out where to look for them. Failure thus far is now simply that we haven't explored all possible targets sufficiently thoroughly. Note here the shift away from distance, in addressing the *elsewhere* dimension, toward specific stars/targets.

But this line of reasoning doesn't really do justice to the full implications of the asymmetry of the situation. In the limit, the identification of a target star to which an ETI *might* transmit a signal comes down to the detection of intelligent life on a known planet orbiting a candidate target star. Earth-based SETI can indeed drive the specification of such targets for scientific attention, but if this is done thoroughly the BQ will be answered as a byproduct of finding targets for SETI. If indeed an ETI has reasoned this way it might well not bother sending signals (reason (a) above notwithstanding). Even if they reasoned that (b) makes more sense, the sociotechnical commitment to support extended transmission efforts might be judged too onerous. Increasingly it looks like "listening" might not be the best way to approach SETI.

7.0. Search Space—III

It was suggested at the outset that SETI is best cast as the (1) *scientific* (2) *search* for (3) *incontrovertible* (4) *evidence* for the (5) *existence* (6) *elsewhere* of (7) *intelligent* (8) *life*. Our earlier discussion of the reasoning behind ETI's reasons for transmitting signals to Earth got us to the point that the *elsewhere* factor in the list above actually reduces to a list of targets, and indeed that the target list will, in the limit, be a list of planets known to have intelligent life on them—thus actually answering the BQ. Notice that reaching this point has involved refining ideas in the *search* dimension in the sense that ideas in the latter domain force us to rework the *elsewhere* dimension.

8.0. Target Planets for SETI

The assumption above is that defining targets for SETI efforts will, in the limit, actually involve the identification of planets with known intelligent life. This line of reasoning is partly based on reworking our ideas about intelligence: the search space is changed by developments in the science deployed. Another change in the search space is also based on scientific development: progress to date with exoplanet discovery. Currently, with around 500 exoplanets known, the assumption is that advances in technology will lead to the detection of smaller, slower planets more probably in the habitable zone for their host stars. Schneider, Léger, Fridlund, White, Eiroa, Henning, Herbst, Lammer, Liseau, Paresce, Penny, Quirrenbach, Röttgering, Selsis, Beichman, Danchi, Kaltenegger, Lunine, Stam, and Tinetti (2010) summarize the technical feasibility for developing advanced systems for detecting exoplanets with detectable life—for example, looking for biomarkers such as the spectral signatures for CO_2 in a warm atmosphere. They argue that the instrument size required for direct imaging of large life forms in reflected light is not feasible, but they do not consider the case of imaging generated light from habitations. Figure 15.1 illustrates the concept that emitted light could be a biomarker, as much as CO_2 or other chemical signatures. This well-known (enhanced) image of Earth points up the need for extra-ordinary technical prowess—but there seems to be no need for any conceptual advances.

9.0. Search Space—IV

We can refashion our response to the multidimensional guidance on "how to succeed at SETI." We, the human race, should develop an international plan for the scientific search for incontrovertible evidence for the existence elsewhere of intelligent life. Billions of dollars/euros can be found for the Large Hadron Collider—a truly international project of tremendous scope and significance—so surely SETI could become such a project?

In this scenario some of the factors discussed earlier become strands in work that would lead to really significant advances in technology and science irrespective of SETI. Searching for microbial life, or its detritus, in the solar

Figure 15.1. Earth's emitted light could serve as a biomarker. Source: http://apod.nasa.gov/apod/image/0512/eunight2_pv_big.jpg.

system is already ongoing. The *Astrobiology Habitability Primer* (Fridlund and Lammer 2010) presents much relevant work, and notes the need for multidisciplinary scientific efforts to be made to tackle some of the issues (e.g., identifying possible biomarkers and schemes for their detection). In a sense the work has started, but perhaps not in a coherent multicultural fashion.

The conclusion is seemingly inescapable—to end Phase II of SETI, the finding of a single additional intelligent life form elsewhere in the universe, we would have to develop on earth a "technologically advanced civilization"—comprising both the science and technology itself and the pan-cultural resolution and commitment to conduct the search. So we end up concluding that maybe a good definition of an "advanced civilization with advanced technology" is one that has ended Phase II of SETI—they know they are not alone. They know this because they went looking for the incontrovertible evidence, rather than searching primarily for signals (my own belief is that we should put the effort into imaging technology, not "messaging" technology, and that we will indeed find ETI using such technology). The latter can of course be part of the effort, especially as exoplanet detection and characterization improve, but it cannot be the whole story because it is focused on direct evidence—nice if you can get it. However, negative outcomes could be very expensive with less scientific payoff than the alternative proposed.

Indirect evidence—markers of one sort or another, or images, are more likely to lead to more interesting and varied science that is worth conducting for its own sake. Ending Phase II of SETI would be a tremendous achievement, and would make us truly an advanced civilization. Phase I—sidestepped for expository convenience—is detection of varied but not *intelligent* life forms elsewhere in the galaxy (exploring dimensions (3), (4), and (8) in the list given earlier), starting close to home of course, perhaps even in our solar system, but moving ever farther as the technology progresses. Phase III would be to communicate with ETIs once we know they exist. Current SETI efforts appear directed at jumping straight to Phase III—great fun, but perhaps increasingly unjustifiable on scientific grounds (cf. Davies 2010).

The dimensions listed at the outset have not all been covered here, but with just a little imagination it can be seen that those not touched on earlier can be reanalyzed, reappraised, realigned in some bigger scheme of SETI endeavor—each contributing value to the whole and shaping our sense of the search space for SETI. For example, current discussions of the emergence of life as we know it are greatly informed by preoccupations in SETI, and in turn reinform SETI endeavors (see the *Astrobiology Habitability Primer* [Fridlund and Lammer 2010]). Note that in this chapter the attempt has been made to show how the dimensions can be used to sharpen our thinking about the SETI enterprise. The effort doesn't remove the need for making assumptions—how could it?—but it does help clarify the nature of those assumptions and thus reveals the effects of exploring some of them explicitly.

Usefully, as the effort devoted to SETI work becomes more sophisticated, and with better instrumentation (cf. the Allen Telescope Array, or ATA), its value to science in general increases enormously, and this serves to shine validity on the general enterprise. Thus we find that reviewing the multidisciplinary or multidimensional effort required to work on SETI takes us to the point where we can see more clearly the role played by valuable contributions in the past. The need now is for identification of the best part of the search space in which to concentrate efforts in the future. Developing technology to the point where we can detect life, especially intelligent life, elsewhere in our galaxy would earn us the status of "advanced civilization with advanced technology." It really is worth striving for, and can be its own reward regardless of whether or not we are alone in the universe.

Works Cited

Berkeley. 2010. The arthropod story. Understanding evolution. University of California Museum of Paleontology. May 23, 2010 <http://evolution.berkeley.edu/evolibrary/article/_0_0/arthropods_toc_01>.

Davies, P. 2010. *The eerie silence: Are we alone in the universe?* London: Allen Lane.

Dick, S. J. 1998. *Life on other worlds.* Cambridge: Cambridge University Press.

———. 2008. The postbiological universe. *Acta Astronautica* 62 (8–9): 499–504.

Edmondson, W. H. 2004. Evolution of intelligence in the context of SETI, Bioastronomy 2004: Habitable Worlds, Reykjavik, July 12–16, 2004. Abstracts published in *Astrobiology* 4 (2), Abstract 60F: 251.

———. 2010a. General cognitive principles: The structure of behavior and the sequential imperative. *International Journal of Mind, Brain and Cognition.* See also http://www.cs.bham.ac.uk/~whe/GCPSOBSI.pdf.

———. 2010b. Targets and SETI: Shared motivations, life signatures and asymmetric SETI. *Acta Astronautica.* doi:10.1016/j.actaastro.2010.01.017.

———, and I. R. Stevens. 2003. The utilization of pulsars as SETI beacons. *International Journal of Astrobiology* 2 (4): 231–71.

Fridlund, M., and H. Lammer. 2010. The astrobiology habitability primer. *Astrobiology* 10 (1): 1–4.

Hoffman, D. D. Forthcoming. Images as interstellar messages. In *Between worlds: The art and science of interstellar message composition,* ed. D. A. Vakoch. Cambridge: MIT Press.

Howard, A., P. Horowitz, C. Mead, P. Sreetharan, J. Gallicchio, S. Howard, C. Coldwell, J. Zajac, and A. Sliski. 2007. Initial results from the Harvard all-sky optical SETI. *Acta Astronautica* 61 (1–6): 78–87.

Oliver, B.M. 1979. Rationale for the water hole. *Acta Astronautica* 6 (1–2): 71–79.

Reed, M. L. 2000. Exosemiotics: An inter-disciplinary approach. *Acta Astronautica* 46 (10–12): 719–23.

Schneider, J., A. Léger, M. Fridlund, G. J. White, C. Eiroa, T. Henning, T. Herbst, H. Lammer, R. Liseau, F. Paresce, A. Penny, A. Quirrenbach, H. Röttgering, F. Selsis, C. Beichman, W. Danchi, L. Kaltenegger, J. Lunine, D. Stam, and G. Tinetti. 2010. The far future of exoplanet direct characterization. *Astrobiology* 10 (1): 121–26.

Shostak, S. 2009. Limits on interstellar messages. *Acta Astronautica.* doi: 10.1016/j.actaastro.2009.10.021.

Stapledon, W. O. 1930. *Last and first men.* London: Methuen.

Vakoch, D. A. 2010. An iconic approach to communicating musical concepts in interstellar messages. *Acta Astronautica* 67 (11–12): 1406–09. doi: 10.1016/j.actaastro.2010.01.006.

Zuckerman, B. 1985. Preferred frequencies for SETI observations. *Acta Astronautica* 12 (2): 127–29.

Part II

Active SETI: Should We Transmit?

Unpacking the
Great Transmission Debate

Kathryn Denning

1.0. Introduction: The Transmission Debate

For many years now, scientists and the public alike have been excited about the possibility of contact with an extraterrestrial intelligence, but also concerned about it. Such an event could be wonderful, and it could be dangerous, and it could be both. We do not know. Accordingly, there is perennial concern about deliberate transmissions from Earth—including both "de novo" transmissions of the sort occurring now, or hypothetical future "reply" transmissions, which human beings might send in response to a signal from an extraterrestrial intelligence. Therefore, there has been a great deal of debate within and outside the SETI community about these issues. Is it wise to attempt to attract the attention of extraterrestrial intelligences? Should transmissions be halted or regulated? Who speaks for Earth? What should we say? How should we say it? Who should decide?

Several documents, produced by the International Academy of Astronautics SETI Permanent Study Group, explicitly address policy concerning transmissions from Earth. These SETI Protocols and papers may be found at http://iaaseti.org/protocol.htm. (See also Shuch and Almar 2007.) Some of these have been under discussion and revision for many years, and represent substantial investments of effort. Yet, they remain contentious. Moreover, times have changed since these debates began, not least because the requisite transmission technology is increasingly widespread, and access to it is easier for nonscientists to obtain, in exchange for payment. Transmissions are undertaken as commemorative acts or public participation projects, by organizations as diverse as national space agencies, retailers, broadcasting corporations, or Internet-based media companies.[1] At the same time, science in astrobiology and SETI is developing, and carefully targeted Active SETI is increasingly feasible, given the growing knowledge of extrasolar planets.

Transmissions show no sign of abating, and neither do the discussions: editorials and articles appear in the popular science press and on the Internet

(Nature 2006; Hecht 2006; Shostak 2006; Grinspoon 2007; Brin 2006; Whitehouse 2007), conferences are held, scientific papers are published, and efforts to catalog the significance of different types of transmissions are made.[2] Some argue for a moratorium or restriction on transmissions (Brin 2006; Nature 2006). Others argue that Active SETI projects should be prioritized and supported (Zaitsev 2006, 2008, and his other works). Others just get involved in transmission projects because they seem like a neat idea.

But is everyone really talking about the same thing? Vakoch (2008) has observed that interstellar transmissions can be considered as scientific experiments, diplomatic action, or artistic expression, and this neatly points to one of the problems; however much we might wish to, we cannot simply define the arena as being one of these domains, and expect others to think the same way. Scientists may have the greatest access to the most significant antennas, but theirs is not the only game in town. Given that the rules of conduct, interaction, and evaluation are different for each domain—and frequently incommensurable—we have a problem. Moreover, attempts to solve the problem of whether or not people should transmit, by using quantitative logic, can take us only so far in answering what is fundamentally a social question about global citizenship.

How can we apply all our joint brainpower most constructively to this issue? Below, I propose that we step back, take a broader look at the cultural context, and then rethink our approach to the transmission debate. I suggest that our collective goal should be the strategic separation of knowable from unknowable risks, a careful focus on the principles that are most severely disputed, though not often specified, and a commitment to learning about ways that have been used to resolve similar debates.

2.0. Some Cultural Substrates

The transmission debate is situated at an unusually powerful locus of human desires. SETI has, of course, taken a highly scientific and technological form, but the question at its heart—Are we alone?—is ancient and passionate. In turn, the question is, arguably, a product of our inborn curiosity about our earthly neighbors: humanity's story is one of constant searching for those not yet met. Human beings are also highly motivated to interact with invisible entities of many kinds. Whether they are gods, ghosts, spirits, ancestors, descendants, pen pals, or aliens, we are driven to attempt to communicate with them, to tell them about our lives and our thoughts, or ask about their worlds and knowledge. Whether they are real or not does not always matter, for they are real in our imaginations. We build monuments and bury time capsules, so that the not yet born will know us. We carve our initials on benches, and paint on rock walls. We write books. We put messages in bottles. We pray. We talk to the dead. We create archives in space. We attach self-portraits of humanity to spacecraft. We mail our names to other worlds. And, of course, we send messages into the sky, hoping that someone, somewhere, someday—perhaps long after we are dead—will receive them . . . and think of us.[3]

One could argue that we should not conflate all these different sorts of communications into one category, and that in a discussion of interstellar transmissions, we should restrict our consideration to signals of significant strength and content. It has been argued by Zaitsev (2008), for example, that the only real Interstellar Radio Messages to date are Arecibo 1974 and the transmissions from Evpatoria in 1999, 2001, and 2003, which rate high on the San Marino Scale. (More on this below.) Presumably the October 2008 "A Message From Earth" transmission also falls in this category (A Message From Earth, www.bebo.com/amessagefromearth).

Other, weaker transmissions are sometimes viewed by Active SETI proponents as mere publicity stunts without significance, or worse, "profanations" of the concept of interstellar messaging, which detract from METI's serious purpose (Zaitsev 2008,1112).

Obviously, there is a quantitative basis for those assertions. However, I am not certain that it helps us to really understand what is going on in our society with transmissions, or to move forward constructively in the debates. If we do view transmissions from Earth within the larger context of the powerful human drive to self-expression, both individual and collective, then it changes the framing of the discussion. We can see interstellar transmissions not as unregulated scientific experiments, or unauthorized diplomatic initiatives, or artistic performance, or PR stunts, but instead as something deeper and bigger which encompasses all of those: a technologically mediated manifestation of our drive to represent ourselves and connect with those unseen.

This reframing dissolves some of the definitional boundaries between science, politics, art, and commerce, and pinpoints why sending transmissions is such a poetically compelling activity. It also suggests that any attempt at a broad-based discussion of transmission projects and their advisability should recognize the multiple driving forces behind them. These projects are popular with the general public because they tap into a positive desire that many people feel, and can easily identify with. Those seeking to limit or regulate transmissions might try to encourage a different positive expression of this desire, instead of its simple repression. For example, as the Voyager record or the recent Operation Immortality initiative shows, people's imaginations can also be captivated by simply packaging up our messages and mailing them to one of our own spacecraft; in terms of self-representation, this is just as good as beaming our thoughts into space, but it doesn't much affect Earth's actual detectability from afar. Similarly, *some* public transmission projects could possibly be diverted from a focus on Others in space, to a focus on Others here on Earth in our own distant future: time capsules instead of transmissions. The current spate of small transmission projects has the hallmarks of a fad that will not last indefinitely.

There is another substrate, of course, to the steady expansion of our names, messages, and memorials into the solar system and beyond; these are the footprints of empires in the process of claiming and colonizing territory. This is a good deal harder to subvert or divert. One might as well ask for reduction in military radars or new land-based laser weapons on the grounds of their detectability from space. That is, there is a logic at work in these projects, with practical impetus behind it, and protests are unlikely to deter them.

There is a complex interplay of motivations at work with transmission projects, of course, and there are those who specifically intend their projects as high-powered Active SETI, and who are specifically intent upon making the Earth more visible. Objections to their efforts tend to focus on its risks. This subject also warrants some unpacking.

3.0. Risks and Benefits

The risk of transmissions has been the focus of much discussion, but I contend that after a point, this becomes a blind alley. Shuch and Almar (2007, 143) astutely observed that the first step in a rigorous risk-benefit analysis of Active SETI must necessarily be the quantification of exposure, that is, "the exposure to which the Earth is subjected by a given transmission." I'd argue, however, that this is the *only* step in a rigorous risk-benefit analysis of Active SETI that we can actually perform. There is no next step we can take.[4]

To be more specific: the first risk to consider is the risk (or likelihood) of *detection* by extraterrestrials. Calculations of signal intensity, transmission duration, and direction can indicate the likelihood that a signal would be detectable by ETI (subject to a variety of assumptions). The San Marino scale takes this kind of information into account—along with the presence or absence of semantic content in the signal—and thus separates out transmissions that are probably too weak to be detectable, from those that could well be detected by ETI (See http://iaaseti.org/smiscale.htm; and Shuch and Almar 2007). The scale's creators (Shuch and Almar 2007) do not argue that it is quantitatively precise—indeed, they label it qualitative and subjective—but the scale does give a pragmatic threshold indicating which transmissions are essentially insignificant, and which are noteworthy and perhaps worthy of regulation (see http://iaaseti.org/smiscale.htm). This is certainly very useful in narrowing the field of concern, and suggests that policy might focus tightly upon the latter. This is practical, but in some ways, it only defers the problem, because such calculations speak only to the risk of detection. They cannot say anything about what ETI would do with the knowledge that we are here, or with specific information about us.

The risk of *contact* (informational/long distance or otherwise) is the ultimate concern, and it is simply unknown. It is true that we have a great deal of historical and anthropological evidence regarding the effects of culture contact on Earth, but actually there is still no detailed, systematic study that extracts patterns from all that data, to precisely identify potential concerns for human-ETI contact. (Such hologeistic studies are very challenging and methodologically fraught.) Even if there was such a study, evidence about intercivilizational contacts on Earth can really only be illustrative of the *range* of potential patterns for human-ETI contact, rather than informative about which consequences are most *likely*.[5] We cannot know how appropriate it is to generalize from Earth experience to the scenario of human-ETI contact. Some very interesting, well-considered speculation has been produced regarding human-ETI contact, but it can only be speculation. It is not verified information upon which we can build a rigorous risk/benefit analysis.

Moreover, even if we could establish, for example, that there was precisely a 5.745 +/− 1.272 percent risk of ETI being hostile and spacefaring, this would not solve our dilemma of whether or not we should expose ourselves to that risk by transmitting. That is a social decision, which depends on the perceived possible benefits of contact, on whether the consequences would be evenly distributed across Earth, and on people's different tolerances for risk. When the consequences of an optional action could be devastating, even if the probability of such consequences being unleashed is vanishingly small, some will always deem the action unacceptable.

Of course, it can still be useful to discuss the possible sequelae of contact, because it can expand our thinking about what we might face someday, if SETI succeeds. But I would argue that we cannot, *by any method*, achieve a realistic quantification of the risks or benefits of contact. Even if we could, it would not solve the problem of whether or not we should send interstellar transmissions. Equations do not necessarily tell us how to live. The relevant question is thus not *"Exactly how risky is it?"* but *"What should be done when risks are unknowable and distributed globally?"* And this question certainly can be subject to rigorous analysis, but analysis of a different kind—about which more will be said below.

As an aside: It is also possible, of course, to analyze the present risks and benefits of simply sending transmissions, without considering whether or not those transmissions eventually result in contact. That is, who benefits today by sending transmissions, and in what ways do they benefit? Who benefits today by explicitly refraining from sending transmissions, and what ways do they benefit? This is amenable to detailed analysis, but that is another discussion. However, financial backing, public perception, and professional positioning are all factors, which surely provide fuel to the debates.

The debates are interesting in their own right because of the form they often take. Taking a closer look at them provides a hint about how we might move the discussion along in constructive ways.

4.0. The Pattern of Debate

The typical pattern goes something like the following, which is a caricatured condensation of many debates I've heard and read on the subject, in the specialist literature, general media, and public comments in a wide variety of Internet venues. Please note: Each position represented here is a fusion of many voices. These do *not* represent specific individuals. Also, in reality, there are more than two positions involved; I have collapsed the diversity into this dichotomy for simplicity's sake, to show the range of topics that get drawn into the discussion.[6]

1: I think sending de novo interstellar transmissions is a good idea; it's a logical extension to SETI.

2: I think it poses a risk to Earth and so you shouldn't do it without my consent.

1: The risk is low and here's why . . .

Both: [Long discussion of calculations and guesses about probable characteristics of ET civilizations.]

2: I still disagree. It's risky.

1: No, really, I did the calculations, and most transmissions probably aren't really going to be detectable, and the ones that are . . . well, don't we want to talk to aliens? Isn't that why we're doing any kind of SETI? Look, you're undermining the whole idea of SETI.

2: No, I support SETI, but come on, contact is risky. They might be hostile. Stop being naïve. Look at the record of contact on Earth.

Both: [Long discussion of episodes of intercivilization or interspecies encounters on Earth.]

1: I still think there's a good chance they won't be hostile, and there might be great benefits to meeting them.

Both: [Long discussion of why an ETI civilization is likely to be older, technologically superior and peaceful . . . or not, i.e., predatory, imperial, or xenophobic.]

2: But they might not be benign. You don't know. Maybe there's a good reason why we haven't detected any signals yet! Maybe the silence is because there's something very hostile out there.

Both: [Long discussion of Fermi Paradox]

1: Well, our leakage and military and planetary radar, etc., are detectable, so if they're out there and looking, face it, they've already found us anyway.

Both: [Long discussion about calculations of our visibility, and subdiscussion about methods of calculating our visibility]

2: I still don't think it's a good idea to *deliberately* increase our visibility, at all. That's different from these incidental/ accidental transmissions, and I think it's risky.

1: Even if I wasn't going to send one, you can't stop other people sending transmissions . . . it can't be enforced. So it's dumb to make up rules about it. Just makes you look silly. And what are you going to do about all that leakage and all the military stuff? You can't regulate that. And if they're not regulated, why shouldn't others send transmissions too?

2: Well, I think we could stop the loudest transmissions and that we should at least try. Anyway, maybe I can't stop people from transmitting, but you could at least join me and say you think it's a bad idea instead of encouraging them. That might have some influence in reducing transmissions. Scientists should use their expertise to offer guidelines for the good of all humanity.

1: No, I really don't think it's my place to tell anyone else what to do. It's not as though I have jurisdiction. That's not my job.

2: Isn't it? Shouldn't you use your influence to promote caution?

1: Not necessarily. Why is caution better? I mean really, if aliens are aggressive and technologically superior, they'll probably find us and obliterate us anyway, so it's not worth worrying about.

2: So you agree that there's a possibility that they might be hostile! Given that, shouldn't there at least be a moratorium on the strongest transmissions? Shouldn't the people who control the equipment refuse to let it be used for these projects?

1: No. I think we have to send transmissions. It's a logical extension to SETI which enhances our chances of contact, and furthermore, it's our obligation as galactic citizens. If no one signals, how will anyone learn about each other? It's unethical to only search for others' signals and not send our own. It shifts the burden of communication to our neighbors.

2: So what? I still think we should just listen for now, to see if there's anything out there, and try to learn about it before we say anything. I'm more worried about protecting Earth than I am about the morality of our theoretical diplomatic relations with aliens.

1: But transmission projects are great. They're so positive and hopeful. Young people get involved, we think together about the best art and achievements of humanity, and we carefully craft messages, and we consider what we'd really say to aliens if we had the chance . . . these projects are profoundly unifying. The process helps us to think about humanity as a collective. And these projects are terrific tools for getting people interested in SETI and in astronomy and science. And it's important practice in case we get a signal and need to send a reply. What's bad about that?

2: There are better ways to get people interested in science and to do nice public participation projects. Transmission is risky, and you're not entitled to take risks that will affect me, and if something bad happens to future generations because of alien contact, it will be your fault.

1: You're trying to silence me, but you have no right to tell me what to do. I have as much right to speak for Earth as you do. Why do you think your right to silence is more important than my right to speak?

2: But you should at least ask my opinion before you do this! And other people's opinions too. Scientists have an obligation to be ethical and to think about the greater public good. You're just following your own interests without thinking of your fellow citizens. If you don't have my consent and the general approval of everyone else, established through democratic processes, you shouldn't transmit.

1: Thousands of people participate in transmission projects, so it's not just *my* interest. And I *am* thinking about the greater public good! You said you support SETI: so you must agree that it would be incredible for all humanity if we found ETI. It would bring our entire planet together and bring us into the galactic community. We should do everything that we can to make this happen. It's cowardly not to. Inaction would be wrong.

2: No it isn't.

1: Yes it is.

2: No it isn't . . .

Again, to repeat, this is a caricature, meant only to sketch the areas of contention: it blends together many separate discussions, and is not a fair representation of any one person's views.

Typically, the discussion gets heated, goes in circles, gets stuck on one of the specific side topics involving speculation about aliens, fizzles when the two parties have different expertise and don't quite understand each other's calculations, or grinds to a halt when it becomes apparent that there is no way to move toward consensus. A great deal of time, effort, thought, and gastric acid has been expended upon these debates, and constructive resolutions that satisfy all parties, even slightly, seem rare.

My point in offering this distillation is, simply, that the real crux of the debate is rarely explicitly addressed, and this causes problems. That crux, I believe, has nothing to do with aliens. Rather, it is something like this: *What is the right way to balance the desires of some people against the concerns of others, and who is entitled to make decisions about the future of the world which we all share?*

The issue, then, becomes one of weighing the rights of some individuals against others, and the rights of the individual versus the rights of the collective. It also involves the subject of experts' responsibilities, and the related matter of how we conceive of the collective: that is, is the collective best represented by a plethora of different individuals, each with their own distinct voice and agency, or a select intellectual elite, or a unified chorus, which achieves consensus through democratic debate or similar means? (Vakoch 2008 also addresses this, from another angle.)

These issues are rarely addressed head-on in debates about transmission. It may be that those involved tend to steer around it because they prefer strictly scientific analyses, and this issue appears to be completely intractable, and simply a matter of opinion and politics. However, I believe we can do better than that; there is an established literature on precisely this kind of dilemma, because although the subject matter (aliens) is unique, it is a common problem.

More precisely, it is a "commons problem."

5.0. The Problem of the Commons

It is noticeable that the transmission debate echoes many debates concerning technology, civil society, and shared resources, about topics as diverse as arms control, land development, acid rain, ozone depletion, and carbon emissions.

This is because the key variables are of the same category: the individual right to action, the accumulated effects of many individual actions upon a society, and appropriate behavior regarding collective resources.

The term *commons problem* derives from Garrett Hardin's famous 1968 article, "The Tragedy of the Commons." Hardin was writing in reference to the issue of population control, echoing a previous analysis of the nuclear arms race, which observed that "[b]oth sides in the arms race are . . . confronted by the dilemma of steadily increasing military power and steadily decreasing national security. *It is our considered professional judgment that this dilemma has no technical solution.* If the great powers continue to look for solutions in the area of science and technology only, the result will be to worsen the situation." (J. B. Wiesner and H. F. York, *Sci. Amer.* 211 (4)(1964): 27, as cited by Hardin 1968)

Hardin argued that, contrary to most scientists' perception, the population problem was one of a set of problems that has "no technical solution"—a technical solution being one "that requires a change only in the techniques of the natural sciences, demanding little or nothing in the way of change in human values or ideas of morality" (Hardin 1968, 1243). Further, he argued that the problem was freedom: in a world where everyone is free to breed without restraint, ultimately, everyone suffers.

More precisely, Hardin discussed how a "tragedy of the commons" unfolds. It is worth quoting at length:

> Picture a pasture open to all. It is to be expected that each herdsman will try to keep as many cattle as possible on the commons. Such an arrangement may work reasonably satisfactorily for centuries because tribal wars, poaching, and disease keep the numbers of both man and beast well below the carrying capacity of the land. Finally, however, comes the day of reckoning, that is, the day when the long-desired goal of social stability becomes a reality. At this point, the inherent logic of the commons remorselessly generates tragedy.
>
> As a rational being, each herdsman seeks to maximize his gain. Explicitly or implicitly, more or less consciously, he asks, "What is the utility *to me* of adding one more animal to my herd?" This utility has one negative and one positive component.
>
> 1) The positive component is a function of the increment of one animal. Since the herdsman receives all the proceeds from the sale of the additional animal, the positive utility is nearly +1.
>
> 2) The negative component is a function of the additional overgrazing created by one more animal. Since, however, the

effects of overgrazing are shared by all the herdsmen, the negative utility for any particular decision-making herdsman is only a fraction of −1.

Adding together the component partial utilities, the rational herdsman concludes that the only sensible course for him to pursue is to add another animal to his herd. And another; and another. . . . But this is the conclusion reached by each and every rational herdsman sharing a commons. Therein is the tragedy. Each man is locked into a system that compels him to increase his herd without limit—in a world that is limited. Ruin is the destination toward which all men rush, each pursuing his own best interest in a society that believes in the freedom of the commons. Freedom in a commons brings ruin to all. (Hardin 1968,1245)

Let me be precise: I am quoting Hardin for a particular reason. I am not arguing that freedom to transmit *must* be curtailed, or that unlimited transmissions *must* inevitably lead to ruin, in the same way as an overcrowded pasture might. I am arguing that Hardin neatly defined a class of problem to which the transmission debate belongs: a class with no simple technical solution, in which the actions of individuals or small groups ultimately affect everyone, and present benefits must be weighed against ultimate results. Since Hardin wrote, this class of problem has received a great deal of attention in economics, the social sciences, and global governance: the literature is vast and useful.

Hardin pointed out that the same kind of logic applies to problematic emissions into the environment, as well as extractions; it applies to pollution as much as to resource depletion. It can suit an individual better to pollute than not, but everyone will end up bearing the consequences of many individuals acting this way. He also noted that this is particularly intractable when it comes to air and water; land can be fenced and either protected or polluted as private or crown property, but emissions into water and air do not stay localized and can only be controlled through "coercive laws or taxing devices." Hardin also observed that the morality of an action depends on the state of the system; that is, behavior that is acceptable in small doses becomes unacceptable in frequent, large doses. One man polluting a river in the wilds is not the same as one man polluting a river in the midst of a city, or the same as one hundred men polluting a river in either place. As conditions change, therefore, rules for behavior should change. Ultimately, Hardin concluded, the problems of the modern world require us to think carefully and constantly about morality and ethics, and to consider that we may need to voluntarily and mutually accept restrictions on rights and freedoms that we hold dear, in order to protect other rights and freedoms that are even more precious.

Again, let me be precise: I am not suggesting that interstellar transmissions should be equated with pollution or resource depletion. I am, however, suggesting that there are shared resources here: Earth, and perceptions of security. And I am suggesting that one of Hardin's general points is applicable; that in complex systems where conditions change over time, societies must reassess individual actions in relation to their use of common resources. Further, Hardin's paper has helped spawn forty years' worth of studies of "commons problems" and "commons dilemmas" and "public goods," which are pertinent to the transmission debate. Such studies outline, for example, strategies through which common resources can be successfully and sustainably managed, and strategies that usually result in failure.

Because we cannot confidently predict the consequences of transmissions, it is unlikely that all participants in the debate will easily agree about whether highly visible transmissions are a good idea. But we can look to the "commons problem" literature to learn about how, in the absence of simple consensus, group decisions can still be made that maximize benefits for all, and how cooperation can be secured. And we might look, for example, to the Kyoto Protocol, initiatives to combat acid rain and ozone depletion, and nuclear arms control, or other projects with voluntary participation from many nations, to learn about what works and what doesn't, in efforts to form collaborative policy of global scope. Much could be learned.

6.0. Conclusions

This chapter has not offered a new scale concerning transmissions, or a new theory about the risks of Active SETI or contact, or a new suggested protocol. What it offers, instead, is a suggestion about a different framing of the transmission debate, a new focus on the underlying sociopolitical questions involved, and an arrow pointing at an existing body of analysis that might contribute to constructive next steps. As unique as the transmission debate appears at the outset, and as scientific as it appears, it is actually one of a known class of social dilemmas, and we can learn from others' engagements with problems like it.

Acknowledgments

My thanks go, as ever, to the scientists in the SETI community for their congeniality, and to my colleagues in the International Academy of Astronautics SETI Permanent Study Group. I am also grateful to participants in discussion groups and transmission projects for sharing their thoughts in various venues. Particularly pertinent voices in the transmission debates have included those of

Michael Michaud, Seth Shostak, John Billingham, Alexander Zaitsev, David Brin, Douglas Vakoch, Guillermo Lemarchand, Paolo Musso, Paul Shuch, and Ivan Almar, among others. I can only hope that no one mentioned in the foregoing three sentences will be offended by my effort to isolate a solvable problem and a workable approach in this debate. For research and travel and moral support, inspiration, and encouragement, I owe thanks, variously, to York University's Faculty of Arts, and the Social Sciences and Humanities Research Council of Canada, Mark Knackstedt, Laurie Goldsmith, and Allen Tough. Despite all this assistance, errors are mine.

Notes

This chapter is reprinted from *Acta Astronautica* 2010, DOI:10.1016/j. actaastro.2010.02.024. Originally presented at the International Astronautical Congress, Glasgow, 2008.

1. Transmissions include:

    the 1974 message from Arecibo to commemorate the dedication of the observatory (www.seti.org/Page.aspx?pid=688);
    several messages from Evpatoria (Zaitsev 2006, 2008);
    the Discovery Channel Canada's "Calling All Aliens" project in 2005 (http://ctvmedia.ca/discovery/releases/release.asp?id=6886& num=3&yyyy=2005);
    Talktoaliens.com, a company that sent public messages for a fee via low-powered transmissions in 2005 (Young 2005);
    Craig's List via Deep Space Communications, billed as the first commercial transmission into space, in 2005 (Than 2005);
    Cosmic Connexion, www.cosmicconnexion.com/static/home-en.html, a broadcast of public-created content from the French ARTE television station, transmitted by the National Centre for Space Studies antenna to commemorate the launch of the Corot satellite in 2006;
    A publicly generated Doritos ad transmitted via EISCAT / Svalbard and Leicester University toward 47 UMa in summer 2008 (Leicester 2008);
    NASA's 2008 broadcast of the Beatles' song "Across the Universe" to commemorate NASA's fiftieth anniversary (NASA 2008; Zaitsev 2008; Shostak 2008);
    The upcoming A Message From Earth transmission of publicly generated content, fall 2008, from Evpatoria (www.bebo.com/ amessagefromearth).

2. At a recent IAA conference in Paris (Searching for Life Signatures, Sept 22–26, 2008), many papers focused upon Active SETI. The topic has also frequently been addressed within the SETI sessions of the International Astronautical Congress annual meetings. There are many papers concerning Active SETI in the proceedings of the International Academy of Astronautics congresses, and in *Acta Astronautica*. Shuch and Almar (2007) discuss the San Marino Scale, for cataloguing the significance of different types of transmissions. See also http://iaaseti. org/smiscale.htm.

3. Ancient civilizations often built for future generations, but we, too, build for eternity—e.g., the projects of the Long Now Foundation, www. longnow.org. Atwood (2002) discusses what a peculiar pastime writing is. The Pioneer Plaque and the Voyager Record are the classic examples of humanity's self-portraits in space (Sagan et al. 1979), but numerous recent initiatives allow individuals to, for example, send their names to the Moon or to Mars (NASA, Send Your Name to the Moon: www. nasa.gov/home/hqnews/2008/may/HQ_08110_Name_To_The_Moon. html; Planetary Society, Send Your Name to Mars: www.planetary.org/ special/fromearth/phoenix). The idea of an archive in space is seen, for example, in Operation Immortality (www.operationimmortality.com, accessed August 27, 2008). The founder, Richard Garriott, ran a public project: participants were invited to contribute their personal messages, their list of humanity's greatest achievements, and their DNA profiles, to be collated and stored on the International Space Station as a digital time capsule. The project was a promotion related to the Tabula Rasa MMORPG video game/virtual world. See also Wired Blog Network: "Why Richard Garriott Can't Play Tabula Rasa in Space," July 30, 2008. http://blog.wired.com/games/2008/07/richard-garriot.html.

4. In saying this, I intend no disrespect to Musso's (2008) delineation of scenarios and risks of Active SETI; I simply mean that the risks of contact with extraterrestrials cannot be assessed with much confidence.

5. The cross-cultural comparison of intercivilizational encounters is under-theorized, and the use of Earth's historical record in SETI literature is often problematic (Denning 2010). Moreover, of course, contact with extraterrestrials could well introduce novel problems, e.g., Carrigan's (2003) "SETI Hacker" hypothesis.

6. There is a spectrum of opinion. From what I have observed, I do not concur with Shuch and Almar's (2007, 142) assertion that in the IAA SETI Permanent Study Group, "participants are neatly divided into two camps; those who believe predation is a possibility in the cosmos, and those who envision benevolent extraterrestrials."

Works Cited

Atwood, M. 2002. *Negotiating with the dead*. Cambridge: Cambridge University Press.

Brin, D. 2006. Shouting at the cosmos. Online at www.lifeboat.com/ex/ shouting.at.the.cosmos.

Carrigan, D. 2003. The ultimate hacker: SETI signals may need to be decontaminated. Presented at IAC Bremen. IAC-03-IAA.8.3.06.

Denning, K. 2010. Social evolution: State of the field. In *Culture and cosmos*, ed. S. Dick and M. Lupisella. Washington, DC: NASA History Press.

Grinspoon, D. 2007. Who speaks for Earth? *SEED* magazine, 12 Dec. 2007, http://seedmagazine.com/news/2007/12/who_speaks_for_earth.php.

Hardin, G. 1968. The tragedy of the commons. *Science* 162 (3859) (Dec. 13): 1243–48.

Hecht, J. 2006. What are the chances of aliens sniffing us out? *New Scientist* online 24 Oct. 2006.

Leicester University. 2008. Press release: Doritos makes history with world's first extra terrestrial advert. 12 June 2008. www2.le.ac.uk/ebulletin/news/ press-releases/2000-2009/2008/06/nparticle.2008-06-12.7228690325.

Musso, P. 2008. Active SETI and its possible dangers: An overview. Presented at First IAA Symposium on *Searching for Life Signatures*, UNESCO, Paris, 22–26 Sept. 2008.

NASA. 2008. NASA beams Beatles "Across the Universe" into space. Press release, 4 Feb. 2008. www.nasa.gov/topics/universe/features/ across_universe.html.

Nature editorial. 2006. Ambassador for Earth. *Nature* 443 (12 Oct.), 606.

Sagan, C., F. Drake, A. Druyan et al. 1979. *Murmurs of Earth*. New York: Random House.

Shostak, S. 2006. The case for transmitting to space. Space.com, 25 May 2006. www.space.com/searchforlife/seti_transmit_060525.html.

———. 2008. Finding them, finding us. Space.com, 28 Feb 2008. www. space.com/searchforlife/080228-seti-finding-them.html.

Shuch, P., and I. Almar. 2007. Shouting in the jungle: The SETI transmission debate. *Journal of the British Interplanetary Society* 60: 142–46.

Than, K. 2005. Craigslist gets beamed into space. CNN/ Space.com. www. cnn.com/2005/TECH/space/03/23/craigslist.space/index.html

Vakoch, D. 2008. Active SETI: Scientific experiment, diplomatic action, or artistic expression? *First International Academy of Astronautics Symposium on Searching for Life Signatures*, Paris, UNESCO, Sept. 22–26, 2008.

Whitehouse, D. 2007. Meet the neighbours: Is the search for aliens such a good idea? *The Independent*, 25 June 2007. www.independent.co.

uk/opinion/commentators/meet-the-neighbours-is-the-search-for-aliens-such-a-good-idea-454511.html.

Young, K. 2005. Hello aliens, this is Earth calling. 10 March 2005, *New Scientist* online: www.newscientist.com/article.ns?id=dn7128.

Zaitsev, A. 2006. Messaging to extraterrestrial intelligence. http://arxiv.org/ftp/physics/papers/0610/0610031.pdf.

———. 2008. The first musical interstellar radio message. *Journal of Communications Technology and Electronics* 53 (9): 1107–11.

Integrating Active and Passive SETI Programs

Prerequisites for Multigenerational Research

Douglas A. Vakoch

1.0. Introduction

One of the challenges of the Search for Extraterrestrial Intelligence, or SETI, is the need to make inferences about appropriate search strategies and the nature of extraterrestrial intelligence in the absence of any direct evidence that such intelligence even exists. In this chapter, we will examine central assumptions that currently guide the dominant search strategy of the international SETI community—passively listening for signals from other civilizations. In this paper we will see that although those assumptions are reasonable starting points, and we should continue to have at the core of our search a firm commitment to listening for signs of technological civilizations through radio waves and laser pulses, as we reflect on those assumptions we should also open ourselves to an alternative approach. Namely, in the coming years and decades, we should supplement traditional Passive SETI programs with active transmissions of our own, in which we transmit intentional signals to other civilizations even prior to detecting direct evidence of extraterrestrial intelligence. Such de novo transmissions represent an appropriate diversification of our search strategies, particularly given the uncertainty of some of the standard assumptions that guide SETI.

Current SETI programs assume an asymmetry of age between humanity and any extraterrestrial civilizations with which we make contact through interstellar communication. Because transmission requires greater resources over sustained periods than passive searches, many assume that longer-lived, more technologically advanced extraterrestrial civilizations will take the initiative in making their presence known through intentional signals. This chapter will examine the possibility that, even if this assumed asymmetry is correct, extraterrestrial civilizations may not be beaming signals to Earth for our benefit. Instead, we may benefit by conducting Active SETI experiments of our own, which would allow us within a few decades to test variations of the Zoo Hypothesis, which suggest that intelligent life is widespread and monitoring our signals, and it may only be willing to respond but not initiate contact. A shift to humans conducting Active SETI, we will see, may also increase the chances that any messages exchanged will be mutually comprehensible. We also consider ways that such an inherently intergenerational research program can be maintained.

2.0. Asymmetry of Age

In the passive search for transmissions from extraterrestrial civilizations, there are two basic approaches. First, we can conduct what are called targeted searches, in which we focus on one star at a time, spending several minutes at each target systematically scanning a wide range of radio frequencies. This was the approach employed by the recently completed Project Phoenix of the SETI Institute, which searched for radio signals from nearly one thousand nearby stars. For such a search to succeed, the other civilizations will need to have been transmitting continuously or nearly continuously, for us to have a reasonable chance of detecting them during the brief time we look their direction.

The alternative approach, an all-sky survey, is projected to be technologically and economically feasible in the next decade or shortly thereafter (Ekers, Cullers, Billingham, and Scheffer 2002). By that time, the intensive computational power needed for analyzing data gathered from the entire sky is projected to be cheap enough to allow continual, real-time processing of data. With the addition of all-sky surveys to the microwave search in the coming decades, we will become capable of identifying even intermittent transmissions.

With either approach, the chances of success increase markedly if extraterrestrial civilizations are much older than human civilization. This asymmetry of lifetimes of alien and human civilizations is commonly assumed to be a prerequisite for SETI to succeed. This assumption that any extraterrestrial intelligence that we detect would be much longer-lived than our own civilization does not presuppose that there is any inherent tendency toward technology fostering survival. It may well be that civilizations attain the capacity for interstellar communication for only a brief period, say one hundred years, before they either annihilate themselves or lose interest in making contact with extraterrestrial civilizations. If that is the case, however, then given the thirteen-plus billion years age of the galaxy and the variability in ages of stars and timescales during which planets with habitable environments might form, it is highly unlikely that an extraterrestrial civilization's century of technology will so closely coincide with our own to allow us to detect their existence, and thus SETI programs will not be successful. (To be precise, the other civilization would need to begin its lifetime as a technological civilization slightly before we do, assuming they are separated from us by more than a few light-years. This difference would allow their transmissions to coincide with our listening, even though they begin transmission before we have the capability to detect their signals, catching up only during the time the signals traverse interstellar space.)

If indeed extraterrestrial civilizations we make contact with have even more slightly advanced technologies than humanity currently does, it seems plausible they would be capable of detecting intentional signals sent from Earth—even if those signals are intermittent as assumed by an all-sky survey, and as proposed for designing cost-effective interstellar beacons that even humankind is now capable of building (Benford, Benford, and Benford 2010).

3.0. Too Young to Transmit?

This line of reasoning about the asymmetry of lifetimes of humanity and of any civilization old enough to make contact with has also been used to argue that the proper role for humanity at this stage of our development is simply to listen, and not to transmit ourselves. An especially clear statement of this assumption that humans are too young as a civilization to mount a sustained transmission program is articulated by Tarter (2004, 18) in her article "What If Everybody Is Listening and Nobody Is Transmitting?":

> During our 1997–99 workshops on the next two decades of SETI research here at the SETI Institute, the workshop participants took the question of an active transmission strategy very seriously. The results of their deliberations have been published in *SETI 2020: A Roadmap for the Search for Extraterrestrial Intelligence.* They concluded that transmission is NOT an appropriate strategy, at least for the next two decades. Humans need to grow up first.

The challenge of using the guideline that "youthful" civilizations should not be expected to transmit is that it may not be clear exactly how old is old enough to transmit. One problem is that it is difficult to anticipate how long our own efforts in SETI will continue, particularly if we take into account that a highly developed culture may continue to *exist* on Earth, even if the motivation to *explore* the cosmos through astronomy disappears.

Ultimately, however, there is a problem that the advocate of a solely passive search strategy for humankind must acknowledge: even considerably longer-lived civilizations than our own may argue that they too are still too young to be responsible for initiating contact. Even if a million-year-old civilization expects to remain intact for another thousand years, it is not obvious that they will ever feel old enough to begin—especially if they believe there are risks associated with transmitting, as some have suggested. As Lemarchand and Tarter (1994, 137) point out, "Every galactic civilization could find arguments to show that there are 'more advanced civilizations' than their own and that it could be dangerous to 'reveal' their existence and position to other beings. If this is so," they continue, "it might be impossible to find a threshold where a galactic civilization could feel themselves completely safe. In this way, every civilization could apply the 'little children' argument and only use a 'passive search strategy' instead of an 'active search strategy.' " Instead, it may be that the younger civilizations are left with the burden of initiating contact—a possibility that would provide support for diversifying the present terrestrial reliance on passive strategies by adding Active SETI as well.

In addition, a *two-way conversation* with extraterrestrial intelligence requires that at least one of the civilizations is extremely long-lived (so that it happens to exist at the time that the younger civilization exists), and even then the younger civilization needs to exist long enough to be around for a reply. Of course, we might imagine variations on a fully two-way conversation. One partner (civilization A) may get a reply, but the other interlocutor (civilization B) may no longer exist (or may no longer be listening) when the reply from the first (civilization A) comes in. The most likely scenario for this asymmetry, in purely statistical terms, would be an exchange that is two-way from the perspective of the older civilization, but one-way from the perspective of the younger. That is, the older civilization may still be listening when a reply from the younger civilization arrives, but the younger civilization may no longer be listening when a reply from the older civilization comes in.

Indeed, it would seem likely that a civilization that has already been engaged in SETI for a million years would to continue to engage in SETI for, say, another millennium. It is much less clear that a civilization such as ours, which has engaged in SETI only sporadically for less than a half-century, will be able to sustain this activity for twenty times that duration. (For a detailed statistical argument for estimating the future lifetime of terrestrial civilization based on the current lifetime of terrestrial civilization, see Gott's [1993] analysis.)

Diamond (1990, 34) has argued that we should expect the civilizations on other worlds to be relatively short-lived:

> Our development of radios was an extremely unlikely fluke; even more of a fluke was our development of them before we

developed the technology that will end us in a slow stew or fast bang. While Earth's history of radio civilizations thus offers little hope that they exist elsewhere, its history of the concomitants of radio civilizations suggests that those that might exist anywhere are short-lived.

Thus, the deafening silence from outer space is not surprising. Yes, there are billions of galaxies with billions of stars. Out there must be some transmitters as well, but not many, and they don't last long.

If Diamond's conclusion is correct, then the norm for interstellar communication, if it exists at all, is one-sided transmission. By transmitting to an extensive list of targets, including many at great distances, a civilization might hope that, if it is lucky, its signal will be detected by another civilization, even if the sender will not be around long enough to receive a reply.

The argument that an asymmetry of lifetimes between human and extraterrestrial correspondents necessarily means humankind should not be transmitting, but only listening, thus makes a critical assumption that long-lived civilizations will take the initiative in transmitting. We will return to this assumption later when we explore the plausibility of believing extraterrestrial intelligence will engage in SETI for the benefit of humankind. In the next section, we will explore some of the technological issues involved in mounting active transmission projects, which on Earth have thus far been primarily symbolic demonstrations that we have the capacity to transmit strong, information-rich signals across interstellar distances (Staff at the National Astronomy and Ionosphere Center 1975; Zaitsev no date).

4.0. Longevity and the Level of Technology

Paired with the assumption that the extraterrestrial civilizations that we detect will almost inevitably be older than ours, is the assumption that with increased age comes increased technological capability. Not only would other civilizations be older than ours, it is argued, but as on Earth, once a critical point is reached in technological development, increased age yields an increased ability to control nature through considerably more advanced technology. This assumption has an important implication for our search strategy. Because the extraterrestrial civilizations will have much greater technological capabilities and greater economic resources than we do, it is often argued, they should be expected to take on the burden of transmitting, while we, the younger civilization, will be expected to take on the easier role of listening.

But as we have seen on Earth, increased technological capacity also brings with it the potential to annihilate the very civilization that brought forth that technology (Diamond 1990). It is not obvious that the sort of extraterrestrial civilizations that humans may make contact with—sufficiently long-lived to be transmitting while we too exist—will necessarily continue to advance in technology far beyond that of humankind. Perhaps the most stable civilizations on galactic time scales are those that are moderately, but not excessively, technologically advanced. Exponential growth may not be a sustainable development pattern for long-lived civilizations (Haqq-Misra and Baum 2009).

This assumption that extraterrestrial civilizations will necessarily have technologies that are markedly advanced relative to humankind's technologies was directly challenged by Oliver as early as the Third Decennial US-USSR Conference in SETI, held in Santa Cruz, California, in 1991. In a published paper based on that conference presentation, titled "Symmetry in SETI," Oliver (1993) uses Kardashev's typology of civilizations to explain his advocacy of initiating Active SETI programs to complement the current emphasis on Passive SETI. Kardashev identifies three types of extraterrestrial civilizations, based on their levels of energy consumption:

I—technological level close to the level presently attained on the earth, with energy consumption at $\approx 4 \times 10^{19}$ erg/sec.

II—a civilization capable of harnessing the energy radiated by its own star . . . ; energy consumption at $\approx 4 \times 10^{33}$ erg/sec.

III—a civilization in possession of energy on the scale of its own galaxy, with energy consumption at $\approx 4 \times 10^{44}$ erg/sec. (1964, 219)

As Oliver applies Kardashev's typology to selecting search strategies, Oliver notes that the energy requirements for a civilization to transmit omnidirectional signals from a single 100 meter dish that would be strong enough to be detected with the best technology available when he was writing presuppose that extraterrestrial civilizations—even those detectable at a range as close as one hundred light-years from Earth—would need to be Type II civilizations. And yet, Oliver (1993, 72) held open the possibility that extraterrestrial civilizations we might expect as interlocutors could well be "merely 'Type I' civilizations."

Oliver therefore considers alternative transmitting strategies, such as powerful directional beacons, in which the burden of transmission remains with the extraterrestrial civilization, as well as a "balanced" system in which both extraterrestrial civilizations and humankind would build SETI systems in which the cost of transmission and reception are equal. It was this latter, balanced system that Oliver most clearly championed toward the end of his life, as we shall see later in this chapter.

The implication of adopting such a balanced approach would be to remove the assumption of extraterrestrial altruism that is implicit in the notion that more advanced civilizations will be disproportionately responsible for the cost of transmission. As Oliver (1993, 72) summarizes this point, "It is interesting that this optimum solution results not because of any assumption of great prowess on their part but simply as a result of reciprocity applied to selfishness. Perhaps," Oliver concludes, "this is the 'golden rule' of SETI."

Subotowicz, Usowicz, and Paprotny (1979, 205) similarly suggest that both terrestrial and extraterrestrial civilizations have a responsibility to transmit as well as to search for signals: "We and every other ETI are morally obligated to realize together *active* and *passive* CETI [communication with extraterrestrial intelligence]), either from orbit or from the surface of a planet. The active transmission of messages should be carried out with the largest possible power, transmitting the most understandable and readable information." This search strategy is founded on what Subotowicz has called the principle of *partnership*: "Every civilization realizes active and passive CETI if only it can do it" (Subotowicz, Usowicz, and Paprotny 1979, 205). As Zaitsev (no date, 9) has stated this position, "[I]f all civilizations in the Universe are only recipients, and not message-sending civilizations, than [*sic*] no SETI searches make any sense."

To be clear, such a call for humankind to engage in Active SETI has not been advocated widely within the international SETI community. When the SETI Institute completed a strategic planning process in the late 1990s to determine its research priorities in SETI for the next twenty years, it concluded that Active SETI should not be among its projects. The participants of the working group that authored the final report, *SETI 2020: A Roadmap for the Search for Extraterrestrial Intelligence*, note that "[t]ransmission is a more expensive strategy than receiving. Within the next two decades, the parameter space explored for signals can be extended by the compounded growth of many technologies. Transmissions could benefit from these same exponential improvements in technology, but with the limited resources likely to be available during this same period, we could not add significantly to the high power of our leakage radiation" (Ekers, Cullers, Billingham, and Scheffer 2002, 244).

In spite of the overall recommendation of the *SETI 2020* report not to initiate a serious program in Active SETI, an appendix to this report, written by Scheffer (2002), describes the designs for six types of beacons that we could build. Among the alternatives is a "low cost personal beacon" that would transmit daily to the one thousand closest stars at a cost of $640 US per year for power (Scheffer 2002, 315). "Compared to leakage," Scheffer notes, "it has a higher EIRP [Effective Isotropic Radiated Power], uniform repetition, and Doppler compensation." Such a design has additional benefits. "No new technology is needed. Except for the ethical issues of 'Should we transmit?,' it would be easy to do and would give us a consistent stand in front of donors. We would not be relying on the ETIs to do anything that we are not doing ourselves." An even more ambitious galactic beacon could be constructed at the cost of other major scientific facilities on Earth (Benford, Benford, and Benford 2010). Additional planning efforts in beacons design and operation seem warranted.

5.0. Scientific Progress

Another standard argument against Active SETI by humankind is that longer-lived extraterrestrials will be better equipped to create intelligible messages, given that their science will have progressed beyond our own. Implicit in this view, typically, is the assumption that these advanced extraterrestrials will be able to anticipate the nature of our current, rudimentary science because they will have gone through a similar path of scientific discovery in their own past. But a closer examination of the nature of scientific change calls this standard view into question.

In this standard account, more advanced civilizations have passed through more or less the same stages as less advanced civilizations on other worlds. If more advanced civilizations want to make themselves understood, it is argued, they will start with the principles that would surely be understood by less advanced civilizations. But, the skeptic might ask, is it so obvious which principles those would be, and even if the principles are widely known, is the conceptual apparatus for describing these principles universal?

Even if humans and extraterrestrials have a common commitment to modeling ever more accurately the nature of physical reality, there is no guarantee that these models of reality will necessarily be obviously commensurable (Vakoch 1998). "Admittedly there is only one universe, and its laws, as best we can tell, are everywhere the same," observes Rescher (1985, 90). "But the sameness of the object of contemplation does nothing to guarantee the sameness of the ideas about it. It is all too familiar a fact that even where human (and thus *homogeneous*) observers are at issue, different constructions are often placed upon 'the same' occurrences." The challenge is to disentangle our descriptions of the world from the nature of the world as it exists independently of our attempts to understand it.

The differing evolutionary histories of independently evolved species may dramatically affect even the fundamental goals that scientists pursue on disparate worlds. As Rescher (1985, 85) argues, "We can hardly expect a 'science' that reflects our parochial preoccupations to be a universal constant. The science of different civilization would presumably be closely geared to the particular pattern of their interaction with nature, as funnelled through the particular course of their evolutionary adjustment to their specific environment."

Similar insights can be derived from the arguments of Laudan (1977), whose characterization of scientific change in his book *Progress and Its Problems* stipulates that research traditions must be evaluated by how effective they are *relative to their competitors*, and not with respect to some absolute, timeless ideal of scientific rationality. Specifically, Laudan (1977, 130–31) notes that the rationality of scientific theories "is partly a function of time and place and context. The kinds of things which count as empirical problems, the sorts of objections that are recognized as conceptual problems, the criteria of intelligibility, the standards of experimental control, the importance or weight assigned to problems, are all a function of the methodological-normative beliefs of a *particular community of thinkers*" (emphasis added). When we take a step beyond Laudan's argument and imagine communities of thinkers living on planets circling other stars, mindful of the fact that the scientific theories of those communities would have developed independently of terrestrial science, the possibility of extraterrestrials possessing progressive scientific understanding that is nevertheless quite different from terrestrial science, and that never passed through a stage akin to contemporary terrestrial science, becomes quite plausible.

6.0. Multiple Paths up the Mountain

Perhaps an analogy of mountain climbing will help clarify the likelihood of divergent scientific practices on other worlds. Science progresses, we might argue, in the same way that a mountain climber progresses toward the peak of a mountain. Not all climbers will progress as far; novice climbers may only make it part way up the mountain. But as these neophytes become more skilled, they will be able to progress to greater altitudes, pointed toward their goal: the highest point of the mountain.

In this analogy, the scientist is akin to the climber, progressing toward ever clearer understanding of the nature of reality as it really is, symbolized by the mountaintop. There may be times when the scientist/climber diverts from the path, but in the long run, the interplay of theory and experiment ensures that the successful scientist—the one who makes progress in ascending the mountain—will find the right path. A more sophisticated scientist/climber, having ascended higher, could look back and even leave pointers for the less experienced scientist/climber, potentially providing clues that might speed up the ascent of the less experienced.

As we apply this analogy to interstellar communication, we typically assume on statistical grounds that we are the less experienced climber. Although one might imagine scenarios in which we make contact with even relatively short-lived civilizations if they are sufficiently prevalent, let us assume for this analogy that the extraterrestrial will have climbed higher up the mountain than we have.

To continue the analogy, it is much more reasonable to think that an extraterrestrial civilization will have approached the top of the mountain from a different point, having evolved in a different niche. As Barker (1982, 82–83) notes, "There is no reason to think that the alien's science must progress through just the same sequence of revolutions as our own." How then could the more advanced civilization point the way up a path it did not take? Instead, it will have taken a different path, up another side of the mountain.

But even this analogy presupposes that both human science and extraterrestrial science seek the same destination (the "mountaintop") representing the most complete understanding of reality. Rescher (1985, 93) cautions us against assuming that all civilizations are headed toward the same scientific destination: "The one-world, one-science argument would only go through if it could be maintained that, while the process or course of scientific development is something variable and contingent, the ultimate product that will issue from these diversified strivings is fixed in preordained uniformity. It would have to be shown that here all different routes lead inexorably and inevitably to the same destination." "However," Rescher (1985, 95) continues, "the development of a science hinges crucially on . . . the sorts of issues that are addressed and the sequential order in which they are posed. And here the prospect of variation arises: we must expect alien beings to question nature in ways very different from our own."

In essence, we must imagine that the human and extraterrestrial scientists/climbers are ascending different mountains—both gaining an increasingly comprehensive understanding of the universe, but each headed toward a different mountain peak, providing a perspective on a different aspect of the universe. If Rescher is right, then science may take varied forms on various worlds.

While the possibility of multiple directions of progress does little to reassure us of easy interspecies communication, it does open the possibility of learning much, if we ever do establish contact. Indeed, the possible plurality of sciences on different worlds may provide a sense of reassurance that even a civilization as young as ours might contribute substantially in an interstellar exchange: our scientific and cultural accomplishments could be of considerable interest on other worlds. If in fact there is not one single path of scientific progress taken by all civilizations, but different paths depending on each species' idiosyncratic environment as well as its unique evolutionary and cultural histories, then even our relatively primitive accounts of the universe may provide novel insights to extraterrestrials. As Rescher (1985, 93) favorably summarizes the position of psychologist William James, "In the course of cognitive evolution, nature, metaphorically speaking, works towards being known in all its various aspects, evolving beings capable of comprehending it in many different and diversified modes. Here we are dealing with a teleology of diversity—a nature striving to be known in a variety of forms and aspects." From this perspective, humans potentially have much to contribute to other civilizations—even ones that may be much, much older than our own—by explaining, as best we can, the idiosyncrasies of terrestrial science.

Just as the anthropologists and historians of Earth are interested in the development of other cultures' ways of understanding the world about them, so too might extraterrestrial intelligence be interested in the specific trajectory our science has taken. Though we tend to value our most recent scientific understanding most highly, assuming this most accurately reflects the nature of reality, historians of science on another world may not be especially interested in learning about the models most widely accepted in the early twenty-first century. Instead, extraterrestrial historians may be more intrigued by the entire history of our science, and in fact it would be more accurate to acknowledge that the many different terrestrial sciences each have their own history, intertwined with other sciences and other cultural practices.

We have made reference to "terrestrial science" in the singular in the preceding paragraphs for the sake of simplicity. But even in the same scientific domain, for example, "biology" or "astronomy," the forms that science has taken in different cultures have been diverse. In addition, the changing conceptualizations within a given culture have been dramatic, for example, the shift from "natural history" to "biology" in Western science (Foucault 1970). We should expect no less of extraterrestrial sciences. As Rescher (1985, 92) observes, "Human organisms are essentially similar, but there is not much similarity between the medicine of the ancient Hindus and that of the ancient Greeks. There is every reason to think that the natural science of different astronomically remote civilizations should be highly diversified."

7.0. Asymmetry of Experience

In spite of the challenge that the species-specific nature of scientific progress poses to the standard SETI view, there remains an essential validity to the core argument that we should expect more long-lived civilizations to have greater capabilities in interstellar exchanges than we do. On purely statistical grounds, if we make contact with another civilization—especially at as early a stage in the search as we are at today—then we can expect they will already have made contact with many other civilizations before us. (If they have not, the argument goes, then extraterrestrial civilizations are so few and far between that we should not expect to make contact with even one such civilization ourselves.)

In the course of those encounters, it would seem, other civilizations should have made contact with beings who have, collectively, ascended an entire range of mountains, approaching their respective mountaintops from varying directions, and at times even descending to better understand the fuller extent of the entire mountain range. While such exposure to other civilizations does not ensure that the more experienced partner will be able to anticipate all of the problems of communication inherent in interspecies contact, it should give the older civilization an advantage. The more advanced civilization should, then, be better than the less advanced civilization at both *encoding* and *decoding* interstellar messages.

The critical question, as we consider whether or not to begin serious Active SETI programs of our own, is whether superior communications abilities are more important when transmitting to another civilization, or when listening for signals from another civilization. Typically, the SETI community has assumed the former: the older civilization is expected to take the burden of creating a message that will be obviously intelligible to any civilization that receives it. If the older civilization succeeds, then not only will *contact* have been made, but there will be a foundation for *mutual comprehension*. That is, not only will we know that they exist, but we will know what they are saying.

To return to our analogy, we suppose that the advanced extraterrestrials will be able to create a message that is so universally comprehensible that it will be understood by anyone climbing any mountain along any path—given the constraints that the only interstellar interlocutors are those with the technology (such as radio) to receive the signal in the first place.

Another way to make use of the greater range of experiences of older civilizations is to call upon them to understand *our* means of representing reality. If a longer-lived civilization is more capable of creating a message that is intelligible to an alien species than we are, then it should also be more capable of *understanding* the messages of other civilizations than we are.

The critical question, then, is which of the following *combinations* yields the greatest chance for mutual comprehension—or least comprehension of one civilization by the other, if it is too much to assume that both will be able to comprehend:

1. an advanced civilization attempting to understand the message from a less advanced civilization, or

2. a less advanced civilization attempting to understand the message from a more advanced civilization?

Unless we can confidently rule out the first option, it seems prudent to diversify our search strategy by including at least some component of Active SETI. Although we are guaranteed of having no message from another civilization in response to a transmission from Earth for years or decades at a minimum, if in fact we have insurmountable difficulty decoding any message we may some day receive in a passive search, we may need to resort to active transmissions—articulating our understanding of the universe in the most accessible manner we can conceive—to move toward mutual comprehension.

8.0. For Whose Benefit?

Following the assumed asymmetry in the lifetimes of humanity and any extraterrestrial civilizations with whom we are likely to come in contact, some argue that our own future is so uncertain that it would be unwise to assume we will exist long enough to receive a reply to any of our messages. This, they argue, should make us reluctant to begin transmitting. Admittedly, it seems more likely that a civilization already one million years old would continue to exist for another century or millennium, than to argue that civilizations such as ours, with the capability of communicating at interstellar distances for less than a century, will continue to live for centuries or thousands of years.

All of this reasoning, however, presupposes that the intended beneficiaries of Earth's SETI programs are ourselves or other humans who will come soon after we do. And indeed, discussions about the benefits of SETI do focus on those benefits that are admittedly short-term in a broader cosmic timescale.

The emphasis on the benefits of interstellar communication for humans, rather than for extraterrestrial intelligence, is also reflected in the Draft Declaration of Principles Concerning Sending Communications with Extraterrestrial Intelligence, developed during the 1990s within the International Academy of Astronautics' (IAA) SETI Committee, and the subject of recent discussion within the IAA Permanent SETI Study Group— the successor of the earlier SETI Committee. This Draft Declaration of Principles notes that if a message is sent to an extraterrestrial civilization, "The content of such a message should reflect a careful concern for the broad interests and wellbeing of Humanity. . . ." No mention is made of the potential benefits of such communication for intelligence on other worlds. Although many have suggested that humankind might benefit from joining a "Galactic Club" of other civilizations, few have suggested that humankind should be expected to pay dues to join, or that we should consider the needs and interests of other members of the club.

One might argue, however, that as we mature as a civilization, we should increasingly take on both a cosmocentric and an intergenerational ethic to guide our search strategies, in which the potential benefit for those humans living now and in the near future is balanced with the potential benefit of extraterrestrial civilizations who may receive our messages, as well as future generations of humans who may receive a reply to any messages we might send. In this line of reasoning, we could potentially make significant contributions to long-lived extraterrestrial civilizations even if we are unable to sustain a transmission project for decades or centuries. While we may be too young to start and finish an Active SETI experiment *as the scientists*, we might still participate in an extraterrestrial's Passive SETI experiment *as the subjects*.

If we are already being monitored by other civilizations, but are too distant from them for our accidental transmissions to be detected, we might provide information of great interest to extraterrestrial sociologists who are attempting to quantify the distribution of lifetimes of civilizations capable of interstellar communication, term L in the Drake Equation. (The Drake Equation is a heuristic that identifies the factors we need to consider when estimating the number of civilizations capable of interstellar communication that currently exists in our galaxy [Drake and Sobel 1992].) The longevity of civilizations is one of the most elusive terms in the Drake Equation, and at this stage in our development, we have only one data point by which to estimate it—our own lifetime. Even if we begin transmitting and fail to survive until we receive a reply, the length of time we have continued to transmit will provide important information to other civilizations conducting a galactic census.

Although some have argued that older civilizations *should* take on the burden of transmitting, while younger civilizations only listen, it is not clear that the longer-lived civilizations *will* take on this burden (Vakoch 2011). Arguably, it is the younger civilizations that have the most to gain from an interstellar exchange, and thus might be expected to take the initiative. Given the challenges of anticipating extraterrestrial motivations, diversifying our search strategies to include Active SETI may increase the long-term chances of success, even if measured solely in terms of benefits for humanity.

9.0. A Galactic Zoo

The preceding sections have laid out a case for Active SETI as a complement to Passive SETI because other civilizations may not be transmitting, and even if they are, we might accelerate the exchange of meaningful information if we take the initiative in transmitting. While these are sufficient reasons to advocate serious Active SETI programs, still another scenario is worth considering.

It is also possible that our existence may already be known to other civilizations, but they have not made themselves known because they do not want to interfere with our independent development as a civilization, as in Ball's (1973) "Zoo Hypothesis." Some variants of this hypothesis would provide little hope for a response to our transmissions, regardless of the content. For example, if the objective of extraterrestrials is to maintain Earth as a pristine laboratory that they can watch develop without external interference (Barrett 1983), we might expect no reply, because interaction would destroy the experiment.

Even under "The Laboratory Hypothesis," however, there may be cases in which we may be able to prompt a response. For instance, one might imagine extraterrestrial research protocols that stipulate silence if only undirected leakage radiation were detected, but that would call for a response to an intentional human attempt to initiate contact. If, after all, humans had designed a laboratory experiment to study communication between mice, but the mice suddenly began trying to communicate with the humans overseeing the lab, we might expect the experimenters to shift focus and to explore the extent to which further communication with the mice could be fostered.

Other variants of the Zoo Hypothesis may also offer prospects of extraterrestrials breaking a silence they currently maintain. If the primary objective of extraterrestrials is to act as caretakers of Earth who are intent "to discourage any overt visitations . . . and to prevent any possible colonisers from moving in" (Deardorff 1987, 375), they may have no objection to responding in a manner they deem appropriate for a civilization at our level of development. Indeed, if extraterrestrial civilizations are attempting to protect us out of concern that knowledge of their existence could "cause severe adverse reactions amongst followers of some of Earth's religions," as Deardorff (1987, 375) has suggested, then a message from Earth detailing the significant extent to which major religions would be able to accommodate such knowledge—if in fact we could legitimately make that case—may help alleviate those concerns. (For a case study showing the diverse range of views toward the possibility of extraterrestrial life within one religious tradition often assumed to be threatened by this discovery, see the author's review of Roman Catholic perspectives at the outset of the Space Age [Vakoch 2000].) While Gerritsen and McKenna (1975, 254–55) have already observed that a transmission project would provide a response to the Zoo Hypothesis, they did not provide a detailed rationale, but merely noted that "[a] special argument in favor of transmission as contrasted with reception is based on Ball's (1973) zoo hypothesis, that advanced civilizations may not try to interact with underdeveloped ones and may wait for the others to initiate communication." Thus, a transmission from Earth may be effective in eliciting a response by signaling our *intention* and *interest* in making contact, and not necessarily because other nearby civilizations are unaware of our existence.

A search consistent with a test of the Zoo Hypothesis was suggested as early as 1972, in the Project Cyclops report, which laid the foundation for NASA's eventual development of a microwave SETI program. Among that report's recommendations was the following: "If our first search of the nearest 1000 target stars produced negative results we might wish to transmit beacons to these stars for a year or more before carrying the search deeper into space" (Oliver and Billingham 1973, 153). This recommendation seems all the more timely in view of the fact that, as noted earlier, recently the SETI Institute completed its comprehensive survey of nearly one thousand nearby stars.

The first author of the Project Cyclops report returned to this topic in the final months of his life. In his draft paper "SETI Scaling Laws and Acquisition Strategies," written less than three months before his death in 1995, Oliver evaluated the reasonableness of shifting from an exclusive reliance on passive search strategies in SETI to a dual strategy of both listening for signals and transmitting intentional signals of our own. He concluded by advocating a blend of search strategies. "It is our contention," Oliver (1995, 4) wrote, "that the best SETI strategy is not just to listen for signals unless our sensitivity is high enough to eavesdrop on their own 'commercial' signals. Failing that, we must listen for interstellar beacons, which, judging by our own reluctance to construct them, reduces our probability of success several orders of magnitude. Another alternative," he continued, "is to arouse their consciousness to the point where they initiate an Active SETI program. Forcing them to explain a persistent signal containing interesting non-natural features is one way of stirring up their interest."

10.0. The Long Now

If in the next few decades or centuries, such a circumscribed test of the Zoo Hypothesis does not prompt a reply from extraterrestrial intelligence, a serious Active SETI program would need to prepare for transmission projects on time scales that far exceed our habitual frames of reference. The mindset needed by those who would transmit messages to other civilizations, with the hope of prompting a response to future generations of humans, is somewhat akin to those designing the Clock of the Long Now, a device designed to keep perfect time over a ten thousand year period. As founding member of the Long Now Foundation, Brand (1999, 2) summarizes concerns that motivate the Clock of the Long Now, "The main problems might be stated, How do we make long-term thinking automatic and common instead of difficult and rare? How do we make the taking of long-term responsibility inevitable?"

As noted above, Tarter (2004) has argued that given our current level of cultural development, we are not in the habit of thinking in the long term. And yet, as she did hold out hope that we might *learn* to think in longer time scales. After noting the paucity of references in her Google search to long-term planning beyond the half-century mark, she noted a spike at ten thousand years, corresponding to the half-life of radioactive waste products, reflecting in part an interest in storage facilities such as those in Yucca Mountain in the United States. "Ten thousand years is also the planning life-time for the Clock of the Long Now Foundation," Tarter (2004, 19) observes. "It is perhaps less important that the clock is still ticking 10,000 years from now than that Yucca Mountain (or some other facility) remains intact, but the clock is a very useful tool for encouraging humans to adopt a long-term approach to the future. It might be worthwhile," Tarter (ibid.) concludes, "to investigate whether the clock could transmit as well as tick."

11.0. Sustaining SETI and Its Community

For SETI to succeed as a multigenerational activity, we need to foster people's expectations that they can contribute to the creation of humanity's "long now." By making a significant commitment to Active SETI, we would also be making a statement of some of our ideals as a civilization: that we hope to continue to exist when a reply arrives, and that even if we do not survive that long, we are willing to expend resources and effort to provide something that may be of benefit to other civilizations.

By integrating Active SETI and Passive SETI programs, we could also establish an institutional framework for sustaining Passive SETI and the scientists who conduct it, even in the face of decades or centuries of silence from the stars. By engaging in a clearly articulated, ongoing, and evolving set of experiments to test various versions of the Zoo Hypothesis, we could build into Passive SETI programs specific dates at which we could expect a first response to messages sent to particular stars. Such a multigenerational activity would need to look not only to the future, but also to the past, recalling the dates, content, and targets of transmissions sent centuries or millennia before.

Although a long-term perspective is vital for a search that could take millennia to succeed, it is equally important to avoid making initial Active SETI projects so ambitious that the prospect of sustaining them becomes daunting. At the earliest stages of developing an ongoing Active SETI program, it is important to consider projects that can be completed within time scales far shorter than ten thousand years.

As a start, a two-phase process targeting nearby stars could profitably engage broader international discussion about appropriate message contents while also addressing concerns that Active SETI poses a risk to humankind by making our existence known to potentially hostile civilizations. While most SETI scientists consider the distances between stars to be a natural buffer that would provide protection, the issue is worthy of additional debate before transmitting to stars that have not yet been targeted with intentional signals from Earth.

In this proposed two-phase process, in the first phase only stars that have previously been targeted by earlier Active SETI projects would be sent messages that had been discussed widely within the international community. In all Active SETI projects to date, such broad-based international discussion has not been sought prior to transmission. For a list of stars within seventy light-years of Earth that have been targeted by past Active SETI projects through November 2009, see Table 17.1.

The second phase, which would target stars within a certain range that had not previously received intentional transmissions, would only occur if the project received broad-based support from the international community after the first phase. By preceding this second phase with a series of transmissions of the same sort proposed for phase two, but sent to previously targeted stars, more widespread engagement from a range of nations and disciplinary perspectives might be achieved than through discussions of hypothetical messages that had never been sent. In addition, the lessons learned in constructing, encoding, and transmitting actual messages in phase one would be considerable. Targets in phase two could include all stars within a certain range, such as the more than 130 stars within twenty light-years (Table 17.2). Alternatively, the list of targets could be restricted to stars deemed most likely to be habitable.

By fostering ongoing community practices, we could increase the likelihood that stars targeted under such a program would actually be reexamined in future years. For example, we might initiate an ongoing series of annual celebrations in conjunction with "first reception days." By specifying in messages to the extraterrestrial recipients the specific date and year that we would hope to receive a first reply, we could even have such "first reception days" fall on the same day of the year on Earth, with different potential "first receptions" coming from different stars in different years. As Sagan, Salzman Sagan, and Drake (1972, 881–82) note when describing the pulsar map onboard the Pioneer plaque, "Since pulsars are running down at largely known rates they can be used as galactic clocks for time intervals of hundreds of millions of years." One of the most critical concepts to communicate in an interstellar message, in this plan, would be our means of reckoning time. Methods have already been developed for using such pulsar maps in interstellar messages that could be transmitted at radio frequencies to describe events over a range of time scales—including time scales appropriate for synchronizing transmission and reception (Vakoch and Matessa 2011).

If the same day of the year were chosen each year for potential "first receptions," this annual celebration could provide a regular opportunity for the SETI community of the time to gather before the reception date, reflect on the nature of the transmission that had been sent to the target stars of interest that year, and discuss current and future SETI projects—always knowing years ahead of time when the celebration would be held, allowing participants to avoid scheduling conflicts. Ideally the date would have a

Table 17.1. Active SETI Transmissions to Targets within 70 Light-Years of the Solar System (adapted and expanded from http://en.wikipedia.org/wiki/Active_SETI).

Name	Designation	Date sent	Arrival year	Message
16 Cygni A	HD 186408	May 24, 1999	2069	Cosmic Call 1
15 Sagittae	HD 190406	June 30, 1999	2057	Cosmic Call 1
	HD 178428	June 30, 1999	2067	Cosmic Call 1
Gliese 777	HD 190360	July 1, 1999	2051	Cosmic Call 1
	HD 197076	August 29, 2001	2070	Teen Age Message
47 Ursae Majoris	HD 95128	September 3, 2001	2047	Teen Age Message
37 Geminorum	HD 50692	September 3, 2001	2057	Teen Age Message
	HD 126053	September 3, 2001	2059	Teen Age Message
	HD 76151	September 4, 2001	2057	Teen Age Message
	HD 193664	September 4, 2001	2059	Teen Age Message
	HIP 4872	July 6, 2003	2036	Cosmic Call 2
	HD 245409	July 6, 2003	2040	Cosmic Call 2
55 Cancri	HD 75732	July 6, 2003	2044	Cosmic Call 2
	HD 10307	July 6, 2003	2044	Cosmic Call 2
47 Ursae Majoris	HD 95128	July 6, 2003	2049	Cosmic Call 2
Gliese 581	HIP 74995	October 9, 2008	2029	A Message from Earth
Gliese 581	HIP 74995	August 28, 2009	2030	Hello from Earth
TZ Arietis	GJ 83.1	November 7, 2009	2024	RuBisCo Message
Teegarden's star	SO 025300.5+165258	November 7, 2009	2022	RuBisCo Message
Kappa Ceti	HIP 15457	November 7, 2009	2039	RuBisCo Message

Table 17.2. Stars within 20 Light-Years of the Solar System (adapted from http://www.solstation.com/stars/s20ly.htm).

Distancde (Light-Years)	Name or Designation	Spectral & Luminosity Type
4.2	Proxima Centauri	M5.5 Ve
4.4	Alpha Centauri A	G2 V
4.4	Alpha Centauri B	K0–1 V
6.0	Barnard's Star	M3.8 Ve
7.8	Wolf 359	M5.8 Ve
8.3	Lalande 21185	M2.1 Vne
8.6	Sirius A	A0–1 Vm
8.6	Sirius B	DA2–5
8.7	Luyten 726-8 A	M5.6 Ve
8.7	UV Ceti	M6.0 Ve
9.7	Ross 154	M3.5 Ve
10.3	Ross 248	M4.9–5.5 Ve
10.5	Epsilon Eridani	K2 V
10.7	Lacaille 9352	M0.5–1.5 Ve
10.9	Ross 128	M4.1–5 Ve
11.1	EZ Aquarii A	M5.0–5.5 Ve
11.1	EZ Aquarii B	M5? Ve
11.1	EZ Aquarii C	M? Ve
11.4	Procyon A	F5 V–IV
11.4	61 Cygni A	K3.5–5.0 Ve
11.4	61 Cygni B	K4.7–7.0 Ve
11.4	Procyon B	DQZ,A4
11.4	Struve 2398 A	M3.0 V
11.4	Struve 2398 B	M3.5 V
11.6	Groombridge 34 A	M1.5 Vne
11.6	Groombridge 34 B	M3.5 Vne
11.8	Epsilon Indi	K4–5 Ve
11.8	DX Cancri	M6.5 Ve
11.8	Epsilon Indi ba	T1 V
11.8	Epsilon Indi bb	T6 V
11.9	Tau Ceti	G8 Vp
11.9	LHS 1565	M5.5 V
12.1	YZ Ceti	M4.5 Ve
12.4	Luyten's Star	M3.5 Vn
12.6	Teegarden's Star	M6.5 V
12.6	SCR 1845-6357	M8.5 V
12.6	SCR 1845-6357 b	T4.5–6.5 V
12.6	Kapteyn's Star	sdM1.5
13.1	Kruger 60 A	M3 Vn
13.1	Kruger 60 B	M4 V
13.2	DENIS 1048-39	M8.5 V

continued

Distancde (Light-Years)	Spectral & Luminosity Name or Designation	Type
13.2	DENIS 1048-39	M8.5 V
13.4	Ross 614 A	M4.5 Ve
13.4	Ross 614 B	M8 V
13.8	Wolf 1061 A	M3.0 V
13.8	Wolf 1061 B?	M?
14.2	Wolf 424 A	M5.5 Ve
14.2	Wolf 424 B	M5.5–7 Ve
14.2	Cincinnati	M1.5–3.0 Ve
14.4	van Maanen's Star	DZ7,F,G
14.6	TZ Arietis	M4.5 V
14.8	BD+68 946 A	M3.0 V
14.8	BD+68 946 B	M?
14.8	LP 731-58	M6.5 V
14.8	CD-46 11450	M2.5-3.0 V
14.8	V1581 Cygni A	M5.5 Ve
14.8	V1581 Cygni B	M6 V
14.8	V1581 Cygni C	M?
15.2	L 145-141	DQ6,A,C
15.3	G 158-27	M5.5 V
15.3	Gliese 876	M3.5 V
15.6	L 143-23	M5.5 V–IV
15.8	BD+44 2051 A	M1 Vne
15.8	WX Ursae Majoris	M5.5 Ve
15.9	Groombridge 1618	K7.0 Vne
15.9	AD Leonis	M3.0–3.5 Ve
16.1	CD-49 13515	M1.5–3.0 V
16.2	DENIS / DEN 0255-4700	L7.5 V
16.3	LP 944-20	M9.0 V
16.4	CD-44 11909	M3.5–4.5 V-IV
16.5	40 (Omicron2) Eridani A	K1 Vne
16.5	40 (Omicron2) Eridani B	DA3
16.5	40 (Omicron2) Eridani C	M4.5 Ve
16.5	EV Lacertae	M3.5 Ve
16.6	70 Ophiuchi A	K0-1 Ve
16.6	70 Ophiuchi B	K5–6 Ve
16.7	Altair	A7 V–IV
17.0	L 722-22 A	M4 V
17.0	L 722-22 B	M? V
17.0	EI Cancri	M5.5 Ve
17.0	Giclas 9-38 B	M5.5 V
17.0	2MASS J09393548-2448279 AB	T8.5 V / T8.5? V
17.5	LTT 17897	M3.5 V

continued

Table 17.2. Stars within 20 Light-Years of the Solar System

Distancde (Light-Years)	Name or Designation	Spectral & Luminosity Type
17.6	G 099-049	M4.5 V
17.6	AC+79 3888	M3.5 V
17.7	LHS 1723	M4.5 V
17.7	Lalande 25372	M1.5 Vne
17.9	LP 816-60	M V
18.0	Stein 2051 B	DC5
18.0	Stein 2051 A	M4.0 V
18.0	Stein 2051 B	M V
18.0	Wolf 294	M3.0 V
18.5	2MASS 1835+3259	M8.5 V
18.6	BD-03 1123	M1.5 Vn
18.7	2MASS 0415-0935	T8 V
18.7	V1054 Ophiuchi	M3 Ve
18.7	Wolf 630 B	M4 V
18.7	Wolf 630 C	M? V
18.7	van Biesbroeck 8	M7.0 V
18.8	Sigma Draconis	K0 V
18.8	Gliese 229	M1–2 Ve
18.8	Gliese 229 b	T6.5 V
18.9	Ross 47	M4 Vn
19.0	L 205-128	M2 V
19.2	Wolf 1055 A	M3.5 Vne
19.2	van Biesbroeck's Star	M8 Ve
19.2	L 674-15	M3.5 V
19.3	Gl 570 A	K4-5 Ve
19.3	Gl 570 B	M1 V
19.3	Gl 570 C	M3 V
19.3	Gliese 570 d	T7-8 V
19.4	Eta Cassiopeiae A	G3 V
19.4	Eta Cassiopeiae B	K7 Vn
19.4	Ross 882	M4.5 Ve
19.4	CD-40 9712	M0-3 V–VI
19.4	BD+01 4774	M1 Ve
19.5	36 Ophiuchi A	K0–1 Ve
19.5	36 Ophiuchi B	K1–5 Ve
19.5	36 Ophiuchi C	K5–6 Ve
19.6	L 347-14	M4.5 V
19.7	HJ 5173 A	K3 V
19.7	HJ 5173 B	M3.5–4.0 V
19.8	82 Eridani	G5 V
19.9	Delta Pavonis	G5–8 V–IV

continued

Distancde (Light-Years)	Spectral & Luminosity Name or Designation	Type
20.0	2MASS 0937+2931	T6 V
20.0	LP 44-113	DQ8,A9
20.0	BD-11 3759	M3.5 V
20.0	Ross 986 A	M4.5 Ve
20.0	Ross 986 B	M6 V

significance that would remain obvious even if calendar systems change over the millennia; for example, each year's "first reception day" could fall on the Summer Solstice, or some other seasonally significant day. Though we would expect that most such annual celebrations would not bring news of the discovery of extraterrestrial intelligence, by connecting intellectual and social gatherings of committed communities of researchers with these potentially critical days, there would be opportunities to build and sustain these communities by reflecting on the current search as part of an ongoing enterprise with a rich history. Such times of community building would provide a means for recognizing tangible progress in developing increasingly sophisticated and ambitious messages and transmission technologies, even if no signal were received.

12.0. To Transmit or Not?

As SETI passes its fiftieth anniversary, it is appropriate to consider whether the passive strategies that have dominated the field are the only legitimate search strategy in the coming decades. To be clear, it is completely unwarranted to argue for the addition of Active SETI to the array of legitimate search strategies by noting that, after nearly fifty years of Passive SETI, no signs of intelligence beyond Earth have been detected, and thus Passive SETI has been found wanting as a basic approach. On the contrary, in spite of the several decades that have passed since Project Ozma in 1960, the total portion of the "cosmic haystack" that has been searched thus far is miniscule, and by many estimates of the prevalence of intelligent life in our galaxy, we should not have expected to have detected them yet, having looked at as few stars as we have to date.

Rather, there are two other reasons that we should begin Active SETI searches at this stage of the development of SETI as a science. First, if the goal of Active SETI is to initiate contact that leads to a response from extraterrestrials to future generations of humans, then the longer we wait to begin a serious transmission process, the longer future generations of humans will need to wait to begin listening for replies that could answer the question of whether there are civilizations out there, ready to reply if only humankind takes the initiative to begin the conversation.

A second reason to begin a serious transmission project in the near future is that there will always be a certain arbitrariness in deciding when to begin the project, just as there was an arbitrariness in deciding that the first SETI project, conducted a half-century ago, made sense. In retrospect, we might argue that it was absurd to expect that we could detect extraterrestrial intelligence by looking for transmissions at a single frequency from only two nearby stars. Would it not be more reasonable, one might have argued, to wait until we had the capacity to search a million or more separate channels, to survey a million or more stars?

There are two problems with this line of reasoning. First, it is possible that the galaxy is so well populated by other civilizations that virtually every star system is inhabited, transmitting at such a wide range of frequencies that even a young civilization just developing the capacity for interstellar communication could pick up the signal. We now know, thousands of times over, that this is not the case. But in April 1960, that remained a distinct possibility, able to be disproved only by conducting the experiment.

More importantly, however, we should not too readily dismiss Active SETI today because the scope of the projects we can now conceive of developing and funding seem inadequate to the task, and because our uncertainty about the future of humankind prevents us from confidently predicting that our successors will be eagerly listening for a response decades, or centuries, or millennia from now.

Instead, we should ask ourselves these questions: "Is it possible that other civilizations are waiting for us to initiate contact?" and "Might we increase our chances of success by adding Active SETI to our portfolio of search strategies?" If the answer to these questions is "yes," we should engage all the more seriously the scientific, technical, ethical, and legal questions involved in mounting a sustained program of Active SETI.

Acknowledgments

The author gratefully acknowledges the following support of this research: The John Templeton Foundation, though its Grant #1840, "Construction of Interstellar Messages Describing the Evolution of Altruistic Behavior"; President Joseph Subbiondo, Academic Vice President Judie Wexler, and Clinical Psychology Department Chair Katie McGovern of the California Institute of Integral Studies for sabbatical and ongoing research leaves; Chairman of the Board John Gertz, CEO Thomas Pierson, and Director of SETI Research Jill Tarter of the SETI Institute for support of research on SETI and Society; Chris Neller of the SETI Institute for administrative support; and Jamie Baswell, as well as Harry and Joyce Letaw, for financial support through the SETI Institute's Adopt a Scientist program.

Note

This chapter is an adaptation of Vakoch, D. A. 2011. Asymmetry in Active SETI: A case for transmissions from Earth. *Acta Astronautica* 68: 476–88. doi:10.1016/j.actaastro.2010.03.008.

Works Cited

Ball, J. A. 1973. The zoo hypothesis. *Icarus* 19: 347–49.

Barker, P. 1982. Omilinguals. In *Philosophers look at science fiction*, ed. N. D. Smith, 75–85. Chicago: Nelson-Hall.

Barrett, J. C. 1983. The laboratory hypothesis. *Speculation in Science and Technology* 6: 373–74.

Benford, J., G. Benford, and D. Benford. 2010. Messaging with cost optimized interstellar beacons. *Astrobiology* 10: 475–90.

Brand, S. 1999. *The clock of the long now: Time and responsibility.* New York: Basic Books.

Deardorff, J. W. 1987. Examination of the Embargo Hypothesis as an explanation for the Great Silence. *Journal of the British Interplanetary Society* 40: 373–79.

Diamond, J. 1990. Alone in a crowded universe. *Natural History* (June): 30–34.

Drake, F., and D. Sobel. 1992. *Is anyone out there?: The scientific Search for Extraterrestrial Intelligence.* New York: Delacorte Press.

Ekers, R. D., D. K. Cullers, J. Billingham, and L. K. Scheffer, eds. 2002. *SETI 2020: A roadmap for the Search for Extraterrestrial Intelligence.* Mountain View, CA: SETI Press.

Foucault, M. 1970. *The order of things: An archaeology of the human sciences.* New York: Random House.

Gerritsen, H., and S. McKenna. 1975. The Luneberg lens and the importance of transmission in establishing contact with extraterrestrial civilizations. *Icarus* 26: 250–56.

Gott, J. R. III. 1993. Implications of the Copernican Principle for our future prospects. *Nature* 363: 315–19.

Haqq-Misra, J. D., and S. D. Baum. 2009. The sustainability solution to the Fermi paradox. *Journal of the British Interplanetary Society* 62: 47–51.

Kardashev, N. S. 1964. Transmission of information by extraterrestrial civilizations. *Soviet Astronomy* 8: 217–21.

Laudan, L. 1977. *Progress and its problems: Towards a theory of scientific growth.* Berkeley: University of California Press.

Lemarchand, G. A., and D. E. Tarter. 1994. Active search strategies and the SETI protocols: Is there a conflict? *Space Policy* 10 (2): 134–42.

Oliver, B. 1993. Symmetry in SETI. *Third Decennial US-USSR Conference on SETI: ASP Conference Series* 47: 67–72.

———. 1995. SETI scaling laws and acquisition strategies. Unpublished manuscript.

Oliver, B. M., and J. Billingham, eds. 1973. *Project Cyclops: A design study of a system for detecting extraterrestrial intelligent life.* NASA CR 114445. Moffett Field, CA: NASA Ames Research Center.

Rescher, N. 1985. Extraterrestrial science. In *Extraterrestrials: Science and alien intelligence*, ed. E. Regis Jr., 83–116, Cambridge: Cambridge University Press.

Sagan, C., L. Salzman Sagan, and F. Drake. 1972. A message from Earth. *Science* 175: 881–84.

Scheffer, L. K. 2002. Appendix B: Beacons, beacons, and more beacons. In *SETI 2020: A roadmap for the Search for Extraterrestrial Intelligence*, ed. R. D. Ekers, D. K. Cullers, J. Billingham, and L. K. Scheffer, 303–20. Mountain View, CA: SETI Press.

Staff at the National Astronomy and Ionosphere Center. 1975. The Arecibo message of November, 1974. *Icarus* 26 (4): 462–66.

Subotowicz, M., J. Usowicz, and Z. Paprotny. 1979. On active and passive CETI from an earth satellite orbit. *Acta Astronautica* 6: 203–12.

Tarter, J. 2004. What if everybody is listening and nobody is transmitting? *Explorer* 1 (2): 18–19.

Vakoch, D. A. 1998. Constructing messages to extraterrestrials: An exosemiotic perspective. *Acta Astronautica* 42: 697–704.

———. 2000. Roman Catholic views of extraterrestrial intelligence: Anticipating the future by examining the past. In *When SETI succeeds: The impact of high-information contact*, ed. A. Tough, 165–74. Bellevue, WA: Foundation For the Future.

———. 2011. Responsibility, capability, and Active SETI: Policy, law, ethics, and communication with extraterrestrial intelligence. *Acta Astronautica* 68: 512–19. doi:10.1016/j.actaastro.2010.01.008.

———, and M. Matessa. 2011. An algorithmic approach to communicating reciprocal altruism in interstellar messages: Drawing analogies between social and astrophysical phenomena. *Acta Astronautica* 68: 459–75. doi:10.1016/j.actaastro.2009.12.012.

Zaitsev, A. No date. Messaging to Extra-Terrestrial Intelligence, arXiv:physics/0610031v1.

Building and Searching for Cost-Optimized Interstellar Beacons

James Benford, Dominic Benford, Gregory Benford

1.0. Minimizing Costs

The usual method of cost optimization is to examine many alternative approaches to building a system, estimating the cost of each, and then comparing them. This is a "bottom-up" approach. We offer a "top-down" method based on analysis and actual experiences of designers.

Cost of large High Power Microwave (HPM) systems is driven by two elements—*capital cost* C_C, divided into the cost of building the microwave source and the cost of building the radiating aperture C_A, and the *operating cost* C_O, meaning the operational labor cost and the cost of the electricity to drive the system (Benford 2007):

$$C = C_C + C_O$$
$$C_C = C_A + C_S \tag{1}$$

One can argue that operating cost of a system is dominated by labor cost, which is in turn proportional to the size and power as well, so that $C_C \sim C_O$, so that $C \sim C_C$.

To optimize, meaning minimize, the cost, the simplest approach is to assume power-law scaling dependence on the peak power and antenna area. Nonlinear dependence is analyzed in the Appendix, but to illustrate the essentials, first we analyze linear scaling with coefficients describing the dependence of cost on area coefficient a($/m^2$), which includes cost of the antenna, its supports and subsystems for pointing and tracking and phase control, and microwave power coefficient ($/W$), which includes the source, power supply, cooling equipment, and prime power cost:

$$C_A = aA$$
$$C_S = \tag{2}$$
$$C_C = aA + pP$$

We neglect any fixed costs, which would vanish when we differentiate to find the cost optimum.

The power density S (W/m²) at range R is determined by W, the effective isotropic radiated power (EIRP), the product of radiated peak power P and aperture gain G,

$$W = PG$$

$$S = \frac{W}{4\pi R^2} \qquad (3)$$

and antenna gain is given by area and wavelength:

$$G = \frac{4\pi\epsilon A}{\lambda^2} = \frac{4\pi\epsilon A}{c^2} f^2 = kAf^2$$

$$W = kPAf^2 \qquad (4)$$

$$k = \frac{4\pi\epsilon}{c^2}$$

where ϵ is aperture efficiency (this includes factors such as phase, polarization, and array fill efficiency) and we have collected constants into the factor k. (For $\epsilon = 50\%$, $k = 7\text{x}10^{-17}\text{s}^2/\text{m}^2$, and $kf^2 = 70/\text{m}^2$ for $f = 1$ GHz.) We carry frequency as a constant here; cost of varying frequency will be treated later. To find the optimum for a fixed power density at a fixed range, meaning fixed W, we substitute W into the cost equation,

$$C_C = \frac{aW}{kf^2 P} + pP \qquad (5)$$

then differentiate with respect to P and set it equal to zero, giving the optimum power and area:

$$\frac{\partial C_C}{\partial P} = -\frac{aW}{kf^2 P^2} + p = 0$$

$$P^{\text{opt}} = \sqrt{\frac{aW}{pkf^2}} \qquad (6)$$

$$A^{\text{opt}} = \sqrt{\frac{pW}{akf^2}}$$

Optimal (minimum) antenna diameter D and cost are:

$$D^{opt} = \left[\frac{16}{\pi^2 f^2} \frac{pW}{ak} \right]^{1/4} \tag{7}$$

$$C_C^{opt} = \sqrt{\frac{aWp}{kf^2}} + \sqrt{\frac{aWp}{kf^2}} = 2\sqrt{\frac{aWp}{kf^2}} \tag{8}$$

$$\frac{C^{opt}_A}{C^{opt}_S} = 1$$

Minimum capital cost is achieved when the cost is equally divided between antenna gain and radiated power. This was first mentioned in 1968 (Brown 1968), independently discovered from cost data on the Deep Space Network (Benford and Dickinson 1995) and stated in the Project Cyclops report (Oliver and Billingham 1996). For a recent example, Kare and Parkin have built a detailed cost model for a microwave beaming system for a beam-driven thermal rocket and compared it to a laser-driven rocket. They find that, at minimum, cost is equally divided between the two cost elements (Kare and Parkin 2006).

1.1. Cost Coefficients

Microwave antenna cost per unit area coefficient (a) can be estimated from astronomical arrays, such as ALMA (National Research Council. 2005). Such systems are driven by tolerances for higher frequencies, ~100 GHz, and are ~4k$/m². Proposed systems such as the Square Kilometer Array (SKA) operate at lower frequencies near the minimum of attenuation, 1–10 GHz. SKA is projected to cost about 1 k$/m² (Benford et al. 2010). Commercial dishes for satellite TV cost about $400/m² and show the cost-lowering effects of huge economies of scale. SETI 2020 estimates ~$350/m² for a large system. We have chosen 1 k$/m² for our examples, but one must be mindful that the optimum area and power depend on the square root of the cost coefficients a and p, so are insensitive to changes in the technology.

Delivered microwave power currently costs ~$1/watt, including everything from the wall plug to the antenna connection. A current point of comparison, the International Thermonuclear Experimental Reactor (ITER) electron cyclotron heating system, heats the Tokamak with 27 gyrotrons of 1 MW continuous power (24 at 170 GHz, 3 at 120 GHz) at a projected cost of $82.5 million, $3/watt (Benford et al. 2010). This includes $26.3 million for power supplies, $14.5 million for microwave sources (gyrotrons) and controls, and \$41.7 million for transmission lines (waveguides). We have chosen 3 \$/W for our examples.

Here we make a distinction between *peak* and *average* power costs. The cost elements are different in pulsed peak power microwave systems, where cost is dominated by the requirement to operate at very high voltage and currents. For peak power systems, costs are in the range of 0.1–0.01\$/W. Then a GW power system costs 10–100 M\$. High average power systems have costs driven by continuous power-handling equipment and cooling for losses. They cost ~1\$/W, so a GW unit costs ~1B\$. The physics for the cost reduction is that the electrical breakdown threshold is much higher for short pulses, so much more energy can be stored in small volumes.

2.0. System Examples

We make an estimate for an HPM system of W = EIRP = 10^{11} W (for example, 10 MW, 40 dB gain), assuming a = $1k\$/m^2$, p = 3\$/W, at 1 GHz. This could be a 1-m^2 antenna driven by a 1.4 GW HPM source, or a 100 m^2 array driven by a 14 MW source. What is the cost-optimum system? (We assume the aperture will be made of an array of elements with aperture efficiency ε = 0.5.) From eq. 5 the cost becomes

$$C_C(M\$) = \frac{1.49}{P(MW)} + 3P(MW) \qquad (9)$$

It's expensive, costing millions of dollars. Figure 18.1 shows the sharp cost minimum. The cost of the system falls rapidly as power increases until power cost equals aperture cost at total cost minimum of 4.14 M\$. Then it increases monotonically. Optimum power is 0.69 MW, optimum antenna area is 2,070 m^2, and diameter is 51 m if the aperture is circular.

The difference between peak and average power cost matters. First, significantly lower costs for pulsed sources drives cost down, as shown in Figure 18.2, which compares the system of Figure 18.1 (3\$/W) with short-pulse systems with 0.3\$/W and 0.03\$/W. This shows an order of magnitude fall in cost, consistent with Eqs. 6 and 8. Optimum power increases by ten to 7 MW, area falls by ten to 207 m^2.

As EIRP increases, optimal area, power, and diameter increase, as Figure 18.3 shows for the constants of Figure 18.1. As EIRP increases 10^{15} W, power rises to 70 MW, area to 0.2 km^2, and diameter to 0.5 km.

2.1. Frequency Scaling

Equations 9 and 12 show cost declining as frequency increases, due to the increased gain. The antenna cost also increases with frequency. But the

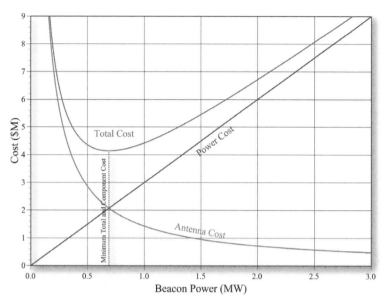

Figure 18.1. Antenna, microwave power and total costs of HPM system with EIRP = 10^{11} W, a = 1 k\$/m², p = 3 \$/W, at f = 1 GHz with aperture efficiency ε = 0.5. Minimum total cost is at the point where the antenna cost and power cost are equal, as in eq. 8.

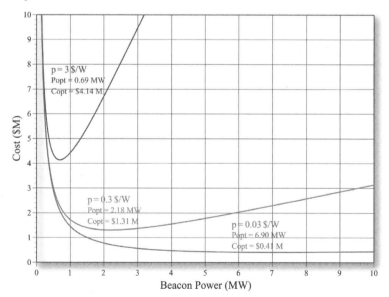

Figure 18.2. Impact of pulsed sources on system cost of systems of EIRP = 10^{11}. Cases of long-pulse sources at 3\$/W (same as Figure 18.1 case) and short-pulse sources at 0.3\$/W and 0.03\$/W.

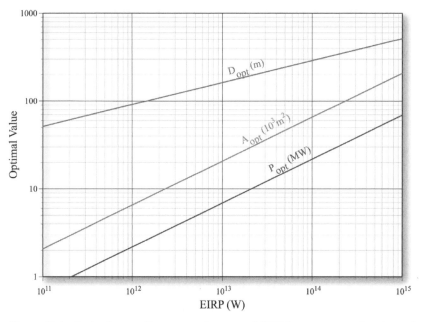

Figure 18.3. Optimal area, diameter, and power of HPM systems as a function of effective isotropic radiated power, an extrapolation of the Figure 18.1 case to higher EIRP.

increase is slow, and the coefficient $a \sim f^{1/3}$ (Benford J. 2010). Microwave sources typically don't depend on frequency (power does, $P \sim f^2$, so more sources will be used to generate a given power at higher frequency). Therefore, the cost scaling with frequency is

$$C_C \propto \frac{1}{f^{5/6}} \tag{10}$$

Figure 18.4 shows frequency dependence of cost for the cases of Figure 18.2. Cost drops a factor of 6.8 from 1 GHz to 10 GHz. Unless the application requires a specific frequency, optimal cost drives the system builder to higher frequencies.

The angular width θ of the optimized Beacon beam is set by the antenna area and frequency:

$$G = \frac{4\pi}{\theta^2}$$

$$\theta^2 = \frac{c^2}{f} \sqrt{\frac{ak}{pW}} \tag{11}$$

284

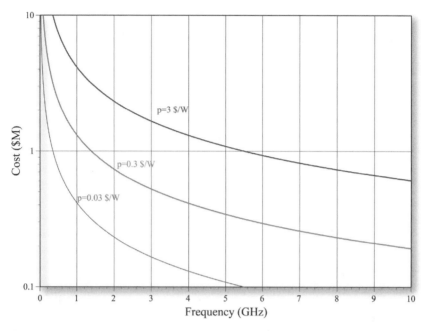

Figure 18.4. Cost vs. microwave frequency for values of the power cost coefficient p for the HPM system of Figure 18.1. Upper end frequencies are favored by almost an order of magnitude.

The beamwidth of a large cost-optimized aperture, such as those discussed in 3.2, is $\theta \sim 10^{-4}$ radians.

To test the cost-minimization approach, consider ORION, a transportable, self-contained HPM test facility first fielded in 1995 and currently in operation (Price et al. 1997) At the heart of the system is a suite of four tunable magnetrons each capable of delivering 400–800 MW peak power, continuously tunable in frequency from 1–3 GHz, fires repetitively up to 100 Hz and can produce 230 MW/m² over a 7m × 15m spot at 100 m range. The EIRP is $W = 3 \times 10^{13}$ W, comparable to Arecibo in peak power, but not average power. ORION fires one thousand pulses in a burst at repetition rates up to 100 Hz. Because of the frequency range, the antenna subsystem also includes a waveguide combiner/attenuator consisting of a hybrid tee/phase shifter and power combiner to sum in waveguide two outputs each magnetron, followed by a hybrid tee/phase shifter attenuator to vary the radiated power over five orders of magnitude.

Using eq. 8, a = 10 k$/ m², p = 0.3 $/W at the mean frequency of 2 GHz, an estimate of the cost is 11.3 M$. This compares well to the 9.8 M$ inflation-adjusted price of ORION.

2.2. Cost-optimized versus Isotropic Beacons

An isotropic beacon broadcasts into all sold angles. If the constraint is that a fixed power density S must be produced at the range of both beacons, then cost-optimized and isotropic beacons differ principally in that they have different beamwidths. Of course C^{iso}, the cost of an isotropic antenna, for example a dipole, is small and C^{opt}_A is half the total cost (eq. 8). So the isotropic beacon cost will be dominated by the of its power P_i. If they use the same technology, the ratio of costs of such Beacons is

$$G_i P_i = 4\pi R^2 S = G^{opt} P^{opt}$$

$$\frac{P_i}{P^{opt}} = \frac{G^{opt}}{G_i} = \frac{1}{\theta^2} \tag{12}$$

$$\frac{C^{iso}}{C^{opt}} = \frac{pP_i}{pP^{opt}/2} = \frac{1}{2\theta^2} \sim 10^8$$

Even if the isotropic beacon broadcasts only into the galactic plane, about 10 percent of the sky, it is far more expensive than a cost-optimized beacon with a scanning broadcast pattern.

3.0. Alien Galactic-Scale SETI Beacons

The above relations have been used to estimate power beaming system costs for the most advanced case: microwave Beacons for communication across galactic distances, for SETI (Search for Extraterrestrial Intelligence) using EIRPs of 10^{17}–10^{20} W (Benford et al. 2010).

Are aliens unknowable in that they are beyond economic arguments? We can call this the Altruistic Alien argument—that aliens of great ability, near-infinite resources and benign intent will transmit to us without taking any consideration to the cost (which would be high in our terms). This argument is seldom directly expressed.

But this argument meets a conceptual danger: If Altruistic Aliens have great resources, they would find it easy to make themselves apparent in our night sky. If so, *where are they?* We now know, from SETI searches of targeted stars conducted largely by the SETI Institute, that within a range of ~400 light years they do not make themselves obvious. Indeed, we see no obvious beacons anywhere in the night sky. Beacons are necessary. Further, no *conversations* occur over several thousand light years; transmissions are announcements or memorials, not letters.

We assume that if they are social beings interested in a SETI conversation (Hetesi and Regály 2006) or passing on their heritage, they will know about tradeoffs between social goods, and thus, in whatever guise it takes, *cost*. But what if we suppose, for example, that aliens have very low cost labor, that is, slaves? (In modern terms we might call them "self-replicating automata.") With a finite number of slaves, you can use them to do a finite number of tasks. And so you pick and choose by assigning value to the tasks, balancing the equivalent value of the labor used to prosecute those tasks. So, choices are still made on the basis of available labor. The only case where labor has no value is where labor has no limit. That might be if aliens may live forever or have limitless armies of self-replicating automata. But such labor costs something, because resources, materials and energy, are not free.

Our point is that *all* SETI search strategies must assume something about the Beacon builder, and that cost is a constraint that may drive some alien attempts at interstellar communication. Instead of the open-ended what-ifs of many SETI discussions, we seek to see what emerges from applying real world constraints, as a guide to smarter searches.

3.1. Beacon Builder Motives

Through most of its history, SETI has assumed a high-minded search for other life forms. But other motives are possible.

What could motivate a Beacon builder? Here we can only reason from our own historical experience. Other possible high intelligences on Earth (whales, dolphins, chimpanzees) do not have significant tool use, so they do not build lasting monuments. Sending messages over millennia or more connects with our own cultures. Human history suggests (Benford 1999*)* that there are two major categories of long-term messages that finite, mortal beings send across vast time scales:

- *Kilroy Was Here* These can be signatures verging on graffiti. Names chiseled into walls have survived from ancient times. More recently, we sent compact disks on interplanetary probes, often bearing people's names and short messages that can endure for millennia.

- *High Church* These are designed for durability, to convey the culture's highest achievements. The essential message is *this was the best we did; remember it.*

A society that is stable over thousands of years may invest resources in either of these paths. The human prospect has advanced enormously in only a few centuries; the lifespan in the advanced societies has risen by 50 percent in each of the last two centuries. Living longer, we contemplate longer legacies.

Time capsules and ever-proliferating monuments testify to our urge to leave behind tributes or works in concrete ways (sometimes literally). The urge to propagate culture quite probably will be a universal aspect of intelligent, technological, mortal species (Minsky 1984).

Thinking broadly, high-power transmitters might be built for wide variety of goals other than two-way communication driven by curiosity. For example:

- *The Funeral Pyre:* A civilization near the end of its life announces its existence.

- *Ozymandias:* Here the motivation is sheer pride; the Beacon announces the existence of a high civilization, even though it may be extinct, and the Beacon tended by robots. This recalls the classic Percy Bysshe Shelly lines,

> And on the pedestal these words appear:
> "My name is Ozymandias, King of Kings;
> Look on my works, Ye Mighty, and despair!"
> Nothing beside remains. Round the decay
> of that colossal wreck, boundless and bare,
> The lone and level sands stretch far away.

- *Help!* Quite possibly societies that plan over time scales ~1000 years will foresee physical problems and wish to discover if others have surmounted them. An example is a civilization whose star is warming (as ours is), which may wish to move their planet outward with gravitational tugs. Many others are possible.

- *Leakage Radiation*: These are unintentional, much like objects left accidentally in ancient sites and uncovered long after. They do carry messages, even if inadvertent: technological fingerprints. These can be not merely radio and television broadcasts radiating isotropically, which are fairly weak, but deep space radar and beaming of energy over solar system distances. This includes "industrial" spaceship launchers, beam-driven sails, "planetary defense" radars scanning for killer asteroids, and cosmic power beaming driving interstellar starships with beams of lasers, millimeter, or microwaves. There are many ideas about such uses already in the literature (Benford and Benford 2006).

- *Join Us*: Religion may be a galactic commonplace; after all, it is here. Seeking converts is common, too, and electromagnetic preaching fits a frequent meme.

We advocate that we know nothing of motives. Whatever the Beacon builders' motives, we should periodically reassess our SETI assumptions in light of how our own microwave-emitting technologies develop. Since the early SETI era of the 1960s, microwave emission powers have increased by orders of magnitude and new technologies have altered our ways of emitting very powerful signals. Given Beacon ranges >1000 ly, EIRPs >10^{17} W are needed. These high powers suggest that all possible motivations will succumb to economics. Is cost/benefit analysis arguably universal? It is certainly a useful hypothesis. As we showed in the previous section, it leads to quantitative methodology for thinking about beacons. We now show how it can be useful in considering unexplained transients seen in the radio sky.

3.2. Beacon Examples

To quantify some classes of possible Earth-based Beacons as we have discussed here, based on modern Earth sources, we present examples of galactic-scale beacons in Table 18.1.

We can estimate the *operating cost* C_O as the cost of electricity to drive the microwave sources, with a cost coefficient p_{ave} ($/W-sec), which at present in the United States is 0.88$/W-yr. (There is also some inefficiency in generating microwave power, about a factor of two.) The operating cost is then p_{ave} times the peak power times the duty factor (product of the pulse length and the pulse repetition rate r) times the operating time τ:

$$C_O = p_{ave}(\text{dutyfactor}) = p_{ave}P^{opt}\text{tr}\tau \qquad (13)$$

These examples show that galactic-scale Beacons can be built for a few billion dollars with our present technology. Such beacons have narrow "searchlight" beams and short "dwell times" when the beacon would be seen by an alien observer in their sky. They cost in the range of 1–10 M$/ly. On a cost basis they will likely transmit at higher microwave frequencies, ~10 GHz. This shows a key advantage to short-pulse technology: much smaller costs, both capital and operating. For a detailed discussion of these examples, see Benford J. et al. (2010).

Note that from eq. 3 R $\sim W^{1/2}$, from eq. 8, $C^{opt} \sim W^{1/2}$, so $C^{opt} \sim R$. For uniform star density, star number N $\sim R^2 \sim C^2$. There are two domains: (1) The near field inside the galactic disk, where stars are a spherical distribution and the number of stars N $\sim R^3$, so number of stars radiated toward increases as the cube of cost: N $\sim C^{opt\ 3}$. Cost per star declines as $\sim 1/N^{2/3}$, a favorable scaling. (2) Even better are more powerful Beacons that radiate into the far field galactic disk, where the number of stars scanned scales as N $\sim R^2$. (The transition is at ~650 ly, the scale height of the galactic disk [Oliver and

Table 18.1. Cost-Optimized Galactic Beacons

Beacon Parameter	Long-Pulse Galactic-Range Beacon	Short-Pulse Medium-Range Beacon	Short-Pulse Galactic-Range Beacon
Range for S/N = 5	6080 ly	1080 ly	10,800 ly
EIRP = W	10^{19} W	10^{18} W	10^{20} W
Peak Power, P_{ppt}	6.9 GW	21.9 GW	218 GW
Pulse Length, t	1 s	1 μs	1 μs
Repetition Rate, r	0.5 Hz	1 kHz	1 kHz
Duty factor (t × r)	0.5	10^{-3}	10^{-3}
Average Power, $P_{ppt} \times (t \times r)$	3.45 GW	0.022 GW	0.22 GW
Antenna Diameter, A^{opt}	5.1 km	0.91 km	2.88 km
Beamwidth, θ	1.2 10^{-4} rad	6.4 10^{-4} rad	2.1 10^{-4} rad
Capital Cost, C_C	41.4 B$	1.3 B$	13.1 B$
Operating Cost of average power, C_O	3 B$/yr	0.2 B$/yr	0.2 B$/yr
Peak Power Cost Coefficient, p	3$/W	0.03$/W	0.03$/W
Dwell Time, τ_d	1.1 s	35 s	1.1 s
Revisit Time, τ_r	1 yr	1 yr	1 yr

(For sky fraction F = 0.1, f = 1 GHz, a = 1k$/m², Beacon and receiver antennas of equal area, $\alpha=1$, $\beta=1$, $\epsilon = 0.5$, $T_{sys} = 7K$, transmitter bandwidth B_t = 1 MHz, average power cost $p_{ave} = 0.88$/W-yr$, S/N = 5.)

Billingham, 1996]). Here, the number of stars radiated toward increases as the square of cost: $N \sim C^{opt\ 2}$. The cost per star scales as $1/N$. That means that more powerful Beacons will have great economies of scale.

3.3. Implications of Cost-Optimized Beacons

3.3.1. IMPLICATIONS FOR MESSAGING FROM EARTH

Galactic-scale Beacons can be built for a few billion dollars with our present technology. Such beacons have narrow "searchlight" beams and short "dwell times" when the Beacon would be seen by an alien observer in their sky. Cost-efficient beacons will be pulsed, narrowly directed, and broadband ($\Delta f/f \sim 0.1\%$) in the 1–10 GHz region, with a cost preference for the higher frequencies. Cost, spectral lines near 1 GHz, and interstellar scintillation favor radiating far from the "water hole." Transmission strategy for such Earth-based Beacons will be a rapid scan of the galactic plane, to cover the angular space. Such pulses will be infrequent events for the receiver, appearing for only seconds and recurring over periods of a month or year.

3.3.2. IMPLICATIONS FOR SEARCH STRATEGY

We conclude that SETI searches may have been looking for the wrong thing. SETI has largely sought signals at the lower end of the cost-optimum frequencies. If there are cost-optimized beacons as we envision them, we argue they can be found by steady searches that watch the galactic plane for times on the scale of years. From Earth, 90 percent of the galaxy's stars lie within 9 percent of the sky's area, in the plane and hub of the galaxy. This suggests a limited sky survey. We will need to be patient and wait for recurring events that may arrive in intermittent bursts. Special attention should be paid to areas along the Galactic Disk where SETI searches have seen coherent signals that are nonrecurring on their limited listening time intervals. Since most stars lie close to the galactic plane, as viewed from Earth, occasional pulses at small angles from that plane should have priority. The strategy that follows is:

- Scan the region pointing directly toward and away from the galactic center.

- Scan the entire plane of the galaxy often throughout the year.

- Since the highest nearby density of stars lies along the Orion Spur we are in, listen in those directions for occasional, transient pulses.

- Scan the region pointing directly toward and away from the galactic center.

- Assume the Life Plane strategy of the Beacon builder; that is, we should concentrate on a narrow range above and below the galactic plane.

To elaborate on the last point: There is a selection pressure for life and therefore civilization that follows from a mechanism proposed by Medvedev and Melott (2007). They point out that extinction events in Earth genera diversity, which is observed to varies with 62 My cycle, may be explained as follows:

- Collision of the galaxy with the intergalactic medium at 200 km/sec produces cosmic rays accelerated by the termination bow shock wave.

- Nonlinear diffusion of these particles due to Alfven waves enhances flux at Earth by factor of 5 at the peak of excursion from the galactic plane (230 ly vs. 30 ly at present).

- Cosmic rays penetrating the atmosphere create muons that propagate deeply into the ocean, causing radiation damage to DNA.

- Therefore genera diversity varies with 62 My cycle, as is observed. This explains both the period of extensions and the timing of their maxima.

The implications for SETI are that life, civilization, and beacons will cluster near the plane. Therefore, SETI searches should look near the galactic plane. *We propose a new test for SETI Beacons, including Cost-Optimized Beacons*:

- Identify coordinates of unexplained transient events, relative to the galactic plane.

- Compare the distribution of these events to the distribution of stars relative to the plane.

- See if they differ. If transients are clustered near the plane more than stars, it is circumstantial evidence for beacons.

- Concentrate SETI searches there, study those locations patiently.

3.3.3. IMPLICATIONS FOR SETI OBSERVING TECHNOLOGY

As a consequence of their area, the large antennas used for radio astronomy see a very small piece of the sky. So to "stare" at the Galactic Disk, one needs a large number of small antennas with each looking at its piece of the plane. Of course, with smaller collecting area comes smaller signals. In particular, steerable phased arrays such as the Allen Telescope Array have the unique ability to produce multiple beams and shaped antenna patterns.

We should revisit the locations of the transient, powerful bursts seen in past surveys in a systematic way. Earlier searches have seen pulsed intermittent signals resembling what we think Beacons may be like, and may provide useful clues. We should observe the spots in the sky seen in previous work for hints of such activity but over year-long periods. Since we know these locations, a search every day or even more often would be inexpensive. We should also scan the region pointing directly toward and away from the galactic center.

Time resolution is key. For a facility well suited to cost-optimized beacon searches the key parameters include collecting area, solid angle, bandwidth, and time resolution. All these parameters are better when maximized. A good figure of merit is thus the survey speed in square degrees per hour to some depth over some bandwidth. However, this doesn't capture the key issue of time resolution, which is a rarely discussed parameter. Some projects specifically want to study fast transients (perhaps ~ms-level timing on the ASKAP, the Australian pathfinder for the Square Kilometer Array [SKA]).

Perhaps newer search methods, directed at short transient signals, will be more likely to see the Beacons we have described (Siemion et al. 2010; Lazio et al. 2009). Likely existing or near-term facilities:

- ASKAP publishes a "continuum" survey speed of around 300 sq. deg. per hour with 300 MHz bandwidth to an intensity of 100μJy.

- MeerKAT (the South African version of ASKAP) has a few alternatives, and would probably be in the range of 250–600+ sq. deg. per hour, perhaps with 512 MHz bandwidth.

- WSRT is the Westerbork telescope, which is supposed to be upgraded with some wideband multi-element receivers for ~1GHz operation, which are called APERTIF. WSRT+APERTIF sounds like it could have something approaching 300 sq. deg. per hour with 300 MHz bandwidth to a depth of 100μJy.

- Fly's Eye, at ATA, looks at 100 deg^2 by pointing each dish of the array in a different direction. This is still only a very small part of the sky, but can see pulses down to 0.6 μsec (Siemion 2008).

- The Astropulse sky survey at Arecibo looks at 1/3,000 of the sky down 0.4 μsec (Siemion 2008).

- The SKA itself won't exist for more than another decade, but its mapping speed should be at least ten times what ASKAP and WSRT+APERTIF advertise.

Signal/Noise is a function of time resolution. Each of these facilities should be able to detect a continuous source easily. However, one that is pulsed, for the same average power, will have a higher signal to noise in a shorter time, so is more detectable, but only *if* the electronics aren't set to average it out. (See the relations in Beacon Examples section, Benford J. et al. [2010].)

So the issue is whether the data systems in the newer systems, and SKA itself, will provide fast-time resolution. These devices point the way for SETI searches in the future.

4.0. How Can We Distinguish Transient Pulsars from SETI Beacons?

Pulsars are clearly radiation from rotating neutron star magnetospheres. There is increasing interest in certain transient phenomena of uncertain origin, but likely due to atypical pulsars. These are occasional observations of repetitive sources, some re-observed, some not.

There is another possible origin: extraterrestrial Beacons trying to attract attention from civilizations such as ourselves. We have recently described such Beacons in terms of how our civilization would build large broadcasting transmitters if we were to undertake announcement of our existence (METI, Messaging to Extraterrestrial Intelligences) or even communication. The approach is cost-optimized beacons: how designing under a likely constraint on cost affects design choices and therefore observable parameters. We found that beacons are likely to be pulsed, both to lower the cost and to make the signal more noticeable. In fact, beacons might mimic a pulsar, because they are likely to be studied. Of course, they would try to distinguish themselves in some way, such as modulations of amplitude, frequency hopping, etc. They would not be isotropic, as pulsars are not isotropic, in order to lower cost. So they might be much like a lighthouse, sweeping over a region of the sky, scanning in a raster pattern, close to the galactic plane. The number of pulses an observer would see must be "enough" to distinguish it: large enough to contain the modulation, but small enough to allow scanning of the beam over the designated target region ("dwell" time), after which it would return ("revisit" time).

There is an increasing interest in shorter transient astronomical sources, so researchers should be aware of the likely properties of beacons. How would observers distinguish such beacons from pulsars or other exotic sources? Here I consider transient observations in light of likely beacon observables, with one example of how to analyze observational data in terms of it's possibly being a beacon, deducing beacon parameters.

Factors to distinguish Beacons from pulsars are:

- *Bandwidth* An obvious difference is that pulsars have large bandwidths, typically peaking at ~1 GHz, falling with about the square of frequency. So a half-power bandwidth would be ~400 MHz. This is much larger than our powerful microwave devices, which have much smaller bandwidths. That is because they are based on conversion of electron beam energy to microwaves by techniques using resonance. For such devices, there is a tradeoff, called the gain-bandwidth product, which means that higher efficiency in microwave generation comes when there is high gain and small bandwidth. (Here "small" means much larger than the 1 Hz signals many SETI listeners have searched for, though.) Efficient Earth sources have bandwidths from 100 kHz to few MHz in the ~1GHz range where pulsars radiate most of their energy, and efficiencies are in the range of 50–90 percent. However, we should not preclude the possibility that more advanced methods of microwave generation could make very broadband emission efficient. In addition, larger bandwidth allows larger data transmission rates. Could ~100 MHz signals be broad transmission channels, a galactic information superhighway? Bandwidth alone is not necessarily a pulsar/Beacon separator. (Note also that some pulsar observation is done with narrower bandwidths, so the full bandwidth is not observed, but is assumed to be broad, but may possibly not be.)

- *Pulse length* Millisecond pulsars were observed later due to observational selection, for instance, integration time gives a natural bias against short periods. Cost-optimized beacons will likely be pulsed to lower cost, with a preference for shorter pulses due to source physics. On Earth, the higher source power, the shorter the pulse, due to breakdown physics, which is universal.

- *Pulse shape* An isotropic beacon will of course be seen whenever one observes it, and a scanning beacon is seen only when it passes by. A difference between Beacons and pulsars is that antennas generally have a gaussian beam shape, so a beacon beam sweeping past Earth will be seen as a gaussian versus time by an observer. In contrast, pulsars do not have a Gaussian shape in general (Lyne and Manchester 1988).

- *Frequency* Pulsar searches cluster in the lower end of the microwave, but Beacons may be more likely to appear at higher frequency. We expect cost-optimized beacons to appear at higher frequency (~10 GHz) due to the favorable scaling of cost with frequency.

4.1. A Specific Case, PSR J1928+15

As an example of analysis for SETI beacon possibilities, consider the transient bursting radio source, PSR J1928+15, which was observed in 2005 just below the galactic plane at 1.44 GHz in an Arecibo two-minute observation and not reobserved in forty-eight minutes of revisits (Deneva et al. 2009). The candidate explanation in Deneva et al. is perhaps an asteroid falling into the neutron star from a circumpulsar disk, perturbing its magnetosphere.

Three pulses were received, the first and third down a factor of ten from the 0.180 Jy central pulse. Separation between was 0.402 sec (= 1/2.48 Hz). The dispersion measure (DM) was 242 pc-cm^3. With density of 3.26 electrons/cm^3 the source is at a distance 24,000 ly, placing it almost as far as the 26,000 ly galactic center (Kraus 2005).

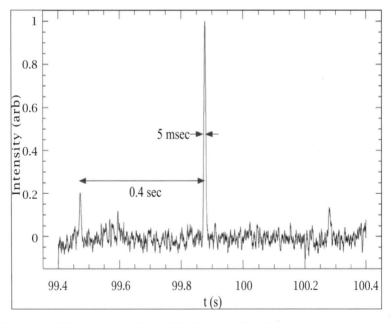

Figure 18.5. Observed pulses from PSR J1928+15: De-dispersed time series showing three pulses, the center pulse with highest amplitude.

4.2. Beacon Analysis of PSR J1928+15

In this section we work exercises in understanding beacon tradeoffs, with a hypothetical PSR J1928+15 beacon. The methods discussed in Section 4.1 allow deduction of beacon parameters. Given a small set of observations, the principal parameters such as power and antenna area can be calculated. Then the time required to listen for a revisit of the scanning beacon can be made if we assume its search pattern. Using the formulation of the Benfords (Benford, J. et al. 2010), parameters of three beacons producing 1.8×10^{-19} W/m^2 at 24,000 ly are given in Table 18.2. We use Earth cost parameters of 1 k\$/m^2, 0.3 \$/W.

To characterize a hypothetical PSR J1928+15 beacon, we make two working assumptions:

1. The beacon is a "lighthouse" scanning the galactic plane. The source is a scanning beacon and, as it swept past, Arecibo caught the central pulse, the true beam. The first and third pulses are at the edges of the antenna's acceptance angle, which is 3.5 arcmin = 1 mrad.

2. The beam bandwidth covers all channels of the 100 MHz span of the detector array. (The channel bandwidths are 0.39 MHz, with total BW 100 MHz.) This assumption drives the beacon power estimate. From the observed power density of 0.18 Jy, the total power across 100 MHz is 1.8×10^{-19} W/m^2.

The scanning beam produces the three pulses 0.4 sec apart. In Figure 18.6, the pulses are fitted to a Gaussian beam pattern of width 0.5 sec, so the observed pulse heights are replicated. As noted in 4.0, antennas generally have a gaussian beam shape, so a beacon beam sweeping past Earth will be seen as a gaussian versus time by an observer. In contrast, pulsars do not have a Gaussian shape in general (Lyne and Manchester 1988). But PSR J1928+15 fits a gaussian.

BEACON A: COST-OPTIMIZED

Using the formulation of the Benfords (Benford, J. et al. 2010), and the Earth cost parameters of 1 k\$/m^2, 0.3 \$/W, the cost-optimized parameters of a beacon producing 1.8×10^{-19} W/m^2 at 24,000 ly are given in Table 18.2. It has a big antenna, high peak power, but costs much less than the two other beacons. The small beamwidth gives a spot size of only 0.25 ly, meaning it's targeted to either ourselves or some target between (or behind) us. So, it is not a scanning beacon. If it were, the revisit time for a complete

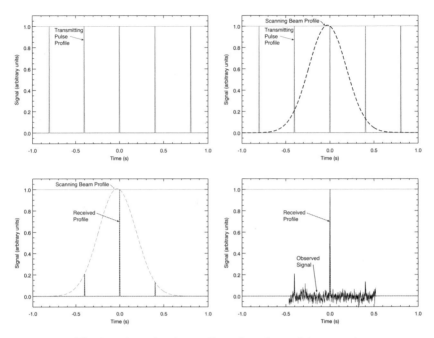

Figure 18.6. The hypothetical pulse profile, pulses observed 0.4 seconds apart, convolved with a Gaussian beam shape 0.5 sec FWHM results in reduced first and third pulses and fits the observed signal.

scan of the disk would be six years (see the calculation for the next example).

Total power in the spot area A_s is then $P = S\,A_s = 1{,}600$ TW, but the duty factor is low (5 ms/0.4s = 2 10^{-3}) so average power is 3 GW, about that of a standard nuclear installation. Note it is the *average* power, not the peak power that matters. The beacon power system will store energy between bursts (or shots) in an intermediate store. Thus, if it's a beacon, it comes from a civilization of our scale. On the Kardashev scale (Kardashev 1964):

$$K = \frac{\log_{10}P - 6}{10} = 0.35 \qquad (14)$$

where P is the beacon power. For comparison, Earth is K = 0.73.

BEACON B: NON-COST-OPTIMIZED, SMALL ANTENNA

As an exercise in understanding beacon tradeoffs, assume the beacon antenna diameter is $D_t = 100$ m. This leads to a small powerful beacon, with a large beamwidth

Table 18.2. Beacon Models Producing the Observed PSR J1928+15 Signal

Beacon	EIRP	Peak Power	Average Power	Antenna Diameter	Beam-width	Capital Cost	Operating Cost	Revisit time	Kardashev Scale (K)
	W	TW	GW	km	radians	B$	B$/yr		
A	10^{23}	1,600	3	25	9μrad	980	3	6 years	0.35
B	10^{21}	190,000	$4\ 10^{14}$	0.1	5mrad	10^{8}	$3\ 10^{14}$	2 hrs	0.86
C	10^{21}	1,900	3,800	1	0.5mrad	6,000	3,000	1 week	0.66

$$\theta = 2.44 \ \frac{\lambda}{D_t} = 5.1 \times 10^{-3} \, \text{rad} \qquad (15)$$

where we use the observed frequency, 1.44 GHz. Then the spot size at our range is $R\theta$ = 122 ly.

The total power is then $P = SA_s$ = 190,000 TW, ten thousand times the total electrical power of Earth. Thus if it's a beacon, it comes from a much more advanced and powerful civilization. On the Kardashev scale K = 0.86, less than the power of an entire planet (K = 1). Therefore, this beacon is an affordable luxury only to a civilization substantially in advance of us.

From the figure, the beam passing Arecibo in an interval 0.5 sec, so $d\phi/dt$ = 5.1 mrad /0.5 sec = 10^{-2} sec. If the pattern of the beacon is scanning the disk of thickness h, which is ~1300 ly at our location range of 24,000 ly, then the spot is moving at $R \, d\phi/dt$ = 250 ly/sec. The time for a cycle around the galactic circumference is $2\pi/[d\phi/dt]$ = 10 minutes. The number of such strips in the scan is 1350ly/122ly = 11. So the beacon will return in 11 × 10 minutes ~2 hours. It's understandable that forty-eight minutes of revisits hasn't seen it again in 40 percent of the revisit time. Of course, it could be scanning a smaller area, so that the revisit time would be sooner.

BEACON C: NON-COST-OPTIMIZED, LARGE ANTENNA

Assume D_t = 1 km. The beam width is reduced by a factor of 10 to 5 × 10^{-4} rad. Spot size diameter falls to 12 ly. Power in the spot falls to 1900 TW, one hundred times Earth's entire power. The Kardashev scale falls to K = 0.66. This is a civilization of planetary scale, commanding the entire energy of civilization of the previous example.

The spot moves at the same rate, 30 ly/sec. But since the spot is smaller, the number of strips in the scan increases to 1350ly/12.2ly = 110. So the beacon will return in 110 × 5 × 10^3 sec = 5.5 × 10^5 sec = 150 hours. Observers have revisited the site for forty-eight minutes, only 0.5 percent of the revisit time, and haven't seen it again.

The beacon nomograph (Benford, J. et al. 2010) can be used to display the above examples. They all cover the same fraction of the sky, the Galactic Disk, but have very different dwell and revisit times.

4.3 Conclusions

Galactic-scale beacons require resources larger than Earth presently has available for radio astronomy or SETI. From the examples, a civilization lower on the Kardashev scale will have a narrower beam, revisit less frequently, and so *will be harder to observe*. But lower power beacons will probably be more numerous. So we should learn how to identify them.

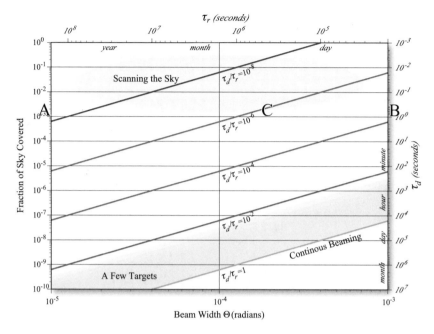

Figure 18.7. The three beacons of Table 18.2 displayed on the beacon broadcast strategy nomograph of the Benfords (Benford, J. 2010). A Beacon builder choosing values of θ (from Figure 18.10) and a sky fraction F to illuminate gives lines of constant duty cycle ratio (dwell time/revisit time, τ_d/τ_r) for the Beacon observer. Then right and top axes give ranges of these times for fixed τ_d/τ_r ratio, and can not be correlated to the lower and left axes. The two relations are independent of each other, except in that they produce the same time ratios. Cost-optimal Beacons lie in the upper region, continuous Beacons targeting specific star targets are in lower region, can be observed with surveys observing for short times.

The discussion here shows a method of analyzing an observed radio transient in terms of a possible beacon. We urge observers to consider SETI beacons as a candidate explanation when perplexing nonrepeating signals are seen in the radio sky.

Appendix: Nonlinear Cost Scaling

On Earth, antenna cost often varies with antenna area more rapidly than linear, i.e., A^α, $\alpha > 1$. The cost of power can also increase more rapidly than linear. If we generalize eq. 2 to give cost elements for aperture and microwave power a power-law dependence of A^α, P^β, the optimum ratio of power cost to aperture cost is β/α:

$$C_C = aA^\alpha + pP^\beta \qquad (16)$$

The optimal aperture and power and total cost are:

$$P^{\text{opt}} = \left[\frac{\alpha}{\beta} \frac{a}{p} \left(\frac{W}{kf^2} \right)^\alpha \right]^{\frac{1}{\alpha + \beta}}$$

$$A^{\text{opt}} = \frac{W}{kf^2 P} = \left[\frac{W}{kf^2} \right]^{\frac{\beta}{\alpha + \beta}} \left[\frac{\beta p}{\alpha a} \right]^{\frac{1}{\alpha + \beta}} \qquad (17)$$

$$C_C^{\text{opt}} = a \left[\frac{W}{kf^2 P} \right]^{\frac{\alpha\beta}{\alpha + \beta}} \left[\frac{\beta p}{\alpha a} \right]^{\frac{\alpha}{\alpha + \beta}} + p \left[\frac{\alpha}{\beta} \frac{a}{p} \left(\frac{W}{kf^2} \right)^\alpha \right]^{\frac{\beta}{\alpha + \beta}} \qquad (18)$$

$$\frac{C^{\text{opt}}_A}{C^{\text{opt}}_S} = \frac{\beta}{\alpha} \qquad (19)$$

Again, the cost ratio depends on only the exponents.

The scaling of aperture is very important for cost, as Figure 18.8 shows: more expensive antenna area ($\alpha=1.375$) drives cost upward. More power is radiated to make up the EIRP. The ratio of costs is

$$\frac{C^{\text{opt}}_A}{C^{\text{opt}}_S} = \frac{1}{1.375} \qquad (20)$$

in agreement with eq. 19. The main point of Figs 18.1 and 18.5 is that when antenna cost increases because α increases to 1.375, the total cost minimum increases at a steep rise. There is a great incentive to keep α close to 1.

The area exponent ranges from $\alpha = 1$ to 1.375 (Benford, J. et al. 2010). The Project Cyclops report concluded $\alpha = 1$ (Oliver and Billingham 1996), and SETI 2020 study gave $\alpha = 1.35$. It is difficult to argue that cost will vary less than linearly with area, so $\alpha = 1$ is probably a minimum. For space-based antennas, with no requirement to support or move the antenna mass against gravity, α will be less than on a planet or moon. For power

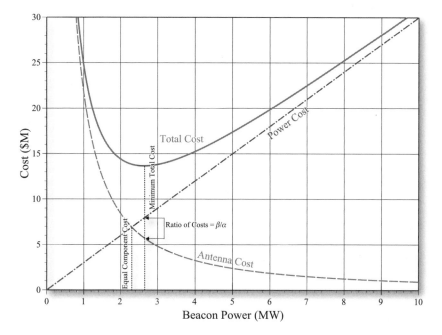

Figure 18.8. System of Figure 1 with $\alpha=1.375$, $\beta=1$. More rapid cost scaling for antenna leads to smaller antenna, higher power, and significantly higher total cost to achieve the same EIRP. Antenna and power cost are in ratio of, β/α, as in eq. 14.

costs, the rule of thumb in industry is $\beta = 1$ and O'Loughlin gives 0.75–1.

Therefore, in the range of experience, the range of the cost ratio is small, about a factor of two:

$$0.75 < \beta < 1$$

$$1 < \alpha < 1.375 \tag{21}$$

$$0.55 = \frac{C^{opt}_A}{C^{opt}_A} < 1$$

The value of α and β have a big impact on cost, as the following figures show. The penalty for not constructing near α, $\beta\sim1$ is high. However, making them <1 is difficult.

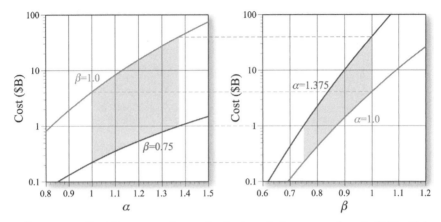

Figure 18.9. Rapid increase of cost of galactic-scale beacons such as in Table 18.1 with α and β. Shaded region is the range of values found on Earth.

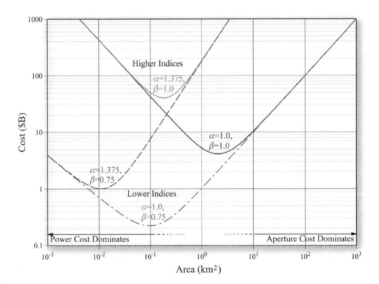

Figure 18.10. Galactic-scale beacon costs as a function of radiating area for various values of α, β and EIRP = 10^{17} W, a = 1 k$/m², p = 3 $/W, at f = 1 GHz with aperture efficiency ε = 0.5. The expected ranges of the antenna and power indices are shown, and hence the acceptable region is the v-shaped region enclosed by them. Cost falls with lower indices. Since the cost scale is logarithmic, the penalty for not constructing near the minimum is severe. Minimum cost is at an antenna diameter of hundreds of meters, and hence is large enough to suggest a phased array approach to keep structures to manageable scales.

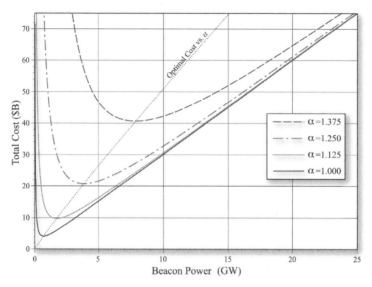

Figure 18.11. Galactic-scale beacon cost vs. power over the expected range of α for EIRP = 10^{17} W, β = 1, a = 1 k\$/m^2, p = 3 \$/W, at f = 1 GHz. Optimal cost increases with α and almost exactly linearly with power as α increases.

Works Cited

Benford, G., 1999. *Deep time.* New York. Harper Collins.

Benford, G., J. Benford, and D. Benford. 2010. Searching for cost optimized interstellar beacons. *Astrobiology* 10: 491–98. Also at arxiv.org/abs/0810.3966v2.

Benford, G., and J. Benford. 2006. Power beaming concepts for future deep space exploration. *J. British Interplanetary Soc.* 59: 104–107.

Benford, J., G. Benford, and D. Benford. 2010. Messaging with cost optimized interstellar beacons. *Astrobiology* 10: 475–90. Also at arxiv. org/abs/0810.3964v2.

Benford J. 2008. Space applications of high power microwaves. *IEEE Trans. on Plasma Sci,* 36: 569–81.

Benford J., J. A. Swegle, and E. Schamiloglu. 2007. *High power microwaves,* 2nd Ed. Boca Raton, FL: Taylor and Francis.

Benford J., and R. Dickinson. 1995. Space propulsion and power beaming using millimeter systems. In *Intense Microwave Pulses III,* ed. H. Brandt, 179–92. SPIE 2557.

Brown, W. 1968. Microwave-powered aerospace vehicles. In *Microwave Power Engineering Vol. 2, Applications,* ed. Ernest C. Okress. 268–76.

Deneva, J. S. et al. 2009. Arecibo pulsar survey using ALFA: Probing radio pulsar intermittency and transients. *Astrophysical Journal* 70: 2259–75.

Hetesi, Z., and Z. Regály. 2006. A new interpretation of the Drake Equation. *J. British Interplanetary Soc.* 59: 11–14.

Kardashev, N. 1964. Transmission of information by extraterrestrial civilizations. *Soviet Astronomy* 8: 217–21.

Kare, J., and K. Parkin. 2006. Comparison of laser and microwave approaches to CW beamed energy launch. In *Beamed energy propulsion—2005*. ed. K. Komurasaki, AIP Conf. Proc. 830, 388–99. New York: American Inst. of Physics.

Kraus J. D. 2005. *Radio astronomy*. Durham, NH: Cygnus-Quasar Books.

Lazio, J. et al. 2009. The dynamic radio sky: An opportunity for discovery. arXiv:0904.0633v1.

Lyne, A. G., and R. N. Manchester. 1988. The shape of pulsar radio beam. *Royal Astronomical Society Mon. Notices* 234: 477–508.

Medvedev, M. V., and A. Melott. 2007. Do extragalactic cosmic rays induce cycles in fossil diversity? *Astrophysical Journal* 664: 879–89.

Minsky, M. 1984. Communication with alien intelligence. In *Extraterrestrial: Science and alien intelligence* ed. Edward Regis Jr., 85–96. Cambridge: Cambridge University Press.

National Research Council. 2005. *The Atacama Large Millimeter Array (ALMA): Implications of a potential descope.* Washington, DC.

Oliver, B., and J. Billingham, eds. 1996. *Project Cyclops.* NASA CR-114445, 83.

Price, D., J. S. Levine, and J. Benford. 1997. ORION—A frequency-agile HPM field test-system. In *Proc. Seventh National Conference on High Power Microwave Technology*, 185–91. Laurel, MD: US Army Research Office.

Siemion, A. et al. 2010. New sky surveys for radio pulses. *Acta Astronautica*, oi:10.1016/j.actaastro.2010.01.016.

Seeking Contact

The Relevance of Human History

Michael A. G. Michaud

If you do not know the history of another people, you will not understand their values, their fears, and their hopes or how they are likely to react to something you do.

—Historian Margaret MacMillan, 2009

1.0. Introduction

In recent decades, we have seen a growing number of speculations about what would happen if we came into contact with an extraterrestrial civilization. Our discussions about the nature and consequences of contact are a vast thought experiment that draws on many different sorts of knowledge, and on many unproven assumptions.

Initially, this discussion was dominated by astronomers (particularly those involved in SETI) who believed that the only possible form of contact was through the transmission and detection of signals. Many envisioned that this would lead to a continuing exchange of messages in what we might call the correspondence model, a vastly expanded example of the invisible college. That form of contact seemed free of serious risks, except for possible cultural disorientation.

Things have changed during the past thirty years as others have questioned the assumptions underlying the correspondence model. One major factor has been the growing credibility of direct contact through robotic interstellar spacecraft. Scientists and engineers have shown that sending such probes to nearby stars would be technologically feasible for a civilization only slightly more advanced than our own (Bond 1978; Gilster 2004).

If there are more technologically advanced civilizations in our galaxy, some may already have this capability. Carl Sagan had argued as far back as 1962 that civilizations eons more advanced than our own must today be plying the spaces between the stars (Sullivan 1994). The possibility of direct contact requires us to widen the range of possible consequences.

2.0. Analogy and Probability

When the best physical scientists study the universe, they report it as it is, not as they wish it would be. For example, scientists are actively researching the probability that our planet will collide with an asteroid large enough to do serious damage. They use scientific methods to estimate what the impact might be. They are doing science, based on facts they can observe in nature.

Other scientists have used the scientific method in their searches for signals from extraterrestrial intelligence, rejecting findings that are not supported by confirmed evidence. Yet some do not observe that standard of objectivity when they predict the nature of an extraterrestrial civilization or the consequences of our coming into some form of contact with such an alien society. We often get opinions rather than systematic analyses based on the only database we have—ourselves.

Obviously, we have a problem in predicting the nature and behavior of intelligent extraterrestrials, as we have no confirmed information about them. Until we do, we have only two methods of analyzing the possible consequences of contact: analogy with ourselves, and probability based on what we know about human history and behavior.

3.0. Optimists

Authors have expressed sharp differences of opinion about the potential consequences of contact; some have been optimistic, others pessimistic. Many optimists support the search for extraterrestrials not just because it is scientifically interesting, but also because they hope that contact will introduce positive factors into human affairs. They *want* contact to have consequences for Humankind—the consequences we prefer, such as the transmission of knowledge that will make us wiser, even save us from ourselves.

Those who are optimistic about the consequences of contact draw on the most positive, least threatening analogies from our history. One analogy, popular during the early years of SETI, was receiving communications from civilizations separated from us in time, as in the Western rediscovery of knowledge from ancient Greece and Rome that helped stimulate the Renaissance. (It is worth recalling that both those admired cultures practiced slavery, and that both engaged in frequent warfare.)

Some optimists have claimed that more advanced civilizations would be generous in sharing their knowledge, even eager to educate us. Frank Drake expected a more advanced civilization to bequeath to us vast libraries of useful information to use as we wish. Sagan imagined that we might have access to an Encyclopedia Galactica. According to one book, all of the important questions in science, engineering, and social science would be answered for us (Drake 1993; Sagan 1980; MacGowan 1966). A related prediction is that a more advanced civilization will help us to solve our current problems, from warfare to cancer, with a kind of interstellar technical assistance program, a grant of useful knowledge by a Prometheus from the stars.

Such predictions were implicitly connected with visions of alien utopias. In the early fictional depictions of intelligent life on other worlds, the alien civilization often was described as more advanced than our own not only in science and technology, but also in morals and ethics, and in social and political behavior (Guthke 1990). Some of those skeptical of traditional Christian concepts of heaven imagined planetary paradises populated by angelic extraterrestrials. That vision was not so much a prediction as a method of social criticism, contrasting humankind with a utopian society. By implication, such a society was a model for our own future.

Now let us move forward to the beginnings of the modern scientific search for extraterrestrial intelligence in the 1960s and 1970s. What we now call SETI was the product of a particular historical moment: the cold war, when many people in industrial nations feared a disastrous conflict.

Again, we find utopian predictions about extraterrestrial civilizations. Sagan imagined alien societies "in excellent harmony with their environments, their biology, and the vagaries of their politics, so that they enjoy extraordinarily long lifetimes" (Sagan 1973, 241). Optimists foresaw that advanced extraterrestrials would share their good fortune with us. Drake wrote that we might learn general rules of civilization that we could apply to our own. The authors of the Project Cyclops report hoped that we would discover social and esthetic forms most apt to lead to self-preservation and a richer life (Drake 1976; NASA 1973).

Instead of seeing the utopian vision of extraterrestrials as a heuristic device, some came to adopt it as an assumption. That assumption gave us an outlet for idealism and hope. It implied the future we would like to have, contrasted with the future that many feared we would have if we did not change our ways.

Regrettably, we have not one shred of scientific or historical evidence supporting the belief that an alien civilization would adhere to these idealized models. A vision to aspire to may not be an accurate prediction of our future, or their present.

Optimists assure us that there will be no risk in contact, either because we are insulated by interstellar distances or because advanced aliens will have benign intentions. Sagan and William Newman claimed that civilizations that do not self-destruct are "pre-adapted" to live with other groups in mutual respect. Astronomer Paul Horowitz thought that civilizations that don't acquire the wisdom to control war will destroy themselves long before they can take to space, so the ones we contact will be, by definition, no longer threatening (Regis 1985; Easterbrook 1988). We have no scientific or historical evidence that supports these assumptions; they reflect hope rather than fact.

4.0. Pessimists

Pessimists draw on less attractive analogies from human history, often invoking examples of more powerful societies disrupting or demoralizing weaker ones. Their concerns range from the cultural disorientation of receiving information from a more advanced civilization to the possible extermination of our species through direct contact.

Nobel Prize–winning biologist George Wald declared that he could conceive of no nightmare as terrifying as establishing communication with a superior technology in outer space. Evolutionist Steven Jay Gould predicted that success in the search would be cataclysmic for our intellectual lives. Even the generally optimistic psychologist Albert Harrison acknowledged that, while intelligent aliens could help us to solve our problems, the introduction of their ideas into our society could backfire and create a nightmare without end (Berendzen 1973; Billingham 1994; Harrison 1997).

Astronomer Robert Jastrow argued that contact between scientifically advanced civilizations and others typically results in the destruction of the less developed culture. International affairs analyst Fareed Zakaria gave us this analogy from our own history: "Within a hundred years of initial European contact, one trend was unmistakable and irreversible: these encounters changed or destroyed the existing political, social, and economic arrangements in non-Western societies." If we have no choice in the matter, warned historian William McNeill, the end of human civilization as we have known it would be an expected consequence—especially in the case of direct contact (Jastrow 1997; Zakaria 2008, 66; Sagan 1973b).

We have no information about how an extraterrestrial society would behave in a contact situation. Our assumptions about alien behavior have not passed the Galilean test. As astrobiologist Christopher McKay put it, the Copernican principle is not established with respect to biology, culture, or ethics (Dick 2000).

We cannot assume that all extraterrestrial civilizations will conform to any standard model, whether that model is utopian or dystopian. It is more likely that separate civilizations will evolve differently. Our cosmic neighborhood might contain a mixture of cultures that would react in different ways to an encounter with us. We have no way of knowing which is which in advance of contact.

5.0. The Present Is Different

Some claim that our time is fundamentally different from the past. According to this school of thought, historical analogies do not apply because we have entered a more peaceful era since the end of the cold war, putting behind us the worst human behaviors. Some have argued that technologically advanced extraterrestrials already will have passed through such a transition, and dismiss worries about the possible negative consequences of contact as paranoia or cold war thinking. Harrison, declaring that war among humans is on the decline and citing computer models showing that "belligerent" societies are likely to collapse, predicted that it is peaceful societies that we are likely to meet (Harrison 1997).

In fact, conflict continues to be part of the human condition. Since the end of the cold war, millions of people have died in wars in the former Yugoslavia and in Africa. Well before the recent conflict in Darfur, an estimated two million people died in Sudan because of the war between the Moslem north and the Christian and animist south. One million people died as a result of the war in Angola that finally ended in 2002.

More people have lost their lives in the Eastern Congo than in any conflict since World War II. A peer-reviewed study found that 5.4 million people already had died in that war as of April 2007; thousands more have died since then (Gettleman 2009; Kristof 2010). The total number of deaths is approximately equivalent to the number of Jewish people who died because of the Nazis, yet this tragedy is virtually ignored in the West. Excluding conflict in Africa from our calculations would be implicitly racist.

U.S. President Barak Obama clearly stated the reality in his Nobel Prize acceptance speech. "We must begin by acknowledging the hard truth: we will not eradicate violent conflicts in our lifetimes. There will be times when nations—acting individually or in concert—will find the use of force not only necessary but morally justified" (Zeleny 2009).

In his massive study of war in human civilization, Professor Azar Gat concluded that violent competition—including intraspecific conflict—is the rule throughout nature. "Within this fundamental reality," he wrote, "organisms can resort to co-operation, competition, or conflict, strategies that they invariably mix, depending on each strategy's utility in a given situation" (Gat 2006, 663). If the biological and evolutionary laws we know on Earth prevail throughout our galaxy, many extraterrestrial societies may have had violent histories like ours. Whether they have moved beyond such behavior is unknown.

6.0. Humans Are Uniquely Evil

Some of those who are most optimistic about the consequences of contact try to get around such facts about our history and behavior by claiming that more advanced beings will not share our faults, even that we humans must be uniquely evil. We have not one shred of scientific evidence to support that belief, which seems at least partly inspired by some human religions.

It is far safer to apply Sagan's assumption of mediocrity, to assume that we are neither the best nor the worst. As astronomer Sebastian von Hoerner put it, assuming ourselves to be average has the highest probability of being right (Hart 1982). The prudent course is to assume that while some intelligent species may be nobler than we are, others may be more brutal. Consider the implications of postbiological societies described by Steven Dick and others (Dick 2003). Would intelligent machines behave altruistically toward biological beings?

Assumptions about alien behavior have not passed the observational test. We have no evidence of what motivates intelligent extraterrestrials, or of how they would behave in a contact situation. If we insist on assigning our best qualities to them, we also must allow them to have our worst.

7.0. The Focusing Issue: Active SETI

In recent years, the debate about the consequences of contact has swirled around the issue of calling attention to ourselves by sending out more powerful and more targeted signals than the Earth normally emits, in the hope of provoking a response. This practice is known as Active SETI, or METI (Messaging to ETI), in contrast to passive listening.

Much of the discussion has focused on the content and format of the messages we send. Those actually are secondary issues, particularly given the probability that our messages will be misunderstood. The larger question is whether we should call attention to ourselves by increasing the power, directionality, and continuity of our deliberate signals. In Active SETI, we are not just seeking to find extraterrestrials, but also to provoke a reaction from an alien society whose capabilities and intentions are not known to us.

Advocates of Active SETI argue that there should be no restrictions on sending high-powered signals because there is no risk. That view rests on implicit assumptions: that interstellar flight is impossible or, if it is possible, that more technically advanced societies will be benign. Both of these assumptions have been challenged forcefully.

Some claim that there is no point in worrying about being detected; they assume that alien civilizations already know that we are here because of the radio, television, and radar signals the Earth normally emits. In fact, most such signals are intermittent and untargeted, or are weak on an interstellar scale. One also may ask why we need to send Active SETI signals if extraterrestrials already know of our existence.

Others have warned that the possible negative consequences are immense and irreversible. "Those astronomers now preparing again to beam radio signals out to hoped-for extraterrestrials are naïve, even dangerous," wrote Pulitzer Prize–winning author Jared Diamond (Diamond 1999, 69). "If there really are any radio civilizations within listening distance of us," he wrote in one of his books, "then for heaven's sake, let's turn off our transmitters and try to escape detection, or we're doomed" (Diamond 1992, 214).

According to Russian astronomer Alexander Zaitsev (an advocate of METI), fifteen real Interstellar Radio Messages have been sent since the first in 1974 (Zaitsev 2010). Many other transmissions have been too weak, and too temporary, to significantly increase our detectability; they are unlikely to be heard by an alien civilization. Some messages have been commercial or advertising ventures with no serious intent to communicate with extraterrestrials (Zaitsev calls them "Pseudo-METI"). Such transmissions are messages to ourselves.

8.0. Principles and Protocols

The so-called First SETI Protocol (actually entitled Declaration of Principles), published in 1989 and adhered to by most of the organizations involved in SETI, emphasized the importance of first confirming any detection of extraterrestrial intelligence, and then making the news public. One of the declaration's other principles stated that no reply to a detected signal should be sent until a process of international consultation has taken place. Adherence to those principles was entirely voluntary; the purpose was not to impose control, but to build consensus. The organizations involved in drafting and approving that text, such as the International Academy of Astronautics and the International Institute of Space Law, have no authority over anyone, not even their own members.

During a later debate within the International Academy of Astronautics about proposed principles for the transmission of messages, disagreement arose over whether international consultations should be required before an Active SETI transmission, that is, a powerful, targeted signal that was not sent in response to a detection. A 2006 editorial in *Nature* described the issue this way: "While the chances of unpleasant consequences may be remote, they must be considered. When technologies offer radical new possibilities, the people who have the privilege of playing with them have an obligation to consult widely about what those possibilities mean" (Nature 2006, 606).

As of this writing, there is no agreed limitation on sending Active SETI signals, though many believe that international consultations should be required before they are sent. Some Active SETI advocates seem to presume the outcome of such a process. Douglas Vakoch was quoted as saying, "I have long held the position that after broad-based international consultation, we should be doing active SETI" (Battersby 2010, 29).

There are other recent examples for such consultations. The geo-engineering community, recognizing that its experiments can affect the public and that the public cannot opt out of their effects, has been developing a statement of principles to guide its work. Like the first SETI Protocol, the draft principles state that all geo-engineering research projects should be made public and their results disseminated openly. Those principles recognize a basic question: Is the proposed project in the public interest? According to the draft, decisions defining the public interest should be made with public participation; governments and the public should work together to decide which schemes are viable, appropriate, and ethical (Kintisch 2010; *Economist* 2010). In Active SETI too, the public would not be able to opt out of the effects of the experiment.

9.0. How History and Social Science Can Help

This debate is an opportunity to get people around the world to think about what is in the best interests of the human species, and to act accordingly. Seeking contact with extraterrestrial intelligence is a species enterprise that should be conducted with our shared interests in mind.

It is time for the participants in this debate to acknowledge the relevance of human history. Historians and social scientists can provide us with the evidence of what actually happened when different human societies first came into contact, both the good news and the bad news. They can describe human behavior as it actually was, not as we would prefer it to be.

Regrettably, history and the social sciences have had only a marginal influence on the debate about the consequences of contact. There appear to be several reasons, including a lack of funding for social science research in this field, and a perception that this kind of work is not respectable. The other side of that coin is that physical and biological scientists have not always welcomed social science involvement (Tough 2000).

Former United Nations Secretary General Kofi Annan observed that separate narratives about the Arab-Israeli problem have become intellectual prisons, paralyzing discourse and hindering understanding. The antidote, he argued, is grounded historical research. Such knowledge of history, such dispelling of public myths about the other, is a precondition to serious dialogue (Ryback 2006). Similarly, we need to apply more firmly grounded history to the debate about the implications of contact.

10.0. What Is To Be Done?

To better understand the consequences of contact with an extraterrestrial civilization, we need a more systematic and objective calculation of the potential benefits and potential risks. We should reflect on the full range of possible outcomes, not just those we prefer. As Zakaria said about foreign policy, this is a matter of costs and benefits, not theology (Zakaria 2008). One starting point might be the Rio scale proposed by astronomers Ivan Almar and Jill Tarter, a framework for classifying the impact of contact (Almar and Tarter 2000).

We also need more thoughtful analyses of the direct contact scenario, which is poorly represented outside of science fiction. Our own history offers many examples of the first direct contacts between different civilizations and different societies. It is time for an agenda-free survey.

We need non-Western perspectives on the implications of contact. In particular, we need more input from Asia. The world's most populous nations—China and India—have played almost no role in this debate, nor has Japan.

11.0. Conclusion

Scientists, perhaps more than all others, should not let belief or preference triumph over evidence. As historian Margaret Macmillan put it, we must continue to examine our own assumptions and those of others and ask, Where's the evidence? (Macmillan 2009).

Science is the best defense against believing what we want to, wrote Jack Cohen and Ian Stewart (Cohen 2002). So is grounded history.

Works Cited

Almar, I., and J. Tarter. 2000. The discovery of ETI as a high-consequence, low-probability event. Paper presented at the International Astronautical Congress in Rio de Janeiro, to be published in *Acta Astronautica*.
Battersby, S. 2010. We're over here. *New Scientist* 205 (2744): 28–31.
Berendzen, R., ed. 1973. *Life beyond Earth and the mind of man*. Washington, DC: NASA SP-328.

Billingham, J., R. Heyns, D. Milne, S. Doyle, M. Kline, J. Heilbron, M. Ashkenazi, M. Michaud, J. Lutz, S. Shostak, eds. 1994. *Social implications of the detection of an extraterrestrial civilization.* Mountain View, CA: SETI Press.

Bond, A., A. R. Martin, R. A. Buckland, T. J. Grant, A. T. Lawton, H. R. Mattison, J. A. Parfitt, R. C. Parkinson, G. R. Richards, J. G. Strong, G. M. Webb, A. G. A. White, P. P. Wright. 1978. Project Daedalus: The final report on the BIS starship study. Special Issue of the *Journal of the British Interplanetary Society.*

Cohen, J., and I. Stewart. 2002. *What does a Martian look like?* Hoboken: Wiley.

Diamond, J. 1999. To whom it may concern. *The New York Times Magazine,* Dec. 5: 68–69.

———. 1992. *The third chimpanzee.* New York: Harper Collins.

Dick, S. J., ed. 2000. *Many worlds: The new universe, extraterrestrial life, and the theological implications.* Radnor, PA: Templeton Foundation Press.

———. 2003. Cultural evolution, the postbiological universe, and SETI. *International Journal of Astrobiology* 2: 65–74.

Drake, F. 1976. On hands and knees in search of Elysium. *Technology Review* (June): 22–29.

———, with D. Sobel. 1993. *Is anyone out there?: The continuing search for extraterrestrial intelligence.* London: Souvenir Press.

Economist. 2010. We all want to change the world. *The Economist,* April 3: 81–82.

Gat, A. 2006. *War in human civilization.* New York: Oxford University Press.

Easterbrook, G. 1988. Are we alone? *The Atlantic Monthly,* August: 25–28.

Gettleman, Jeffrey. 2009. A wound in the heart of Africa (review of Prunier's *Africa's World War). The New York Times Book Review,* April 5: 16.

Gilster, P. 2004. *Centauri dreams: Imagining and planning interstellar exploration.* New York: Springer (Copernicus).

Guthke, K. 1990. *The last frontier: Imagining other worlds, from the Copernican revolution to modern science fiction.* Trans. Helen Atkins. Ithaca: Cornell University Press.

Harrison, A. 1997. *After contact: The human response to extraterrestrial life.* New York: Plenum.

Hart, M. H., and B. Zuckerman, eds. 1982. *Extraterrestrials: Where are they?* New York: Pergamon.

Jastrow, R. 1997. What are the chances for life? (Review of Dick's *The Biological Universe). Sky and Telescope,* June: 62–63.

Kintisch, E. 2010. Asilomar 2 takes small steps toward rules for geoengineering. *Science* 328: 22–23.

Kristof, N. D. 2010. Orphaned, raped, and ignored. *The New York Times,* January 31.

MacGowan, R. A., and F. I. Ordway III. 1966. *Intelligence in the universe.* Englewood Cliffs, NJ: Prentice-Hall.

Macmillan, M. 2009. *Dangerous games: The uses and abuses of history.* New York: Modern Library.

Michaud, M. A. G. 2007. *Contact with alien civilizations: Our hopes and fears about encountering extraterrestrials.* New York: Springer (Copernicus).

NASA. 1973. *Project Cyclops: A design study of a system for detecting extraterrestrial intelligent life* (revised edition). CR114445.

Nature. 2006. Ambassador for Earth. *Nature* 443: 606.

Regis, E. Jr. 1985. *Extraterrestrials: Science and alien intelligence.* New York: Cambridge University Press.

Ryback, T. 2006. Enter the historians, finally. *International Herald Tribune*: November 24.

Sagan, C. 1980. *Cosmos.* New York: Random House.

———. 1973. *The cosmic connection: An extraterrestrial perspective.* New York: Anchor.

———, ed. 1973b. *Communication with extraterrestrial intelligence.* Cambridge: The MIT Press.

Sullivan, W. 1994. We are not alone: The continuing search for extraterrestrial intelligence. New York: Plume.

Tough, A., ed. 2000. *When SETI succeeds: The impact of high-information contact.* Bellevue, WA: Foundation for the Future.

Zaitsev, A. 2010. Email to SETI transmission group, February 12.

Zakaria, F. 2008. *The post-American world.* New York: Norton.

Zeleny, J. 2009. Accepting peace prize, Obama offers "hard truth." *The New York Times*: December 11.

Pragmatism, Cosmocentrism, and Proportional Consultation for Communication with Extraterrestrial Intelligence

Mark L. Lupisella

1.0. Introduction

The primary task given to me by the SETI (Search for Extraterrestrial Intelligence) session organizers was to address philosophical and policy challenges of communicating with extraterrestrial intelligence (ETI), with an emphasis on "cosmocentric" thinking and "Active SETI" (transmissions that are sent to putative extraterrestrial intelligence prior to having received one, sometimes also called METI—Messaging to Extraterrestrial Intelligence). Such questions regarding communicating with ETI, either as individuals, groups, or more collectively as a species, pose more near-term practical policy challenges than it may first appear. And like many policy challenges, this issue ultimately connects to, and perhaps rests on, matters of philosophy and ethics.

This chapter has essentially two parts: The first part (sections 2–4) briefly explores potentially relevant philosophical views with an emphasis on a hybrid view combining pragmatism and cosmocentrism—namely, "cosmocultural evolution" or "cosmoculturalism"—which acts as a source of guidance for the second part of the chapter (section 5), which emphasizes the importance of attempting to communicate with ETI and makes practical suggestions for "proportional consultation" when attempting to do so.

Given the cosmic dimensions of this issue, including the assumption that we presumably have the universe in common with ETI, these questions lend themselves to what might be called "cosmocentric" thinking, which generally invokes cosmological perspectives and makes the universe a kind of priority in a worldview, perhaps with other co-priorities. Cosmocentrism is a broad notion that can have a wide variety of meanings and implications, with details ranging from empirically scientific to spiritually divine.

This chapter explores a number of philosophical views within the context of a cosmic perspective and examines their applicability for informing issues regarding communication with extraterrestrial intelligence, with an emphasis on Active SETI since that arguably poses the most pressing policy challenges. Because Active SETI and communications with ETI more generally are practical policy challenges with potentially unusually significant implications, the relationship of "pragmatism" to "cosmocentrism" is particularly relevant and is explored by examining a number of views such as anthropocentrism, ratiocentrism, cosmocultural evolution, teleology, and pantheism—with an emphasis on "bootstrapped cosmocultural evolution," which has elements of both pragmatism and cosmocentric views, as shown in Figure 20.1. With general guidance from a bootstrapped cosmocultural evolutionary perspective, I then suggest a few practical examples of how we might strike balances and compromises regarding communicating with ETI by appealing to "proportional consultation."

2.0. Pragmatism

As shown in Figure 20.1, anthropocentrism and ratiocentrism can be thought of as occupying the pragmatic end of a continuum of philosophical views that increase in their scope and prioritization from human beings to the universe.

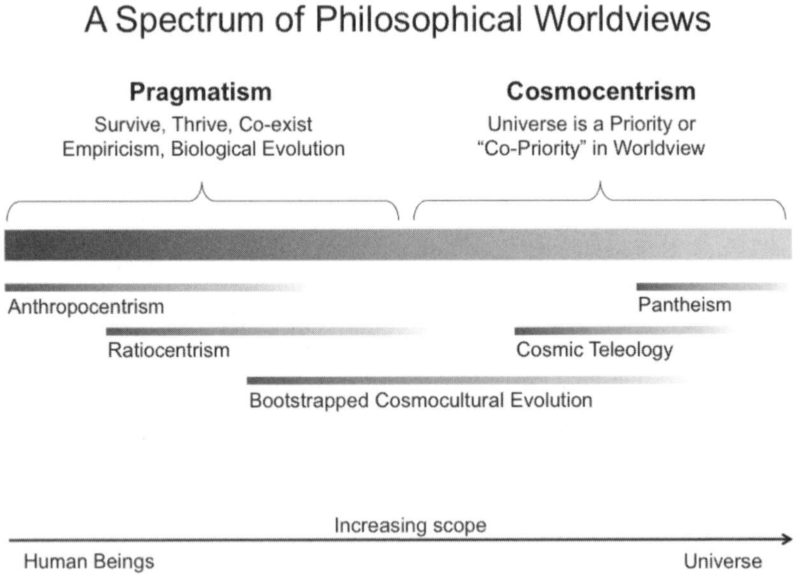

Figure 20.1. A spectrum of philosophical worldviews.

The philosophical tradition of pragmatism can be traced to Charles Peirce in 1878 and later to William James in 1907 when he published a series of lectures on "Pragmatism: A New Name for an Old Way of Thinking," which popularized the philosophy and coined the term *pragmatism*.

The original purpose of pragmatism was to help clarify disputes by clearly articulating practical relevance of terms, arguments, implications, etc. But for the purposes of this chapter, the pragmatist tradition (which is broader than the original conceptions of Peirce and James) can be generally associated with empiricism, secular humanism, and a weak form of relativism regarding truth (stemming in part from a reliance on "collective knowledge")—resulting in a worldview that prioritizes practical relevance to human beings and effectively rejects that which is beyond the test of human experience.

Two broad views that are consistent with (but not equal to) the pragmatist tradition are *anthropocentrism*, which makes human beings the priority in a worldview, and *ratiocentrism*, which broadens that prioritization to all rational beings. With these views, morality and ethical obligations lie primarily with human beings, or more generally with any nonsupernatural rational agent, where "rational" is defined loosely for the purposes of this essay to imply at least (1) the potential and/or ability to understand relatively complex rules and norms of social behavior (Smith 2009) as well as perhaps complex bodies of social knowledge, and (2) to have a relatively long-term level of awareness about the future that can, in part, result in conscious suffering due to anticipating future adverse circumstances. "Complex" might be characterized minimally as that which involves a large amount of intentionally created abstract symbols (Deacon 1997) to manage large amounts of information in both space and time. Pragmatism, anthropocentrism, and ratiocentrism rely primarily on notions of value, meaning, and purpose that are created by, or in service to human beings or other rational beings.

Pragmatism, and more specifically, anthropocentrism and ratiocentrism have a relatively narrow scope of applicability in that nonrational entities are generally devalued (often resulting in the unwarranted adverse treatment of nonrational entities—e.g., animal abuse), and long-term perspectives are often deemphasized (intentionally and unintentionally), sometimes resulting in longer-term self-destructive tendencies. However, pragmatist views, particularly ratiocentrism, need not necessarily result in short-termism or unfortunate outcomes for nonrational entities—particularly if those nonrational entities are valuable to rational entities (Smith 2009).

3.0. Cosmocentrism

Figure 20.1 shows cosmocentrism at the other end of the worldview continuum, signifying cosmic scope and prioritization of the universe. At its core, cosmocentrism makes the universe a priority in a worldview, perhaps along with other priorities. It ascribes value to the whole of the universe and hence to cosmic evolution, and in some cases, may attempt to ground degrees of cosmic value, and perhaps hence degrees of intrinsic value rooted in the nature of the universe (Lupisella and Logsdon 1997; Lupisella 2009a). There can be many kinds of cosmocentric views, two of which are *teleological cosmocentrism*, which suggests that the universe has intrinsic directionality and/or perhaps an ultimate "purpose," and *pantheism*, which ascribes a kind of spirit or divinity to the universe, essentially equating it with many conceptions of God.

There are a number of ways to think about teleology, and while the idea has fallen out of favor among many scientists, it still receives attention (Manson 2003). Cosmic teleology suggests that the universe has natural directionality and/or fundamental cosmological trends, or perhaps even cosmic "imperatives" or cosmic purpose. Forms of teleology have been implicitly or explicitly suggested by a number of scientists, suggesting, for example: trends toward increasing self-organization and complexity (Chaisson 2005); life and intelligence as "cosmic imperatives" or inevitable cosmic phenomena (Lloyd 2006; Davies 2007); "multiverse" and/or "anthropic" worldviews that suggest our particular universe is made for life (Smolin 1997; Rees 1997); and more explicit eschatological treatments that have pantheistic themes (Teilhard 1955; Tipler 1994).

Pantheism generally equates God with the universe and rejects the notion of a personal and/or transcendent God. There are numerous conceptions of pantheism, including Taoism, some mystical versions of Western religions, and purely naturalistic views based on biology and cosmology that focus on the physical realities of the natural world, the universe, and cosmic evolution (Dick 2000). Pantheism is a form of metaphysical and religious views where unity, reverence, sacredness, and divinity play important roles (Levine 1994; Harrison 1999).

Cosmocentrism implies the broadest scope of applicability possible, but may inappropriately confer value and prioritization too broadly and perhaps inconsistently with interests of rational beings.

4.0. Bootstrapped Cosmocultural Evolution

A view occupying a middle ground between pragmatism and cosmocentrism, and that may have particular relevance for SETI, is what might be called *cosmocultural evolution* or *the cosmocultural principle*, which suggests that the cosmos and culture co-evolve and will increasingly co-evolve with culture playing an important role in the overall evolution of the universe—ranging from creating and manifesting value in an otherwise valueless universe to perhaps eventually exercising control over the universe itself (Lupisella 2009b). This kind of view may seem fanciful, irrelevant (since it is so long-term), and perhaps hubristic given what we know of our universe today (e.g., the second law of thermodynamics) and the sometimes unfortunate consequences of human arrogance. However, the cosmocultural principle or cosmocultural evolution does not necessarily make specific claims about how the significance of cultural evolution will ultimately be realized in the universe. It suggests primarily that it is sufficiently plausible that cultural evolution can be significant in the overall evolution of the universe—and that the details of that significance may ultimately simply be choices for cultural beings to make.

Despite the notoriously difficult challenges in defining culture (Dick and Lupisella 2009; Traphagan 2011), it is still helpful to attempt an operational characterization, however incomplete or imperfect it may be. I have in mind a characterization of culture as something like "the *collective manifestation of value*—where 'value' is that which is valuable to 'sufficiently complex' agents, from which meaning, purpose, ethics, and aesthetics can be derived" (Lupisella 2009b, 322). "Sufficiently complex" may rest primarily on the ability to intentionally create collective symbolic abstractions—in part to assist with social life, conflict management, meaning, purpose, etc. With this characterization of culture in mind, "cultural evolution," then, is the variance of that complex collection over time—which may or may not involve directionality or "progress."

Bootstrapped cosmocultural evolution more specifically suggests that culture is not necessarily inherent in the universe, but simply arose from the emergence of successful replicators that eventually led to cultural beings. On this view, the universe has in some sense "bootstrapped" itself into the realm of value via the emergence of replicators that drove the subsequent evolution of cultural beings who are now sources and arbiters of value, meaning, and purpose—including potentially for the whole of the universe (Lupisella 2009b).

In this way, bootstrapped cosmocultural evolution is both pragmatic and cosmocentric: (1) it does not appeal to anything supernatural or intrinsically teleological and (2) places primacy on both rational cultural beings and the cosmos by emphasizing the unlimited practical potential and philosophical significance that cultural beings can have for the evolution of the universe as a whole.

5.0. Application to Communication with ETI

Bootstrapped cosmocultural evolution makes rational cultural beings and the universe "co-priorities" and "co-creators" in a worldview. It places rational cultural beings in a special role, along with the universe more broadly—and that has implications for how we might apply this view to practical ethical policy challenges associated with communicating with ETI. There are a number of key questions: Should we transmit before receiving a signal? Should we transmit after receiving a signal? What should be communicated? What, if any, consultation should be pursued before transmitting to ETI?

On the question of message content, cosmocentric thinking should at least be considered as potential content for what might be communicated to ETI, in part for the practical reason that we likely have the universe in common with ETI, including perhaps what may be our ultimate shared cosmic origins (Vakoch 2009).

Regarding policy questions of process associated with communicating with ETI, a bootstrapped cosmocultural evolutionary view implies high uncertainty regarding ETI because cultural evolution and anything associated with it (such as altruism, stable social structures, philosophies such as cosmocentric views, etc.) are assumed to be at least partially, if not completely situational, and not inherent in the universe or in intelligence. Culture simply arose due to the emergence of replicators that sought strategies for optimal replication, leading to social cultural rational beings. So views such as bootstrapped cosmocultural views, or any other philosophical views, are not assumed to be universal, but are instead merely intellectual interpretations and philosophical choices made by reasoning cultural beings—which may or may not be reflective of some sort of universal reality. These kinds of worldviews may vary dramatically across extraterrestrial civilizations—as they have throughout human history, resulting in a high level of uncertainty regarding the nature of putative ETI and their motives.

While the uncertainty regarding ETI is clearly high, so is the potential value of communicating with ETI. Bootstrapped cosmocultural evolution would ascribe a very high level of significance to communication with ETI, in part because of the potential cosmic significance of cultural interaction contributing to and enhancing cosmocultural evolution. It may be reasonable to suppose that most long-lived intelligent civilizations would exhibit a significant level of sensitivity regarding other rational beings—in part because that sensitivity is so important for social beings to co-exist and thrive for extended periods of time. However, biological evolution, and arguably much, if not most cultural evolution, often selects for moral behavior to the extent that it satisfies precariously balanced cost-benefit outcomes—which can be complex, uncertain, imprecise, and unstable. The knife-edge of moral evolution suggests it doesn't take much to end up on the wrong side of the blade.

Given the high uncertainty and extremely high significance of communication with ETI (Michaud 2007), and given the more pragmatic obligations to rational beings and our clear obligation to our fellow human beings (who we know exist for sure!), it seems reasonable to engage in some degree of proportional consultation before taking significant steps to communicate with ETI. Pursuing some degree of collective understanding and readiness before attempting to communicate with ETI is presumably a healthy approach. However, depending on the details of what constitutes "consultation," "collective understanding," and "readiness," it could be unduly confining, time-consuming, and perhaps ultimately misguided in terms of process and outcomes. Individuals and groups around the world may find it to be an unjustified infringement of freedom if the international community was too slow or prevented communication attempts with ETI without approval resulting from significant international consultation.

Others may claim that consultation is not relevant since (1) we have already sent signals to ETI (intentionally and unintentionally), (2) if ETI wanted to, they could harm us anyway, or (3) they can't harm us because of the large astronomical distances involved. The fact that we've already transmitted doesn't mean those transmissions have been received or understood. Arguments 2 and 3 are reasonable speculation, but given the uncertainties and potential implications, they are not necessarily grounds on which to base confident policy positions at this time. So adopting a precautionary principle seems reasonable, and it would arguably suggest making minimal assumptions and not relying on the assumptions of a limited number of individuals who may choose to engage in substantial attempts to communicate with ETI.

If the potential hazard of an attempt to communicate with ETI could be assessed with a reasonable level of confidence to be relatively small, that might serve as justification to relax a certain level of required international consultation. This is clearly a big "if," but it may be worth pursuing.

A partial approach to assess what constitutes "significance steps" and/or potential hazards regarding communicating with ETI could be to use something like the "San Marino Scale" to help assess the potential hazard of communication attempts with ETI. The San Marino scale was adopted by the IAA (International Academy of Astronautics) SETI Permanent Study Group as an accepted tool to assess transmissions from Earth) and is defined as the sum of signal strength plus the characteristics of the transmission (e.g., content, intention, direction, duration). The resulting index can be placed on an overall scale of 1 to 10 where 1 poses an "insignificant" potential hazard and 10 poses an "extraordinary" potential hazard (Almár and Shuch 2007).

Assessing the second term of the index, namely the characteristics of the transmission—and specifically the *content* and *intent* of any given transmission attempt—is the most difficult part of this kind of assessment—but there have been attempts to do it. As an example, Almár and Shuch assessed the Arecibo Message of 1974 to have a transmission characteristic value of 3 (on a scale of 1 to 5), which corresponds to "a special signal in a preselected direction at a preselected time in order to draw attention" (Almár and Shuch 2007, 58). Once its transmission strength was added, the overall resulting index value was 8, or "far-reaching" in terms of the potential hazard levels defined by Almár and Shuch.

Perhaps for efforts that are assessed to be above a certain level on the San Marino scale, international consultation would be warranted. The details of the particular kind and level of international consultation is obviously a complex matter, but to serve merely as a simple example as shown in Table 20.1 (a "Binary Scale"), if the potential hazard was assessed to be above a threshold level of, say, 5, it could be determined to require some degree of international consultation or perhaps even approval or agreement. Below level 5 on the San Marino Scale, there might be a "disclaimer guideline," or perhaps a regulation, that might call for (1) the message origin and process to be clearly communicated, specifically noting (2) whom the message is from, (3) what kind of consultation was involved, and (4) that the message does not necessarily "speak for Earth."

International consultation and agreements would be complex of course, and one approach could be to use the United Nations Committee on the Peaceful Uses of Outer Space (COPUOS) for consultation and perhaps consensus-level agreement as appropriate. Like the UN more generally, COPUOS tends to operate on a consensus basis, which can make the process difficult and even prohibitive, so other mechanisms should be considered. But given the potential significance of communicating with ETI, working for COPUOS consensus (or a similar mechanism) would seem to be prudent. Further rigor would be obtained by going through the UN Security Council or General Assembly (the latter being more representative of all humanity).

Another approach could be to pursue international consultation that is more commensurate with, or proportional to the potential hazard—as examples Table 20.2 and Table 20.3 indicate. This approach arguably rests on

Table 20.1. Example of "Binary Scale" for Proportional Consultation

Potential Hazard	Consultation Level
10. Extraordinary	International Consultation
9. Outstanding	International Consultation
8. Far-reaching	International Consultation
7. High	International Consultation
6. Noteworthy	International Consultation
5. Intermediate	Disclaimer Guidance/Regulation
4. Moderate	Disclaimer Guidance/Regulation
3. Minor	Disclaimer Guidance/Regulation
2. Low	Disclaimer Guidance/Regulation
1. Insignificant	Disclaimer Guidance/Regulation

the assumption that there is an even higher level confidence in the potential hazard assessment (higher confidence than for the "binary" approach) to justify the greater precision of additional proportional consultation levels. Indeed, the San Marino Scale might benefit from an associated confidence-level assessment. This could be done by allowing for ranges of the terms of the Index and perhaps expanding the signal characteristic scale to be 1–10 to better allow for a range for that term to be captured. A confidence level could also be separately assessed through another, perhaps "independent" evaluation process. Regardless, if the level of international consultation were proportionally aligned with the potential hazard threat assessment, confidence in the potential hazard threat assessment would presumably have to be relatively high. This is obviously a significant challenge and may not be possible. Having sufficiently precise confidence in assessments of potential hazard threat levels 1–10 regarding communicating with ETI would presumably require a level of information we won't likely have in an Active SETI approach. If the communication attempt were in response to a received and understood signal, then presumably our confidence in our threat assessments could go up given the assumed additional information we would have from that initial communication from ETI.

But if the potential hazard assessment confidence level could be high enough, we might then be able to more precisely assess the level of international consultation that would be appropriate prior to a communication attempt. For example, as shown in Table 20.2, Levels 4–7 ("moderate" to "high" potential hazard) might call for international consultation but not necessarily approval or agreement, and Levels 8–10 ("far-reaching" to "extraordinary") might require international agreement (perhaps via majority or consensus). An additional distinction or level of rigor could result from considering agreement through

Table 20.2. Example of Three-Level Proportional Consultation Scale

Potential Hazard	Consultation Level
10. Extraordinary	International Approval/Agreement
9. Outstanding	International Approval/Agreement
8. Far-reaching	International Approval/Agreement
7. High	International Consultation
6. Noteworthy	International Consultation
5. Intermediate	International Consultation
4. Moderate	International Consultation
3. Minor	Disclaimer Guidance/Regulation
2. Low	Disclaimer Guidance/Regulation
1. Insignificant	Disclaimer Guidance/Regulation

COPUOS versus the UN General Assembly versus some other body—as indicated in Table 20.3, which attempts to provide a "multilevel" example.

Such efforts at international consultation would almost certainly be difficult and may inappropriately result in curtailing communication attempts with ETI for no good reason. While international consultation and consensus attempts are justified, it is also conceivable that after some duration of "failed" consultation, communication attempts should proceed. Again, the kinds and amount of consultation that constitute that threshold is a complex matter, but the complete prevention of transmission attempts, especially for a long period of time, may be unwarranted. Indeed, a bootstrapped cosmocultural philosophy suggests not only that it is important to seek contact with other rational beings, but also that we should consider implications that encompass all rational/cultural beings, including potential benefits to ETI that may result from our initiatives. Douglas Vakoch writes: "[W]e could potentially make significant contributions to long-lived extraterrestrial civilizations even if we are unable to sustain a transmission project for decades or centuries. While we may be too young to start and finish an Active SETI experiment *as the scientists*, we might still participate in an extraterrestrial's Passive SETI experiment *as the subjects*" (Vakoch 2011b, 483). While this may sound like a b-grade science fiction movie, it is nevertheless a plausible and worthy consideration. Our transmissions would tell ETI what we ourselves are very interested to know—that someone else is out there. And more specifically, as Vakoch notes, it would tell ETI something about how long civilizations survive—a question we are also interested in. Isn't this the kind of consideration we might like from ETI?

Indeed, humanity should attempt to transmit to ETI—and this chapter does not necessarily advocate for requiring international agreement before

Table 20.3. Example of Multilevel Proportional Consultation Scale

Potential Hazard	Consultation Level
10. Extraordinary	International—e.g., UN General Assembly
9. Outstanding	International—e.g., UN General Assembly
8. Far-reaching	International—e.g., UN Security Council
7. High	International—e.g., UN Security Council
6. Noteworthy	International—e.g., COPUOS Majority/Consensus?
5. Intermediate	International—e.g., COPUOS Majority/Consensus?
4. Moderate	Disclaimer Regulation
3. Minor	Disclaimer Regulation
2. Low	Disclaimer Guideline
1. Insignificant	Disclaimer Guideline

doing so. Such agreement could be a healthy pursuit, but it may not be possible or even necessary in the end. I argue instead for at least attempting proportional consultation for significant communication efforts (which may or may not result in agreement), allowing for the possibility that while the motivations and outcomes of such consultation could ultimately be misguided, it should be attempted nonetheless as we do with so many other matters of social importance. Such consultation may be complex and burdensome, as it often is, but that does not obviate its need or usefulness. Indeed, consultation can contribute to truthful messages that reflect human diversity (Vakoch 2011a). Not only is the attempt to communicate with ETI a diplomatic act of sorts (Michaud 1995; Ekers et al. 2002)—it is also a pursuit that calls for diplomacy among humans. It is not simply a matter of what kind of threat ETI might pose, or our readiness as a species to cope with such a step; it is a matter of basic obligation to consult with our fellow human beings when engaging in an endeavor of such unprecedented and uncertain global consequence.

6.0. Summary

This chapter briefly covers a number of pragmatic and cosmocentric worldviews and suggests the potential applicability of a "bootstrapped cosmocultural evolutionary" view that has aspects of both pragmatism and cosmocentrism. It is a view that suggests that rational cultural beings and the universe co-evolve and will increasingly co-evolve into the future, with cultural evolution playing an important role in the overall evolution of the universe—making rational cultural beings and the universe "co-priorities" in the overall evolution of the cosmos.

The pragmatic aspects of this philosophical view provide guidance suggesting obligations to our fellow human beings (and rational beings more broadly) to engage in "proportional consultation" before making significant attempts to communicate with ETI. The cosmocentric aspect of bootstrapped cosmocultural evolution suggests that the potential cosmic significance of cultural evolution makes actively seeking out and communicating with other cultural beings a worthy pursuit. It suggests we may be part of a grand cosmic story, one that may be increasingly of our own making, and one that we may share and co-create with other cultural beings that may exist throughout the cosmos.

Acknowledgments

Many thanks to Michael Michaud, Iván Almár, Paul Shuch, and Doug Vakoch for their feedback on this chapter.

Works Cited

Almár, I., and H. P. Shuch. 2007. The San Marino Scale: a new analytical tool for assessing transmission risk. *Acta Astronautica* 60: 57–59.

Deacon, T. 1997. *The symbolic species: The co-evolution of language and the brain.* New York: W. W. Norton.

Davies, P. 2007. *Cosmic jackpot: Why our universe is just right for life.* New York: Houghton Mifflin.

Dick, S. J. 2000. Cosmotheology: Theological implications of the new universe. In *Many worlds: The new universe, extraterrestrial life, and the theological implications*, ed. Steven J. Dick, 191–210. Philadelphia: Templeton Foundation Press.

———, and M. L. Lupisella, eds. 2009. *Cosmos and culture: Cultural evolution in a cosmic context.* NASA SP-2009-4802.

Ekers, R. D., D. K. Cullers, J. Billingham, and L. K. Scheffer, eds. 2002. *SETI 2020: A Roadmap for the Search for Extraterrestrial Intelligence.* Mountain View, CA: SETI Press.

Harrison, P. 1999. *The elements of pantheism: Understanding the divinity of nature and the universe.* London: Element Books. Later via self-publishing site of Taramac FL: Llumina Press.

Lupisella, M. L. 2009a. The search for extraterrestrial life: Epistemology, ethics, and worldviews. In *Exploring the origin, extent, and future of life: Philosophical, ethical and theological perspectives*, ed. Connie Bertka, 186–204. Cambridge: Cambridge University Press. Based on American Association for the Advancement of Science workshops.

———. 2009b. Cosmocultural evolution: The coevolution of culture and cosmos and the creation of cosmic value. In *Cosmos and culture: Cultural evolution in a cosmic context*, ed. Steven J. Dick and Mark L. Lupisella, 321–59. NASA SP-2009-4802.

———, and J. Logsdon. 1997. Do we weed a cosmocentric ethic? Paper IAA-97-IAA.9.2.09, International Astronautical Congress. American Institute of Aeronautics and Astronautics, Turin, Italy.

Levine, M. 1994. *Pantheism: A non-theistic concept of deity.* London: Routledge.

Lloyd, S. 2006. *Programming the universe: A quantum computer scientist takes on the cosmos.* New York: Random House.

Michaud, M. A. 1995. SETI and diplomacy. In *Progress in the search for extraterrestrial life*, ed. Seth Shostak, 551–54. San Francisco: Astronomical Society of the Pacific Conference Series, Volume 74.

———. 2007. *Contact with alien civilizations: Our hopes and fears about encountering extraterrestrials.* New York: Springer.

Manson, N. A., ed. 2003. *God and design: The teleological argument and modern science.* New York: Routledge.

Rees, M. 1997. *Before the beginning: Our universe and others.* New York: Perseus Books.

Shuch, H. P., and I. Almár. 2007. Quantifying past transmissions using the San Marino Scale. Paper IAC-07-A4.2.04, International Astronautical Congress, Hyderabad, India.

Smith, K. C. 2009. The trouble with intrinsic value: A Primer for astrobiology. In *Exploring the origin, extent, and future of life: Philosophical, ethical and theological perspectives*, ed. Connie Bertka, 261–80. Cambridge: Cambridge University Press.

Smolin, L. 1997. *The life of the cosmos.* New York: Oxford University Press.

Teilhard De Chardin, P. 1955. *The phenomenon of man.* Trans. Bernard Wall. New York: Harper and Row, 1959. Originally published as *Le Phenomene Humain.* Paris: Editions du Seuil.

Tipler, F. 1994. *The physics of immortality.* New York: Doubleday.

Traphagan, J. W. 2011. Culture, meaning, and interstellar message construction. In *Communication with extraterrestrial intelligence (CETI)*, ed. Douglas A. Vakoch. Albany: State University of New York Press.

Vakoch, D. A. 2009. Encoding our origins: Communicating the evolutionary epic in interstellar messages. In *Cosmos and culture: Cultural evolution in a cosmic context*, ed. Steven J. Dick and Mark L. Lupisella, 415–39. NASA SP-2009-4802.

———. 2011a. Responsibility, capability, and Active SETI: Policy, law, ethics, and communication with extraterrestrial intelligence. *Acta Astronautica* 68 (3–4): 512–19.

———. 2011b. Asymmetry in Active SETI: A case for transmissions from Earth. *Acta Astronautica* 68 (3–4): 476–88.

SETI and International Radio Law

Francis Lyall

1.0. Introduction

SETI researchers and students need to be more active in promoting and publicizing their interests in the modern world. Unless they stand up for their legitimate interests in the use of the radio spectrum they stand in acute danger of being overlooked or ignored amid the demand of the commercial and military communities for their uses of the radio spectrum. Wake up, and be active.

There are two ways in which SETI may employ the radio spectrum, the first being passive (listening) use and the second what is known as Active SETI. Of these the passive use of the radio spectrum is the more common. I will therefore in section 2 note its problems and only turn to the international rules and procedures as to radio in section 3 because these make sense only in relation to the problems they are designed to solve. Then in the light of these rules I comment on Active SETI and radio in section 4. Section 5 sketches the responsibilities of states nationally to implement the relevant international agreements. Section 6 contains comment on other related SETI matters.

2.0. Passive Use of the Radio Spectrum

While some SETI research is carried out optically (Ekers, Culler, Billingham, and Scheffer 2002; Schwartz and Townes 1961) the scrutiny of extrasolar radio frequencies for evidence of an artificial signal remains the main method by which the search is carried out. The radio frequencies involved are extremely weak. Some of the "best" frequencies are also of interest to active users of the spectrum, particularly for newly emergent radio services. That extrasolar radiations might be masked by signals generated locally by our own ever-increasing use of radio is therefore a problem for the SETI community. This is an unusual aspect of interference between radio signals. Technically, the control of the use of radio remains an aspect of the sovereign right of a state to regulate what is done within its boundaries. However, radio knows no state boundaries and hence international rules are required. Radio interference figured in the discussions of the very first international conference on radio (Berlin, 1903, see Appendix I), and rules as to its avoidance are in the subsequent treaty of 1906 (Berlin, 1906, see Appendix I). The avoidance or at least the minimization of harmful interference remains highly important, the International Telecommunication Union being now the vehicle through which that is pursued internationally (Kahlmann 1992; Lyall and Larsen 2009).

3.0. The International Telecommunication Union (ITU)

3.1. The International Organization

As of May 2010, 191 states were members of the International Telecommunication Union (ITU), one less than the United Nations itself. Despite its modern importance the ITU is the second-oldest international organization still functioning (the oldest is the Rhine Commission, which dates to 1831). The roots of the ITU go back to 1865 when the International Telegraph Union was created to cope with international telecommunications by wire. Originally separate arrangements were made for radio (Berlin, 1906, see Appendix I), the wired and wireless services coming together as the ITU in 1932 (Madrid, 1932, see Appendix I). After World War II the ITU was revised (Atlantic City, 1947, see Appendix I), arrangements that in essence persisted until major structural alterations were adopted at Geneva in 1992 (Geneva, 1992, see Appendix I), which split the basic documents of the Union into a Constitution and a Convention. The ongoing work of the Union was entrusted to three new "sectors," the Radiocommunication Sector (ITU-R), the Standardization Sector (ITU-T) and the Development Sector (ITU-D).

The ITU Constitution and Convention have both been subsequently modified in detail but not in substance by quadrennial plenipotentiary conferences and the ITU has regularly published a "clean text" of the result (Collection, see Appendix I).

3.2. The Radiocommunication Sector (ITU-R)

The Geneva reforms entrusted the bulk of radio matters to a Radiocommunication Sector (ITU-R). Although some standardization matters are dealt with by the Standardization Sector, ITU-R is the major authority through which radio SETI can be helped. Article 45 of the ITU Constitution requires ITU members to avoid causing harmful interference to the telecommunication services of other members. This, of course, does not prohibit causing interference to a SETI search since that is not a telecommunication service. However, the regulations and procedures administered by ITU-R provide some, if limited, protection through the Radio Regulations (RR).

3.3. The Radio Regulations (RR)

The Radio Regulations are adopted and modified by world and regional administrative radio conferences (RR, see Appendix I). The RR are not merely the expression of hope; they have the same international treaty status as the ITU Constitution and Convention. To the extent they shelter radioastronomy they serve SETI activities.

RR Art. 29 makes special provision for the radioastronomy service. There are two elements. First, states are required to select appropriate sites for the location of relevant antennae so as to avoid incoming interference, and to ensure that the stations are constructed so as to reduce their susceptibility to interference. Second, states are to take account of the need for the protection of radioastronomy in their licensing of transmitters (whether fixed or mobile) including as to the elimination of spurious transmissions, site shielding, and periods of radio silence. Radioastronomy stations are notified to the ITU Radiocommunication Bureau and their data is published to ITU member states. The Radiocommunication Bureau also publishes recommendations as to how states can reduce interference with radioastronomical observations. SETI can benefit from such provisions.

The other way in which the RR can protect radioastronomy and SETI is through rules as to the use of radio frequencies. The RR lay down the procedures through which frequency assignments are recognized internationally through the Master International Frequency Register (see 3.4). Article 5 of the RR comprises the Table of Allocations (Lyall and Larsen 2009). In it the whole of the currently usable radio spectrum is divided into nine spectrum bands and their use is allocated on a primary or secondary basis for use for particular services on a worldwide or a regional basis. On occasion an exception may be made for a state or group of states in the use of a particular waveband. This is incorporated in a footnote, which shares the treaty status of the rest of the text.

Most of the Table of Allocations deals with the active use of the radio spectrum for broad- and narrowcasting. This also includes higher-frequency uses such as for medical imaging, microwave cookers, etc. However, the Table also recognizes that a passive use may be made of the radio spectrum. Particular frequency bands are noted as used by radioastronomy and as indicated above states should notify the establishment of radioastronomy sites and the frequencies to be employed there (Spoelstra 1997; Stull and

Alexander 1977). No transmissions in these designated bands should take place if there is the possibility of interference with that scientific inquiry. SETI is a similar passive use of the radio spectrum but it does not itself presently receive any protection. Mention is made here and there in relation to particular bands that "passive research is being conducted by some countries in a programme for the search for intentional emissions of extra-terrestrial origin." However, portions of these bands are also allocated to active spectrum use for a variety of purposes and that means that SETI is not protected. In various bands fixed and mobile services, space operations, maritime and aeronautical mobile satellite, meteorological satellite, radio-location and other services may be clamoring for active use of spectrum areas that SETI would wish to be free of terrestrial signals.

The RR will be reconsidered at a World Radio Conference (WRC) to be held in Geneva in January, 2012. As can be seen from the ITU Web site, preparatory discussions are to be held in various regions to discuss, and, where possible, to come to a common mind on particular matters. In addition national telecommunications administrations are also preparing for the WRC (FCC 1, see Appendix I). Those with radioastronomy and SETI interests should make their concerns known in these national and international fora so that proper recognition is given to their requirements. This should include the protection of particular frequencies from local interference and, where necessary, the moving of national frequency assignments where the existing ones are not quite SETI satisfactory. Make no mistake about it, the pressure from commercial and other interests for spectrum is growing. A spirited defense of the passive use of the radio spectrum will be required.

And, whatever the outcome of the 2012 Geneva World Radio Conference, SETI people should remain alert to developments within the international use of the radio spectrum.

3.4. The Master International Frequency Register (MIFR)

The other element in the international regulation of the use of radio is the Master International Frequency Register (MIFR). Maintained by the Bureau of ITU-R this is a listing of the assignments by states of the grant of a right to a national entity to use a frequency (and in the case of satellites an orbit), that have been notified to ITU-R. Operated under procedures laid down in RR Chapter III (Arts. 7–14) an incoming notification is logged, collated with the RR Table of Allocations and checked to see it will not cause harmful interference to a previously notified assignment. If "harmful interference" might so be caused the notifier is required to coordinate with the prior notifier to resolve the matter. In the ultimate, the notifier can insist on activating the use of the assignment that it has notified, but the laws of physics, unalterable by law or agreement, usually compel compromise. All notifications are circulated regularly to ITU members. However, since both radioastronomy and radio SETI are passive uses of the radio spectrum they can be effectively invisible to the procedures of the MIFR, albeit that the publication of a notification could alert a national administration of an incipient problem for national SETI efforts.

4.0. Active SETI

Some forms of Active SETI require the transmission of radio signals. Irrespective of the merits and demerits of such efforts, that use brings the RR into play. No radio frequencies have been allocated for use by Active SETI in the RR Art. 5 "Table of Allocations." That does not mean that it is unlawful under international law to send a SETI radio message, although a national law may prohibit such. However, the general obligation laid on states to avoid causing "harmful interference" to other radio users could come into play. An Active SETI signal is likely to be powerful and will punch through the orbital location of at least some of the many satellites that now exist. A transmitted signal that affects a satellite service either directly by use of an assigned, registered, and operational satellite frequency, or indirectly through generating spurious emissions, distorting polarization, or in any other way, would be unlawful. Were a satellite to be affected, disabled, or killed by an Active SETI signal it is my view that a civil or international claim might be brought against either the generator of the signal or the relevant licensing state or both. An unlicensed transmission would, presumably, be unlawful and potentially criminal under national laws.

5.0. National Action

As stated above, states have the right to control the use (including the nonuse) of the radio spectrum by stations within their boundaries. While the international obligations and procedures outlined above impact on SETI, they do so largely only when implemented through national action. In the United States this is largely a matter for the Federal Communications Commission (Stull and Alexander 1977 [note: some FCC procedures have altered since 1977: cf. Spencer 2009]; Spoelstra 1997; see also FCC 2, Appendix I). In the UK it is the responsibility of the Office of Communications although by the time this is published and subsequent to the UK General Election of 2010 the UK arrangements may have been altered (Ofcom, see Appendix I). Other states have other constitutional arrangements.

5.1. Assignment

In the assignment of the right to use a radio frequency states should have regard to the interests of passive users of the radio spectrum, within the parameters of the ITU RR, though departing from these when desirable. It is up to the radio astronomers and SETI users to make their requirements plain to the relevant licensing authorities. For Active SETI, appropriate frequencies must be selected to the satisfaction of the licensing authorities so that no "harmful interference" is caused to other radio users including those in other countries. The main users likely to be affected are the satellite services, and their operators may oppose the grant of a licence.

5.2. Licensing of Sites

In the licensing of relevant transmitters and radioastronomy and SETI-apt receivers, states should have regard to the requirements of the latter. Again, the radio astronomers and SETI users need to make their requirements plain to the relevant licensing authorities.

6.0. Other Matters

The various SETI communities should consider how raise the profile of the importance of what they do both nationally and internationally with regard to the use and nonuse of radio frequencies. The opportunity given by the 2012 Word Radio Conference is indicated above (Sec. 3.3, *ad fin*), but the effort should continue; otherwise, SETI will be overlooked.

I am aware of the SETI Detection Protocol (SETI 1, see Appendix I) and the draft Reply Protocol (SETI 2, see Appendix I). These also relate, albeit more indirectly, to the use of radio. Paragraph 7 of the Detection Protocol calls for radio frequencies surrounding a possible detected signal to be cleared so it may be further studied. The Reply Protocol involves questions as to the content of a message and, naturally, the frequencies to be used for its transmission will have to be chosen. National administrations should be pressed to consider whether, and if so how, to give legal effect internationally and within their jurisdictions to the Detection Protocol. Work should continue on the Reply Protocol.

Appendix I

Key Telecommunications Conventions and Regulations

Atlantic City, 1947: International Convention on Telecommunications, Atlantic City, 2 October 1947; 193 UNTS 188, 194 UNTS 3; 1950 UKTS 76, Cmd. 8124; (US) 63 Stat. 1399, TIAS 1901.

Berlin, 1903: Preliminary Conference at Berlin on Wireless Telegraphy, Procés Verbaux and Protocole Finale, 194 CTS 46; (UK) (1903) Cd. 1832,. Final Protocol only: http://earlyradiohistory.us/1903conv.htm.

Berlin, 1906: Radio-telegraphic Convention, Final Protocol and Regulations, signed at Berlin, 3 November 1906, 1909 UKTS 8, Cd. 4559; http://earlyradiohistory.us/1906conv.htm. The Convention and Final Protocol but not the Service Regulations are at 203 CTS 101; (US) 37 Stat. 15665, TS 568; (1906) 3 AJIL Supp. 330-40.

Collection: *Collection of the basic texts of the International Telecommunication Union adopted by the Plenipotentiary Conference*, 3d ed. 2007, (Geneva: ITU, 2007). A new edition will be proceeded to take account of the Final Acts of the ITU plenipotentiary conference to be held in Guadalajara, Mexico, in the autumn of 2010.

FCC 1: See http://www.fcc.gov/ib/wrc-12/.

FCC 2: See *In the Matter of the 4.9 GHz Band Transferred from Federal Government Use*; 2002 17 FCC Rcd 3955; 26 Comm. Reg. (P & F) 50, (various sites inc. Goldstone and the new Allen Telescope Array); *In the Matter of The 4.9 GHz Band Transferred from Federal Government Use*, 2003 18 FCC Rcd 9152; *In the Matter of Amendment of the Commission's Rules to Establish a Radio Astronomy Coordination Zone in Puerto Rico*, 11 FCC Rcd 1716; (Arecibo); *In the Matter of Amendment of the General Mobile Radio Service (Part 95) and Amateur Radio Service (Part 97) Rules to establish procedures to minimize potential interference to Radio Astronomy Operations*. 1981 85 FCC 2d 738.

Geneva, 1992: Constitution and Convention of the International Telecommunication Union: Final Acts of the Additional Plenipotentiary Conference, Geneva, 22 December 1992 (Geneva: ITU, 1993). Constitution, 1825 UNTS 331; Convention, 1825 UNTS 390; 1996 UKTS 24, Cm. 3145; US Tr. Doc. 103-35.

Madrid, 1932: Telecommunication Convention, General Radio Regulations, Additional Radio Regulations, Final Protocol to the General Radio Regulations, an Additional Protocol (European) to the Radio Regulations, Telegraph Regulations and Telephone Regulations, Madrid, 9 December 1932; 151 LNTS 4; (US) 49 Stat. 2393, TS 867; 6 Hudson 109.

Ofcom: The Wireless Telegraphy (Automotive Short Range Radar) (Exemption) (No. 2) Regulations 2005 (2005 SI 1585) as amended, restrict potentially interferent radio usage in designated areas round the Jodrell Bank, Cambridge, Darnhall, Pickmere, and Knockin radio telescopes.

RR: The Radio Regulations (Geneva: ITU). The Final Acts of the 2007 World Radio Conference are separately available, but a more useful source is the consolidated *Radio Regulations* of 2008 which is a "clean text" of the current rules.

SETI 1: Declaration of Principles Concerning Activities Following the Detection of Extraterrestrial Intelligence. 1990. *Acta Astronautica* 21:153–54; http://iaaseti.org/.

SETI 2: SETI Reply Protocol: http://www.setileague.org/general/reply.htm.

Works Cited

Ekers, R. D., D. K. Cullers, J. Billingham, and L. K. Scheffer. 2002. *SETI 2020: A roadmap for the Search for Extraterrestrial Intelligence.* Mountain View, CA: SETI Press.

Kahlmann, H. C. 1992. SETI and the radio spectrum. 1992. *Acta Astronautica* 26: 213–17.

Lyall, F., and P. B. Larsen. 2009. *Space law: A Treatise.* Farnham, Surrey, UK: Ashgate.

Schwartz, R. N., and C. H. Townes. 1961. Interstellar and interplanetary communications by optical masers. *Nature* 190: 205–208.

Spencer, R. L. Jr. 2009. State supervision of space activity. *Air Force Law Review* 63: 75–127.

Spoelstra, T. A. Th. 1997. Radio astronomy in telecommunication land: The ITU and radio astronomy. *Air and Space Law* 22: 326–33.

Stull, M. A., and G. Alexander. 1977. Passive use of the radio spectrum for scientific purposes and the frequency allocation process. *Journal of Air Law and Commerce* 43: 459–534.

What the World Needs Now

Identifying the Relative Degree of Specific Maslovian Needs and Degree of Species-Level Self-Identification in Interstellar Messages Submitted by a Multinational Sample

Timothy A. Lower, Douglas A. Vakoch, Yvonne Clearwater, Britton A. Niles, John. E. Scanlin

1.0. Introduction

The SETI Institute's *Earth Speaks* project (http://earthspeaks.seti.org) was initiated to encourage global discussion about the content of interstellar messages that may some day be transmitted to other civilizations (Vakoch 2011). As we have discussed elsewhere (Vakoch et al. 2010), we can gain significant insights into what people would want to communicate to an extraterrestrial intelligence through this Web-based project. Past interstellar messages have often relied on input from small groups of experts (e.g., Sagan, Salzman Sagan, and Drake 1972; Staff at the National Astronomy and Ionosphere Center 1975). In this chapter, we continue to explore the most prevalent themes of messages submitted to *Earth Speaks* as a step to make more inclusive ongoing discussions about content of actual interstellar messages that may some day be transmitted from Earth. In addition, we examine a few of the potential social scientific and global health benefits of including the general populace into the dialogue about interstellar message construction. These benefits include the increasing ability to identify, describe, and assess species-level self-identification in humans, as well as to assess the global climate of human need as expressed through Abraham Maslow's theory of motivational needs.

Previously we detailed the themes arising from initial analyses of the messages people submitted to the *Earth Speaks* Web site (Vakoch et al. 2010). As we reported in this earlier work, the results we found were indicative of the presence of very strong themes in the current messages. These themes, however, were somewhat surprising. Though many types of messages exist in the data, the main themes the participants spoke to expressed rather exigent and practical ideas. The dominant themes were:

We are humans of the planet Earth.
You are alien to us, but you have know-how.
Hello and welcome.
Please help.
Peace, love, and friendship.
Mathematics and binary expressions.
We feel alone and we are fearful, primarily of our own propensity for violence.
Our gods and religions are influential in our lives.
We recognize our cultural heritages and the civilizations they produce.

We found that many people currently submitting messages to the *Earth Speaks* Web site were asking for help (Vakoch et al. 2010). Instead of submitting lists of our historic achievements and goals, or attempting to sensibly encode our biological structure, or providing instructions on navigating our solar system, many people are choosing to ask for help. They are doing so frankly, and are holding out hope that those who hear them will respond kindly. In this current project we demonstrate how messages from the *Earth Speaks* Web site can reveal the state of human needs across the globe, and suggest an established psychological framework for conceptualizing these needs.

2.0. Participants

At the time of data analysis, 995 submissions were received from participants. We excluded submissions consisting exclusively of photographs and sounds. We also eliminated repetitive, nonelaborative submissions. We eliminated submissions without a reported age, or those that reported "0" or "99" as their age. We also eliminated submissions from those reporting no gender, or reporting "other" for their gender. This left 699 submissions, all of which were included in this analysis.

The participants ranged in age from eight to seventy-five years old. The mean age of participants was thirty-five and a half years old, while the median age was thirty-four years old. The modal age of the participants was twenty-one years old.

Of the messages, 556 (79.5%) were submitted by males and 143 (20.5%) were submitted by females. Submissions came from participants in sixty-eight different nations. The top ten nations are listed in Table 22.1.

3.0. The Instrument and Its Function

The *Earth Speaks* Web site (http://earthspeaks.seti.org/; see Figure 22.1) is a free site that allows users to sign up and submit interstellar messages that they would like to see sent to extraterrestrial intelligent species. The site confidentially records participants' sex, age, and geographic location. Participants can also supply their own "tags" or labels that summarize the content of each message. The format of the site allows participants a great amount of variability in the topics they write about, and the manner in which they write them. Participants are largely free to write whatever comes to mind when they think about the message they would want to send to an intelligent extraterrestrial species.

We believe that because of the way the messages are submitted to the *Earth Speaks* Web site, they are amenable to textual analysis similar to other projective psychological assessments. Thorndike and Hagen (1969) assert that a projective psychological assessment is based on the hypothesis that people insert meaning in unique (or somewhat unique) ways when presented with stimuli. They write that projective techniques generally involve three characteristics: (1) the participant is provided with "a series of fluid, weakly structured stimuli," (2) instructions provided to the participant "emphasize freedom of response," and (3) responses are analyzed "for insight into [the client's] basic personality dynamics" (Thorndike and Hagen 1969, 495). It

Table 22.1. The Top Ten Nations from Which Messages Were Received.

Nation	Submissions
United States	332
Great Britain	51
Netherlands	32
Australia	27
Canada	21
France	20
Mexico	17
Indonesia	16
Turkey	16
India	13

Figure 22.1. A sample of messages from the *Earth Speaks* Web site.

would seem that the *Earth Speaks* Web site complies with these characteristics and is supportive of the projective hypothesis.

As Vakoch et al. (2010) demonstrated, the theme *We are humans of the planet Earth* was the largest theme found in the messages submitted to the *Earth Speaks* Web site at the time of analysis. The seventeen word-concepts representing this theme comprised approximately 9.1 percent of the total words in the more than twenty-five thousand word database available for us to analyze. On average these seventeen words were represented thirty-two times more frequently than their baseline frequency in common written British English.

Given the prevalence of this species-level form of group identification, we explored the extent to which messages reflect group identity that is either: (1) global and at the level of the species (identifying with being "human"), or (2) group-specific (e.g., belonging to a particular nationality, ethnicity, political party, social class, or religion). We then compared the frequency of references to being human to the frequency of these other group references, providing a measure of degree of affiliation at the species level, versus other forms of affiliation and identification. The difference in frequency between words reflecting nationality or other group membership was rather large. At most, other forms of within-species group identity were represented at a small fraction of a percent of the total volume of word-concepts present (usually in the small hundredths of a percent, indicating it was mentioned less than five times out of the twenty-five thousand word database created from the 699 submissions). When the messages are used as a projective psychological assessment, the *Earth Speaks* Web site appears to serve primarily as a projective assessment of species-level self-identification.

Furthermore, the relationship between the word-concepts that constitute the participants' self-identification as members of the human species is observable in the messages. We can map the cognitive network of the most prevalent and overrepresented word-concepts used to refer to the theme, *We are humans of the planet Earth*.

To determine both the relative frequency of themes in the messages and the cognitive map of these themes, we used the British National Corpus. The British National Corpus is a consortium project led by Oxford University. It contains one hundred million words drawn from written (90%) and spoken (10%) British English. Sources are designed to be representative, and include a full spectrum of writing samples, such as memoranda, letters, published books (fiction and nonfiction), newspaper articles, magazines, and academic essays by students. We used Leech, Rayson, and Wilson's (2001a) frequency table of words from the British National Corpus, drawing upon the freely available frequency table from this work as published on the Internet (Leech, Rayson, and Wilson 2001b). This table reports raw frequency per one million written words. For our analysis, we used the column within the published table that lists the written word frequency. We then converted this word frequency into a proportion.

We compared the rank order and relative proportion (ES/BNC) of the most frequently found word-concepts constituting the theme *We are humans of the planet Earth*. The ES/BNC score represents the proportional overrepresentation of the word-concept in the *Earth Speaks* (ES) submissions, as compared to the base proportion of that word in the British National Corpus (BNC). For instance, the word-concepts *We* and *Earth* in the examples presented in Table 22.2 were proportionally represented in the *Earth Speaks* messages at a rate that is more than eleven times and seventy-three times respectively more frequent than the base frequency for these word-concepts as found in the British National Corpus. Though some word-concepts, such as the words *and* or *a*, are frequent in the *Earth Speaks* data, their relative proportions of .93 and .80 respectively indicate that they are found at about the same frequency as they are in the British National Corpus. Because their frequency in the *Earth Speaks* messages is not particularly different from their frequency of usage in everyday British English, they and words like them fall into the background noise for this projective assessment and are not evaluated.

Table 22.2.

Earth Speaks Rank (ES)	Word	ES Freq.	ES Proportion	Relative Proportion (ES/BNC)	British National Corpus Rank (BNC)	Word	BNC Freq.	BNC Proportion
1	we	673	0.0316929	11.383965	1	the	64420	0.06442
2	to	616	0.0290087	1.7643055	2	of	31109	0.031109
3	you	612	0.0288203	6.0610607	3	and	27002	0.027002
5	and	537	0.0252884	0.9365394	5	in	18978	0.018978
7	a	377	0.0177537	0.8080151	7	is	9961	0.009961
8	are	359	0.0169060	3.5871104	8	to	9620	0.00962
10	our	273	0.0128561	14.065791	10	it	9298	0.009298
11	us	270	0.0127148	22.189978	11	for	8664	0.008664
19	earth	160	0.0075347	73.869905	19	s	4945	0.004945
27	please	125	0.0058865	55.533095	27	have	4416	0.004416
39	know	86	0.0040499	5.5175988	39	been	2756	0.002756
45	help	77	0.0036260	13.138003	45	there	2354	0.002354
47	but	74	0.0034848	151.51360	47	all	2297	0.002297
48	peace	74	0.0034848	36.682240	48	can	2211	0.002211
50	welcome	70	0.0032964	117.73016	50	who	2086	0.002086
52	how	66	0.0031080	3.3968046	52	do	2016	0.002016
53	love	65	0.0030609	20.138054	53	what	1936	0.001936

Examples drawn from the rank ordering and comparison of word concepts in the Earth Speaks messages and British National Corpus.

Therefore, by looking at exponential differences in the proportional representation of a word in the *Earth Speaks* data, as compared to the British National Corpus, we can differentiate words that are common because they are important responses to the stimulus being presented (i.e., the *Earth Speaks* Web site) from those words or themes that are simply used frequently in our everyday communication with each other. In essence, such a procedure allows us to identify the verbal signals among the background noise.

A cognitive semantic network of word-concepts was created for the most frequent themes using the *Mathematica* software program's synonym network function. This software program includes the British National Corpus in its WordData calculations, such as calculating a synonym network. We visually inspected the synonym networks that *Mathematica* returned from each word entered. We then overlaid the graphs and drew connections between each word that was both proportionally overrepresented in our data and present with the first-word association of other word-concepts in a given theme. This procedure provided qualitative construct validation for the derived themes, as well as assisted in elaborating on the internal relations between the word-concepts that together form a theme.

The resultant cognitive map of the word-concepts that form the theme *We are humans of the planet Earth* is illustrated in Figure 22.2.

All connections between word-concepts depicted in Figure 22.2 are first-order synonyms in the *Mathematica* synonym network. The unidirectional arrows indicate that the word from which the arrow originates serves to modify the word to which the arrow points (e.g., the word-concept *planet* when used by participants for this theme, serves to modify or elaborate on

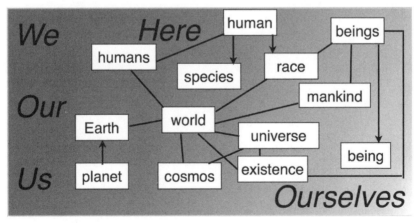

Figure 22.2. The cognitive map of the relations between word-concepts constituting the theme *We are humans of the planet Earth*.

the word-concept *Earth*). Bi-directional lines indicate that both connected words were found in each other's first-order synonym network. As depicted in Figure 22.2, the centering word-concept in this theme appeared to be the concept *world*. Our physical Earth and humans as an intelligent (but flawed) species are intimately tied through our word-concept *world*. The rank order, frequency, and relative proportion of the word-concepts constituting this theme are reported in Table 22.3.

4.0. Revealed Human Needs

Using a widely known organization and differentiation of human needs proposed by Abraham Maslow (1943), we examined the needs expressed in the message content. Maslow identifies five universal areas of motivational human needs and generally presents these in the shape of a triangle. The base motivational need area contains the physiological needs, which represent the basic needs required to sustain human life. Examples of physiological needs include breathing, food, sexual activity, air, water, defecation and elimination, and the overall maintenance of a homeostatic state of being.

Table 22.3. The Rank-Order, Frequency, and Relative Proportion of the Word-Concepts Constituting the Theme *We are humans of the planet Earth.*

Earth Speaks Rank	Word	Raw Frequency	Relative Proportion
1	we	673	11.38396542
10	our	273	14.06579184
11	us	270	22.18997827
19	earth	160	73.86990586
22	planet	144	0 in BNC
55	here	62	4.948657677
61	universe	55	92.50227051
64	species	53	23.54603249
75	world	44	3.294198505
78	humans	42	89.90303316
104	human	30	7.099306279
109	race	29	17.28696056
124	beings	24	0 in BNC
129	ourselves	23	25.18877895
132	being	22	35.72501482
149	we're	20	0 in BNC
266	mankind	10	0 in BNC

The second level, referred to by Maslow (1943) as our safety and security needs, represent the human desire to have control over one's personal self as well as one's situational and environmental experiences. Safety and security needs are met when humans procure personal safety, financial security, personal mental and physical well-being, stability, and structure. Specific examples of safety and security needs include (but are not limited to) such aspects of life as are often revealed through psychological attention to job security, education, and financial well-being, such as savings accounts and retirement plans.

Maslow (1943) refers to a third need area that he calls our love and belonging needs. These represent the human desire to experience a sense of belonging and acceptance from others. Examples of love and belonging needs include the following: friendship, emotional connection, sexual intimacy, family, participation in professional organizations, cultural identification, and participation in sports.

The fourth level of motivational needs consists of esteem needs, which according to Maslow (1943) represents the human need—either at the individual level or group level—to be respected by oneself and others. The perception of esteem or respect from others toward oneself often results in higher levels of self-value. Examples of esteem needs include (among others) assertiveness, emotional competence, intellectual competence, and self-awareness.

The fifth and final motivational need is referred to by Maslow (1943) as our self-actualization need area. This need area encompasses the human desire to become and maintain one's ideal self, consistently using all the potential one possesses. Some examples of self-actualization include acceptance, problem-solving mastery, strong ethics, and less desire for social interaction.

We conducted exploratory analyses to determine the most efficient method for identifying and elaborating on themes within the messages submitted through *Earth Speaks* and then placing them into their respective Maslovian need areas. At least two research members (one female and one male) were actively involved in all stages of data analysis. As with the method presented previously, we identified the most frequently cited themes in the messages by: (1) comparing the internal relative frequency and rank order of word-concepts found in the submitted messages, and (2) comparing the proportional representation of word-concepts among the messages to the baseline proportional representation of the same word-concepts in the British National Corpus.

Figure 22.3 illustrates the most frequent and overrepresented word-concepts that express an identifiable need and are present in the *Earth Speaks* messages. The numbers in parentheses next to the words in the triangle are the number of times that word-concept is overrepresented as compared to

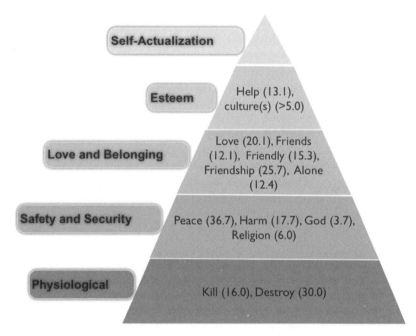

Figure 22.3. The most frequent and overrepresented word-concepts revealed in the Earth Speaks messages and their corresponding placement on Maslow's hierarchy of needs.

the word's baseline proportion in the British National Corpus. As can be seen from the figure, the esteem, love and belonging, and safety and security need areas are currently the most frequent and proportionally overrepresented need areas in the messages. Self-actualization needs are apparently rare. The needs reported in Figure 22.3 can be given greater context by the cognitive maps of the theme areas reported in Vakoch et al. (2010).

5.0. Discussion

The following conclusions appear warranted, based on the results of this first attempt to measure species-level identification and the global climate of need via the messages submitted on the SETI Institute's *Earth Speaks* Web site:

- The *Earth Speaks* Web site functions not only as a means of acquiring potential messages to be sent to the stars, but also as a projective assessment of species-level identification and expressed need. The *Earth Speaks* Web site appears to be one of the first viable instruments used to assess such identification and need.

- Scientific study regarding the construction of interstellar messages can facilitate greater knowledge about and ability to assess (and therefore respond to) the global climate of human need here on Earth.
- Our method, when applied to the messages, efficiently reveals not only the main themes in the messages, but also the cognitive map of the thoughts constituting these themes in our participants.
- The current messages, when taken as a whole, indicate a population attempting to cope with needs that range from physiological to esteem. What the world needs now—according to the current participants—is primarily a sense of esteem, love and belonging, safety and security, and a cessation of human-caused threats to its immediate existence.
- The world, as it is currently revealed through those submitting messages to the *Earth Speaks* Web site, feels somewhat helpless, but it maintains a hope that some intelligent life form will hear its message and will provide it with the knowledge necessary to overcome the contemporary global climate of human-caused fear, loneliness, and violence.

In this report we focused on a particular benefit of researching interstellar message construction as a component of a larger and active search for extraterrestrial intelligence. By contemplating what we would want to say to an intelligent, extraterrestrial species, we simultaneously express our own fundamental needs, our cognitive processes, and our common identity as human beings. The *Earth Speaks* Web site serves as one of the first viable projective psychological assessments of species-level human identification, and appears to be particularly revealing of contemporary human needs as they are expressed across the globe. As more submissions arrive, we will be able to assess these needs with greater accuracy and specificity.

When applied to the messages, the method reported in this chapter greatly enhances the efficiency and (we believe) objectivity of textual analysis. It provides, as well, a cognitive map of the thoughts comprising the themes arising from the messages that were submitted. As a side benefit beyond the search for extraterrestrial intelligence, the ability to identify expressed needs within the context of the cognitive map of the thoughts that surround and support those needs can help us to address those human needs with accuracy and in a comprehensive manner. By supplying us with the terms and ideas used most by the participants themselves to conceptualize their needs, it also facilitates communication with those in need.

Given increased participation over the coming years, we believe we will be increasingly able to assess clearly and rapidly a large portion of human need as it occurs across the globe. We will, furthermore, be able to identify variation or disparity between certain segments of our species. These should include at least gender, age, and geographic variables. However, given the projective psychological nature of the project, we should theoretically be able to distinguish disparity in needs according to other psychological phenomena, such as culture, personality, and intellectual or academic variables.

In this project we examined the range of human needs as well as forms of self-identity that were reflected in the text messages submitted by people from around the world through the SETI Institute's online project *Earth Speaks*. There is certainly much more work to be done in the area, and more participants are needed prior to generalizing the results with certainty to specific populations who are not represented in the data at this time. Future research will soon be able to draw conclusions to increasingly specific populations. Such research could therefore provide results that tell us what is needed, how that need is manifested, where that need occurs, in whom it is most prevalent, and what words to use to speak to those in need so that they understand that we empathize with them.

This project encouraged the people of the world to reflect on and report their thoughts about the active search for extraterrestrial intelligence. Currently, the *Earth Speaks* Web site contains submissions from locations in sixty-eight nations. The diversity and global reach of this study are encouraging. Nevertheless, as of right now, the typical message was submitted by a young adult male from the British Commonwealth or its former colonies. As one step to increase the cultural diversity of messages submitted, the SETI Institute has recently launched a Spanish version of the *Earth Speaks* Web site called *La Tierra Habla* (http://latierrahabla.seti.org) (Vakoch 2011).

Some have theorized that our cultural divides prevent the identification of human universals. Our initial empirical evidence would suggest otherwise, at least as it pertains to involving everyday people from around the world in the discussion of interstellar message construction. Our participants do maintain a common identity as *We are humans of the planet Earth*, and this by far exceeds the other explicit group identities found in the data; they are indeed all more human than not.

Acknowledgments

The first author would like to acknowledge Alaska EPSCoR NSF award #EPS-0701898 and the state of Alaska, which provided partial financial support for his involvement in this project. The second author gratefully acknowledges the following support of this research: The John Templeton Foundation, though its Grant #1840, "Construction of Interstellar Messages Describing the Evolution of Altruistic Behavior"; the Astronomical Society of the Pacific (www.astrosociety.org) through a SEED grant, "La Tierra Habla (Earth Speaks): Reaching Spanish-Speaking Audiences through an Interactive Museum Exhibit Based on the Search for Extraterrestrial Intelligence"; President Joseph Subbiondo, Academic Vice President Judie Wexler, and Clinical Psychology Department Chair Katie McGovern of the California Institute of Integral Studies for sabbatical and ongoing research leaves; Chairman of the Board John Gertz, CEO Thomas Pierson, and Director of SETI Research Jill Tarter of the SETI Institute for support of research on SETI and Society; Chris Neller of the SETI Institute for administrative support; and Jamie Baswell, as well as Harry and Joyce Letaw, for financial support through the SETI Institute's Adopt a Scientist program.

Works Cited

Leech, G., P. Rayson, and A. Wilson. 2001a. *Word frequencies in written and spoken English: Based on the British National Corpus.* London: Longman.

———. 2001b. List 2.3: Rank frequency order: written English (not lemmatized). Lancaster University. http://ucrel.lancs.ac.uk/bncfreq/lists/2_3_writtenspoken.txt.

Maslow, A. 1943. A Dynamic theory of human motivation. *Psychological Review* 50: 370–96.

Sagan, C., L. S. Sagan, and F. Drake. 1972. A message from Earth. *Science* 175: 881–84.

Staff at the National Astronomy and Ionosphere Center. 1975. The Arecibo message of November 1974. *Icarus* 26 (4): 462–66.

Thorndike, R. L., and E. P. Hagen. 1969. *Measurement and evaluation in psychology and education.* 3rd ed. New York: John Wiley and Sons.

Vakoch, D. A. 2011. A taxonomic approach to communicating maxims in interstellar messages. *Acta Astronautica* 68 (3–4): 500–11. doi:10.1016/j.actaastro.2010.03.017.

———, T. A. Lower, Y. Clearwater, B. A. Niles, and J. E. Scanlin. 2010. *Earth speaks: Identifying common themes in interstellar messages proposed from around the world.* Poster presented at the Astrobiology Science Conference 2010, League City, TX, USA.

Part III

Interstellar Message Construction: Can We Make Ourselves Understood?

Limits on Interstellar Messages

Seth Shostak

1.0. Introduction

While SETI experiments designed to find radio and optical signals intentionally sent from other worlds have been conducted for more than four decades, there has never been a sustained, deliberate broadcasting effort. Arguments against a transmitting project range from the practical (cost) and the philosophical (our communication technology is less than a century old, so we should listen first), to the paranoid (it might be dangerous to betray our location with a signal). A summary of some of the arguments for and against terrestrial broadcasting efforts is given by Ekers, Cullers, Billingham, and Scheffer (2002).

Although we are not beaming signals to other star systems ourselves, there has nonetheless been considerable thought given to how we might reply to, or even initiate, any extraterrestrial communication. Various suggestions (Vakoch 1998) have included encoding our message with mathematics, music, or graphics. The last has some precedent in the pictorial messages affixed to the Pioneer 10 and 11 spacecraft, as well as the analog records carried by Voyager 1 and 2. In addition, a simple graphic was also used in the first, deliberate high-powered transmission to deep space, the Arecibo broadcast (Staff of the Arecibo Observatory 1975).

Irrespective of coding schemes, one might argue that exercises in message construction for replies are superfluous, given that mankind has long been transmitting information to the stars inadvertently. Indeed, high-powered broadcasting at frequencies above ~100 MHz will traverse the ionosphere and continue into space. The earliest television signals have already reached several thousand star systems. However, the strength of TV signals at light-years' distance will be low, given the small gain of the transmitting antennas. For VHF broadcasts, the maximum effective radiated power is between 100 and 300 kilowatts, and for UHF is 5 megawatts. At 100 light-years, these will produce signals of flux density no more than 10^{-33}–10^{-31} watts/m^2-Hz, even in the very narrow parts of the band where the carriers are located. The best SETI experiments today are seven orders of magnitude too poor to be able to detect a comparable signal. And note that retrieving the video components of a TV broadcast would require ~10^4 greater antenna collecting area than required to find the carriers.

Our military radars, thanks to their higher-gain antennas, produce more detectable emissions, but cover only a fraction of the sky at any one time. Finally, it has been pointed out that high-powered terrestrial broadcasting is likely to be a transitory activity, as improved technology will soon encourage us to use either optical fibers or low-power, highly targeted transmissions to disseminate information and entertainment.

In other words, to assume that leakage automatically generates a "reply from Earth" to any SETI signal we might receive is unrealistic. Consequently, it's useful to seriously consider the general nature of signals intended for deliberate communication between star systems, as these might (1) elucidate the construction of any future replies to extraterrestrial transmissions, and (2) help to gauge what sort of signals our SETI experiments might discover. In this chapter, we consider some realistic limits on information content that can be easily sent across interstellar distances via light or radio, and suggest what might be reasonable signaling strategies.

2.0. Information Content

Leaving aside for the moment the ostensible content of an interstellar message, be it a photograph, mathematics, music, or plain text, one can ask how much information can be sent in a reasonable time between galactic star systems that are separated by hundreds, or possibly thousands of light-years' distance. In the case of electromagnetic signaling (radio or light), this depends on (1) distance, (2) transmitter power, (3) transmitting beam size and receiver collecting area, and (4) the chosen frequency. In this chapter, we do not consider the bodily transmission of information, although as pointed out by Rose and White (2004) actually rocketing highly compressed inscribed data (which in the case of genetic material can reach densities of ~10^{24} bits/kg) to deliberately chosen recipients could convey a great deal of information at low cost. Physical conveyance of data also has the advantage that the signal is not transient—it does not require that the recipient be monitoring the communication when it arrives. On the other hand, electromagnetic signaling is fast, and—if the information conveyed is limited—can be an inexpensive way to reach very large numbers of target star systems, as will be shown below.

We are interested in estimating a reasonable *maximum* data rate for interstellar communications, on the assumption that a society only modestly more advanced (a few centuries) than ours would have the technology to construct the requisite transmitting apparatus. We can then compute the likely size of messages, a parameter that will directly influence the type of information that is sent.

2.1. Data Communication at Microwave Frequencies

Since most SETI is conducted at microwave frequencies—both because the Galaxy is highly transparent in this part of the band, and also because natural "marker" frequencies such as that of neutral hydrogen (1,420 MHz) and the hydroxyl radical (1,612, 1,665 and 1,667 MHz) delimit this spectral region—it is instructive to compute the amount of information, and the requisite power, that can be conveyed in this spectral regime.

Common radio practice is that a broadcast will use a bandwidth that is ~5 percent of the carrier frequency, or in this case, ~70 MHz. The amount of information C (bits/second) that can be conveyed with a channel of bandwidth W is given by Shannon (1948)

$$C = W \log_2 (1 + P/N), \qquad [1]$$

in which P/N is the signal power-to-noise ratio at the receiver = T_A/T_S, where T_A and T_S are respectively the antenna temperature produced by the source and the receiving system temperature. So for circumstances in which this ratio is 1, we have C = W, or 70 megabits/second.

Note that for most SETI experiments, W~1 Hz, and P/N is less than one, but these efforts are intended to find carriers or very slowly pulsed signals, both of which have extremely low information transmission rates. If sufficient transmitter power and/or antenna gain are available to produce a P/N ~1, and if the entire 70 MHz can be recorded by those receiving the signal with high temporal resolution (~10 nanoseconds), then in the course of a day, 750 gigabytes of information could be received, and in a year, 270 terabytes.

To get some idea of how feasible this is, consider the transmitter power required to produce $T_A/T_S = 1$. We have

$$T_A = A_R S_v/2\kappa \qquad [2]$$

where A_R is the collecting area of the receiving antenna, κ is Boltzmann's constant, and S_v is the incident flux density (watts/m^2-Hz).

If we define P_T as the transmitter power over a band W, and additionally assume that this power is uniformly distributed over that band, then equation [1] becomes

$$C = W \log_2 (1+ [P_T\, A_R A_T\,]/[\pi\kappa W\lambda^2 D^2 T_R]), \qquad [3]$$

where A_T is the transmitting antenna area, T_R is the system temperature of the receiver, D is the distance between sender and receiver, and λ is the wavelength. As example, consider Arecibo-sized (7 $\times$ 10^4 m^2 area) transmitting and receiving antennas separated by D = 100 light-years, with λ=21 cm, and T_R = 5 K. To achieve a signal-to-noise P/N = 1 requires a transmitter power density of 1 kilowatt/Hz, or 70 gigawatts over the entire 70 MHz band. The latter figure is considerable, approximately 0.5 percent of the total energy generation on Earth today, but permits a data rate of ~10 megabytes/sec, roughly comparable to that of a high-definition television signal. There are, however, several ways that an extraterrestrial transmitting society could reduce the power demand, by using (1) a larger transmitting antenna, (2) a reduced bandwidth, or (3) a reduction in the data rate, either by sending less information, or by taking longer to send it. Note that scheme (3) has some small benefit from the slow, logarithmic dependence on power implied by equation [3].

We have assumed Arecibo-sized transmitting and receiving antennas. This is, of course, merely an anthropocentric guess. There are already plans to build a radio telescope on Earth whose maximum dimension is 1 km. At the transmitting end, an Arecibo-sized antenna used at 21 cm wavelength has a beam that is ~4,000 AU in size at 100 light-years. This is enormously larger than the zone within which one expects to find Earth-like worlds. It seems more reasonable to suppose that an advanced society bent on sending deliberate signals would focus its transmissions to cover no more than the habitable zone of the target star, defined to be of radius R_H. If we postulate that they have optimized their broadcasting effort in this fashion, equation [3] then becomes independent of distance, and

$$C = W \log_2 [1 + P_T A_R/(2\pi\kappa W R_H^2 T_R)],$$

or, solving for P_T

$$P_T = 2\pi\kappa W R_H^2 T_R /A_R (2^{C/W} - 1). \qquad [4]$$

Assuming we have a 1 km diameter receiving antenna, and using other parameters as in our example above, the power requirement drops so that only 4 kilowatts is required to transmit ~10 Megabytes/sec, assuming R_H = 2 AU, which delimits a zone larger than the orbit of Mars in our own solar system. While this optimization is logical, and rewarded by a dramatic drop in power cost, the transmitting antenna is now impressively large (a filled aperture of 300 km size for targets at 100 light-years). Nonetheless, this optimized approach gives us an estimate of the lower limit on the required power.

We have assumed that only a small fraction of the microwave band would be modulated, as this is typical practice. However, Jones (1995) has pointed out that optimal encoding could make use of the entire free-space microwave window from 1–10 GHz. If this is done, the gain in information transfer rate would be ~10^2 over our example, with a concomitant increase by the same factor in required power. Note also that interstellar scattering will smear high-frequency signal components, and this will require special transmission schemes if broadcasts are made over long distances in the galactic plane (Shostak 1995).

2.2. Data Communication at Optical Frequencies

Optical communication using pulsed light is both feasible and is being looked for by SETI practitioners. For distances extending over many hundreds of light-years or more, scattering by the interstellar medium argues for the use of infrared wavelengths. However, at wavelengths $\geqslant 1$ micron, dispersion will broaden individual pulses, limiting the maximum number of pulses that can be sent per second. The amount of the dispersion is (Taylor and Cordes, 1993)

$$\Delta t = 4.1 \times 10^{15} \text{ DM } \lambda^2 c^2 \text{ sec,} \qquad [5]$$

where a typical value for the dispersion measure DM = 30 sec^{-1} over ~1 kpc distance. This limits pulse repetition rates to 10^{12} sec^{-1} at 1 micron wavelength, and 10^{10} sec^{-1} at 10 microns. We will assume one bit per pulse because of the slow increase in data rate with power implied in the Shannon formulation [1] and its derivative, equation [4]. Clearly, if energy cost is no object, higher bit rates than those we consider could be achieved.

Suppose, as above, an optimized system that targets a star's habitable zone. We further assume that the incoming flux of photons in one pulse width τ (seconds) at the receiving end must be ~10 times that produced during time τ by the transmitting society's own star, of luminosity L_*. If we send C binary bits/sec (one bit per pulse), with a duty cycle for "on" bits of 50 percent, the required power is

$$P_T = C\tau \, L_* \, R_H^2/D^2 \qquad [6]$$

The necessary power *decreases* with distance D because in this simplified calculation we have assumed that the only noise source is light from the transmitting society's star. This has the counterintuitive result that targeting distant stars requires less power, which is true if the habitable zone strategy is used. Using [6], with $\tau = 10^{-10}$ sec, an information rate of C = 10^{10} bits/sec, D = 100 light-years, R_H = 2 AU, and a solar-type star with infrared luminosity L* ~10^{26} watts, then P_T = 10^{13} watts. Reducing the bit rate proportionately reduces the power required, and if only one bit per second is sent, only 1 kilowatt is necessary to signal stars at 100 light-years.

There is a minimum power requirement for signaling in the optical: the level necessary to produce one photon per pulse ("on" bit) in the receiving device, assuming that there is no noise introduced by the transmitter's home star. This might be the case, for instance, if the distance D is great enough so that the star delivers $\ll \tau^{-1}$ photons/sec to the receiver's mirror, or if the transmitter is located far enough from the home star to be cleanly resolved by the receiver.

The minimum power required is

$$P_{Tmin} = \pi C \ R_H^2 \ hc/(A_R \lambda), \qquad [7]$$

where h is Planck's constant and A_R is, once again, the area of the receiving mirror. Assuming λ = 10 microns, R_H = 2 AU, and a 100 m diameter receiving mirror, P_{Tmin} = 7 $\times$ 10^9 watts for C = 10^{10} bits/sec.

Note that from [6] and [7], we can deduce that beyond a distance of

$$D = [\tau \ L_* \ A_R \ \lambda/(\pi hc)]^{1/2}, \qquad [8]$$

the power requirement given by [6] reaches the minimum level, and no further decrease occurs. For a data rate of C = 10^{12} bits/sec at λ = 1 micron, and our example parameters, this happens at D = 120 light-years. For C = 10^{10} bits/sec at λ = 10 microns, the minimum power applies for distances greater than D = 3,700 light-years.

2.3. Radio versus Optical

In Figure 23.1 we plot the required power for radio (both 5% bandwidth, and full microwave window) and two optical regimes as a function of distance.

The straightforward considerations above have shown that data rates from 10^7 to 10^{12} bits/sec can be straightforwardly achieved in the radio and optical. The power requirements as given in Figure 23.1 are considerable, particularly for optical, although one should bear in mind that these values are somewhat dependent on our assumptions about the size of the transmitting and receiving apparatus. For all but the nearest stars, the required powers are less than the current energy production on Earth (1.5 $\times$ 10^{13} watts), and much less than the Earth's insolation (2 $\times$ 10^{17} watts, above the atmosphere). These facts suggest that advanced societies could muster the energy required for transmitting at the given bit rates.

One aspect of Figure 23.1 worth noting is that full spectrum microwave transmissions have a bit rate comparable to 10 micron optical, but achieve this with four to eight orders of magnitude less power. However, against this energy efficiency, one must weigh the greater instrumental and decoding

Power for Max Bit Rates

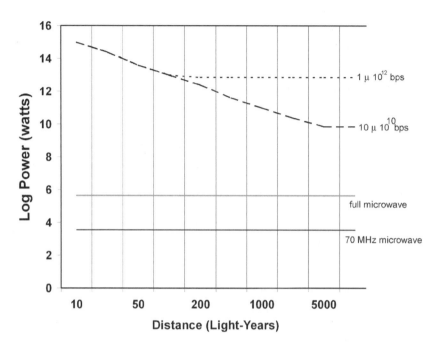

Figure 23.1. Minimum power levels required for high bit-rate transmissions. There are four regimes plotted, and all assume that the senders are beaming to a circle of radius 2 AU centered on the target star: (a) Microwave with a fully modulated 70 MHz bandwidth, having a data rate of 7 10^7 bits/sec, (b) microwave using the spectrum from 1–10 GHz, conveying 10^{10} bits/sec (c) a 10 m pulsed laser with 10^{-10} sec pulses sending 10^{10} bits/sec, and (d) a 1 m laser with 10^{-12} sec pulses and a data rate of 10^{12} bits/sec. The radio examples assume a 1 km diameter receiving antenna. Note that the power required in the optical regime flattens with distance once the influence of noise introduced by the luminosity of the home star (assumed to be L. = 10^{26} watts) drops to less than 1/10th photon per pulse width collected by a 100 m diameter receiving mirror (assumed in these examples). For comparison, terrestrial power generation is currently ~1.5 10^{13} watts.

challenges of the microwave scheme. To put information into 9 GHz bandwidth requires ingenuity both in the design of hardware and in the encoding scheme to minimize the effects of dispersion. In addition, we have assumed that all transmitters will target a star's habitable zone. At 1,000 light-years, the necessary transmitting antenna aperture in the radio, which will vary across the 9 GHz spectrum, is 5 × 10^3 km (nearly one-half

Earth's diameter) at the lowest part of the band. For 10 micron optical, the transmitting mirror would only need be 160 m in diameter to properly target the habitable zone of a star 1,000 light-years' distant. These differences in implementation requirements, in situations where energy is cheap, might easily favor the use of optical.

3.0. Possible Messages

With data rates in hand, we can trivially compute the total information conveyed for any broadcast length. The types of messages we routinely encounter in daily intercourse are usually formulated, decoded, and understood in a few days at most. However, such short messages may not be appropriate for a deliberate transmission intended to reach other societies. If there are 1,000–100,000 civilizations (Dick 1996) spread throughout the Milky Way, then their average separation is hundreds to a few thousand light-years. The round-trip message times will probably lead senders to assume they are engaged in one-way communication. They will want to send everything at once, since interaction with the communicant may not occur. This encourages long messages.

But messages shouldn't be *too* long, as virtually no one will chance to pick up the signal just as the message begins, but will "tune in" somewhere in the middle. Consequently, repetition will be necessary, and should occur at intervals that are short compared with the time that the recipients will be devoting to a single listening project. On Earth, this time is usually less than a human lifetime, suggesting message lengths of no more than a few years.

In considering what message might be encoded by interstellar signaling, it is worth remarking that our own efforts at interstellar communication have been extremely modest, a fact that may influence how we have thought about message content in general. The Pioneer 10/11 and Voyager 1/2 messages contained pictorial information about our appearance, culture, and location, as well as some music and verbal greetings in the case of the Voyager probes. The Voyager message was inscribed on a mechanical record, of ~10 megabytes carrying capacity. The earlier Pioneer plaques were engraved, and a bit map rendition of that graphic is ~600 kilobytes in size. The 1974 Arecibo message, which was sent digitally at about 1 byte/sec, comprised 210 bytes.

The data that our current radio SETI experiments could receive are also very limited. Project Phoenix, for example, a highly sensitive radio scrutiny of ~750 nearby star systems, could recognize narrow-band signals that pulsed every few seconds, and could collect no more than ~10 bytes in a single observation of a target star system. On the other hand, today's optical SETI experiments could, in principle, record short (ten minute) bursts of nanosecond pulses, at 10^8 bytes/sec, making for a total recordable message of ~75 gigabytes.

The limitations of the radio efforts, in particular, have seduced us into assuming that highly efficient encoding schemes would be necessary for interstellar messaging. However, this might be analogous to extrapolating Samuel F. B. Morse's carefully weighed message in 1844, "What hath God wrought?"—the first telegraphic message between cities—to present-day communication. Given the very large size of messages that could be sent by a modestly advanced civilization, as calculated above, it may be unnecessary to be overly concerned about either encoding schemes or specifics of content, but rather rely on redundancy in the transmitted information to provide the key to understanding by the recipient.

In the face of the fact that large information transfers are possible, let us consider what sorts of messages might be sent, based on current human activity. Among SETI researchers, it is occasionally (and usually offhandedly) said that altruistic societies will transmit their "Encyclopedia Galactica." Our own encyclopedias could be sent in a matter of seconds. A more ambitious project would be to transmit the contents of a major library. The content of the Library of Congress is ~1.4×10^{14} bytes (U.C. Berkeley School of Information Management and Systems, 2003) and could be sent in under an hour with the highest speed optical signaling link. The iconic information repository in contemporary times is the World Wide Web, and the estimated amount of immediately accessible data on servers in 2003 was comparable to that in the collections of the Library of Congress (U.C. Berkeley School of Information Management and Systems 2003), although the Web is growing rapidly (doubling time in the 1990s was less than six months).

These and other possible "messages" are listed in Table 23.1. What we see is that, even at radio wavelengths, we can send content as extensive as the Library of Congress in a half-year's time or less. On the other hand, the amount of *new* data currently being stored on magnetic media is about 5×10^{18} bytes per year (U.C. Berkeley School of Information Management and Systems 2003). While this may be only a temporary phenomenon, there's little doubt that information is growing at a prodigious rate. Our fastest channel, a 10^{12} bits/sec optical link, could just barely keep up with this new information flow. In a few years' time, it will obviously be unable to do so. The implication is that, while we might be able to broadcast a comprehensive "Encyclopedia Terrestria," the annual updates will require editing, and the editing will become more severe with time.

Nonetheless, Table 23.1 encourages us to think that, with transmitting times of a year or less, a society could send enormously rich content, with enough redundancy to facilitate decoding (in the same way that anyone

Table 23.1. Size of Sample Messages and Transmission Time for Various Modes.

Message	Size (bytes)	Microwave Radio Narrow-band	Microwave Radio Wide-band	Optical 10^{10} bps	Optical 10^{12} bps
Arecibo 1974 Message	2×10^2	2×10^{-5} sec	2×10^{-7} sec	2×10^{-7} sec	2×10^{-9} sec
Pioneer plaque	6×10^5	6×10^{-2} sec	5×10^{-4} sec	5×10^{-4} sec	5×10^{-6} sec
Voyager record	10^7	1 sec	0.01 sec	0.01 sec	10^{-4} sec
Human DNA	10^9	2 min	1 sec	1 sec	8×10^{-3} sec
Encyclopedia Britannica	5×10^9	10 min	5 sec	4 sec	4×10^{-2} sec
Library at Alexandria	~7×10^{11}	1 day	10 mins	10 mins	6 secs
A human memory	3×10^{12}	4 days	40 mins	40 min	25 sec
New books in 1 year	5×10^{12}	1 week	1 hour	1 hour	40 sec
Library of Congress (17 million books)	2×10^{14}	6 months	1 day	1 day	20 min
World Wide Web (2003)	2×10^{14}	7 months	2 days	2 days	20 min
E-mail in 1 year	4×10^{17}	1,500 yrs	10 yrs	10 yrs	1 month
New information generated in one year (2002)	5×10^{18}	20,000 yrs	140 yrs	130 yrs	1 yr
Memory content of all human beings	2×10^{22}	70 million yrs	0.6 million yrs	0.5 million yrs	5,000 yrs

Some table entries are based on information found in U.C. Berkeley School of Information Management and Systems (2003), and in Reupke (1992).

receiving the Library of Congress would eventually be able to figure out English). However, there is the serious problem of knowing where to broadcast the information, especially since, as we have seen, the power requirements for high bit-rate transmissions are substantial, making multiple targeting expensive. It is well and good to say that the Web can be sent in two days, but if one is compelled to sequentially target every star in the Galaxy, the transmitting project will last a billion years, and the chances that someone is listening when the broadcast reaches their planet is small.

One means to ameliorate this unfavorable approach is to use advanced astronomical information. Extraterrestrial societies that are only a century or two beyond our own will possibly have long lists of planets whose atmospheres give evidence for biology, as Earth's has for ~2 billion years. This would tell them which target stars have worlds with life. However, technological societies might only inhabit a planet for a tiny fraction of its biological history, and thus even for those transmitting societies with lists of fecund worlds, the number of possible signaling targets could still be large ($>10^6$).

These facts incline us to suggest that only three strategies seem appropriate for a transmitting society:

1. Develop very large energy sources (10^{17} watts or more, for optical transmissions) and construct a device capable of *simultaneously* targeting millions of likely worlds. If the energy of a star (~10^{26} watts) can be harnessed (e.g., via a Dyson sphere), then continuous broadcasts in the optical could be made in all directions.

2. Transmit only to those star systems from which signals have already been heard, confirming the presence of intelligent recipients. In this case, it is highly unlikely that the Solar System is now on anyone's target list, as Earth's leakage signals have only reached to ~60 light-years.

3. Transmit short messages sequentially, but repeatedly. As example, imagine a society that "pinged" a million star systems once a day (~0.1 seconds per ping). That would be adequate to daily convey the equivalent of the Encyclopedia Britannica to each of these systems using 1 micron infrared as the carrier and one bit per pulse. Needless to say, the message could be different each time.

The relative ease and economy of approach (3) suggests that it is a good candidate for the type of transmission that might be made to star systems for which the senders have no direct proof of technically competent listeners;

in other words, systems like our own, as judged by any extraterrestrials more than ~60 light-years distant. It also suggests that SETI researchers should consider looking for stars whose infrared luminosity regularly spikes.

In conclusion, we have seen that instrumentation that is technologically feasible, coupled to power sources that an advanced society will surely command, would be able to transmit, in a year's time or less, quantities of information comparable to the largest collections on Earth. However, unless the transmitting society is so advanced that it can afford both the instruments and the energy necessary to broadcast simultaneously to vast numbers of stars, it will most likely adopt the strategy of sending short, frequently repeated messages. In the case of optical signaling, these could be gigabytes in size. It seems reasonable to suspect that, until we make our presence known to others, we will have to make do with the fact that other worlds might be sending us only an encyclopedia's worth of information daily.

Note

This chapter is an adaptation of S. Shostak. 2010. Limits on interstellar messages. *Acta Astronautica*, doi:10.1016/j.actaastro.2009.10.021.

Works Cited

Dick, S. J. 1996 *The biological universe: The twentieth-century extraterrestrial life debate and the limits of science.* Cambridge: Cambridge University Press.

Ekers, R. D., D. K. Cullers, J. Billingham, and L. K. Scheffer, eds. 2002. *SETI 2020: A roadmap for the Search for Extraterrestrial Intelligence.* Mountain View, CA: SETI Press.

Jones, H. W. 1995. Optimum signal modulation for interstellar communication. In *Progress in the search for extraterrestrial life*, ASP Conference Series, Vol. 74, ed. G. Seth Shostak, 369–78. San Francisco: Astronomical Society of the Pacific.

Reupke, W. A. 1992. Efficiently coded messages can transmit the information content of a human across interstellar space. *Acta Astronautica* 26: 273–76

Rose, C., and G. Wright. 2004. Inscribed matter as an energy efficient means of communication with an extraterrestrial civilization. *Nature* 431: 47–49.

Shannon, C. E. 1948. A mathematical theory of communication. *Bell System Technical Journal* 27: 379–423, 623–56.

Shostak, G. S. 1995. SETI at wider bandwidths? In *Progress in the search for extraterrestrial life*, ASP Conference Series, Vol. 74, ed. G. Seth Shostak, 447–54. San Francisco: Astronomical Society of the Pacific.

Staff at the National Astronomy and Ionosphere Center. 1975. The Arecibo message of November 1974 *Icarus* 26 (4): 462–66.

Taylor, J. H. and J. M. Cordes. 1993. Pulsar distances and the galactic distribution of free electrons. *Astrophysical Journal* 411: 674–84.

U.C. Berkeley School of Information Management and Systems 2003, http://www.sims.berkeley.edu/research/projects/how-much-info-2003/execsum.htm.

Vakoch, D. A., 1998. Signs of life beyond Earth: A semiotic analysis of interstellar messages. *Leonardo* 31: 313–19.

Communication among Interstellar Intelligent Species

A Search for Universal Cognitive Maps

Guillermo A. Lemarchand and Jon Lomberg

1.0. Introduction

"Take Me to Your Leader" are the classic first words spoken by a prototypical visitor to Earth. But how will she/he/it know English grammar and vocabulary? How could we possibly communicate with an extraterrestrial intelligence?

Since the early-nineteenth-century speculations made by prominent scientists such as Karl Gauss, Charles Cros, Francis Galton, or Konstantin Tsiolkovsky to the present day, several scholars have put forth solutions to the question of how intelligent life on different planets might signal one another. The proposed methods range from giant burning symbols in the desert to the manipulation of viral DNA to contain an encoded message.

This problem invites us to think about language, communication, and intelligence at the deepest levels. Do we perceive the real world or just models of it in our brain? Is there anything in physics and mathematics that we can consider truly universal? How about art, ethics, or culture? How alien can the aliens be?

The aim of the present chapter is to call the attention of scholars to the need for a systematic exploration of *cognitive universals* that might be used to develop new strategies for the Search for Extraterrestrial Intelligence (SETI) research programs and eventually to seek for new ways to design schemes for communication among hypothetical galactic civilizations. We think that this work should be focused on four types of cognitive universals: (1) physical-technological, (2) aesthetic, (3) ethical, and (4) spiritual. All previous SETI observational projects have been designed taking into account only the assumed universality of the laws of nature, along with some technological considerations. Our working hypothesis assumes that one way to improve the odds of finding extraterrestrial intelligent life (ETIL) is by making conjectures about the methodology needed to transform *human cognitive maps* into cognitive universals.

We introduce the term *cognitive map* to define more accurately the conceptualization taken from the epistemology of science known as a paradigm or worldview (Kuhn 1962). Here we conceive a cognitive map as the process by which "intelligence" makes representations of its environment in its own processing system (in our case the human brain).

Features of the physical environment are only one of the significant parts of human cognitive maps. We live in an environment where other people act and where information linguistic in nature (signs, orders, instructions, descriptions, etc.) operates. Therefore, our cognitive maps must represent objects, living beings and their behavior, linguistic abstractions, and perceptions about the external world. To these we would also add perceptions of aesthetic, ethical, and spiritual values. How universal and ubiquitous are these human representations? Is it possible to find cognitive maps within the physical, aesthetic, ethical and spiritual dimensions that could be considered "universal"? These are the types of questions that we would like to introduce to a broad interdisciplinary academic community.

2.0. Communication Assumptions

SETI researchers believe that the basic principles of our science and the science of extraterrestrial beings should be fundamentally the same, and we should be able to communicate with ETIL by referring to those things we share in common: the principles of mathematics, physics, chemistry, and so on (Shklovskii and Sagan 1966; Lemarchand 1992; Lemarchand 1998).

The latter statement assumes that there is only one way to perceive and understand the laws of nature and because of that, the language of science and mathematics must be universal. At this point we may call attention to some other epistemological considerations. Kuhn (1962) envisioned the possibility of situations in which two theoretical conceptions are so different that, because of the irremediable paradigm dependence of observations and other methodological features of science, there is no rational method acceptable to defenders of each conception that could serve to ground the resolution of the dispute between them. In such situations, the two traditions are said to be *incommensurable*.

The question here is whether the standards of evidence, interpretation and understanding dictated by the terrestrial "cognitive map" on the one hand and by the ETIL's "cognitive map" on the other are so different that a "mutual understanding" (communication) between them cannot be interpreted as having been dictated by any common standards of rationality. Since there are no significant theory-independent standards of rationality, it follows that the communication in question is not a matter of rationality. It needs to involve the adoption of a wholly new conception of the world, complete with its own distinctive standards of rationality. This argument incorporates the claim that the semantics of the two cognitive maps differ to such an extent that those terms that they may have in common should not be thought of as having the same referents in the two cognitive map systems.

At this point, we found two different conceptualizations of the term *incommensurability* that may be used to define the impossibility of exchanging intelligible information among galactic cultures.

1. Methodological incommensurability: a relation claimed to hold between different theoretical frameworks such that, because of the deeply cognitive map–dependent character of observation and scientific method, no rational method for exchanging intelligible information can be found.

2. Semantic incommensurability: a relation claimed to hold between instances of the same term as it occurs in the confrontations between two different cognitive maps.

The evolutionary paths of extraterrestrial life will have strong dependence on the characteristics of their planetary environments. Alien intelligences will be very different organisms, with different needs, senses, and behaviors. They could inhabit environments in which neither science nor technologies are needed for survival. The science of an extraterrestrial civilization would reflect the way they perceive nature, as funneled through the course of their particular evolutionary adjustment to their specific environment. It may be impossible for us to distinguish any of their possible intelligent manifestations if we are using different cognitive maps. In the significantly reduced context of our planet, every human language maps the environment of its native speakers and the information accumulated by culture over many generations. Each language therefore reflects the particularities of the environment in which its speakers live. The Inuit people have many words for snow, while people living in the tropics can get along with one or none.

A more serious problem in differing cognitive maps in human societies is the way different cultures express the experience of seeing color (Jameson and Lomberg 2010). Physiologically all humans use the same eye-brain neurology to perceive color, but the description of color is determined by the number of color words in each language. There are some languages that have only three color words: black, white, and red. People in these cultures see and can distinguish between different colors, but feel no need for more terms than the three cited. Communication within the culture is unaffected, but communication about color with other cultures is severely impaired. In many other cultures there is only one color word used to describe blue and green. Even in English, which has hundreds of words to describe different colors, we use the same word *red* to describe an apple, a brick and a red-haired person. Context determines the meaning, but a literal translation into another language might result in confusion because of different human cognitive maps.

Vakoch (1998) goes beyond and shows several examples of incommensurability in the interpretation of pictorial interstellar messages by hypothetical galactic cultures. However, his alternative proposal to use iconic representations, in which the sign bears a physical similarity to that which it represents, still requires some "basic common theoretical framework" to avoid misinterpretation.

SETI pioneer Sebastian von Hoerner (1974) explored in detail the boundary conditions needed for the propagation of sound that are imposed by different planetary atmospheres. He arrived at the interesting result that there is only a very limited number of possible musical scales, inferring the conclusion that music might be universal. Even so, there is no doubt that different planetary environments around other stars may present a far stronger incommensurability problem for exchanging information using different cognitive maps.

This work was followed by Lomberg (1974, 1978). Particularly during the selection of the *"Sounds of Earth"* for the Voyager I and II disks, he had also explored the ubiquity of "life sounds" in order to determine hypothetical cognitive universals (see box). Lomberg proposed to start with the sound of a human heartbeat. It is interesting to note that recently, the Russian biologist Valery Tsevetkov (1997) found that if we take the middle blood pressure in the aorta as the measurement unit, then the systolic blood pressure is 0.382 and the diastolic pressure is 0.618. Their ratio corresponds to the Golden Section (GS), a well-known proportion that appears in pure mathematics, nature, living forms, and the arts. This is not a rare coincidence, taking into account that the GS also works as a self-organizing principle in nature. Inadvertently, the original proposal to include a human heartbeat involved a deep aesthetic principle.

All intelligent problem solvers are subject to the same ultimate constraints—limitations on space, time, and resources, including thermodynamic boundaries for physiological and information processing systems (Dyson 1979). In order for intelligent life forms to evolve powerful ways to deal with such constraints, they must be able to represent the situations they face and must have processes for manipulating those representations (cognitive maps and languages). Thus, every galactic civilization must develop symbol-systems for representing objects, causes, and goals and for formulating and remembering the procedures it develops for achieving those goals. During its evolutionary process, each species would eventually encounter certain very special ideas—for example, about arithmetic, causal reasoning, and optimization processes—because these particular ideas are very much simpler than other ideas with similar uses. In the same way, we can speculate about the need to develop aesthetic, ethical, and spiritual values. We need to begin new interdisciplinary research programs to focus on these other dimensions of the search for intelligent life in the universe. Eventually, these new windows will be very useful for the redesign of SETI observational strategies.

It is a reasonable hypothesis to assume that any advanced intelligent species would have to establish some of these principles in order to survive its evolutionary process, to develop technology and, in the long term, to evolve in harmony both with its planetary environment and as a society and with other species. The last is a requirement for the development of ethical restrictions. Eventually, some ETIL could have evolved thought processes and communication strategies that match our own to a degree that will enable us to comprehend them.

Because the only way we have to find evidences of ETIL is through the detection of artificial signals generated by technological activities, many sorts of cognitive universals can be surmised by considering various topics, including pragmatic requirements—what the extraterrestrials must know to build sending and receiving equipment. We can also introduce some assumptions about aesthetic and ethical principles that will define some characteristics of the contact strategies.

3.0. Exploring the Wholeness and Its Implicate Order

Within the scientific community there is a general consensus that mathematics contains "cognitive universals" that cut across cultural and linguistic boundaries among humans, which can be used to communicate with ETIL. Are there as well any universal aesthetic principles that could also solve the problem of incommensurability? Is it possible also to derive or find spiritual and ethical cognitive universals?

Obviously, no matter how careful we are in establishing the universality of these concepts, we will have biases due to our culture (ethnocentrism) and our human nature (homocentrism). In fact it may be impossible to avoid these biases, and we may never know how severely they limit our ability to recognize, decode, and respond to communications from extraterrestrial intelligences.

Our traditional cognitive maps usually divide the whole into parts in order to study them and to make the universe intelligible for us. Bohm introduced a different epistemological approach to develop a new notion of order that may be appropriate to a universe of unbroken wholeness (Bohm 1980). He defined this notion as implicate or enfolded order. In the enfolded order, space and time are no longer the dominant factors determining the relationships of dependence or independence of different elements. Rather, an entirely different sort of basic connection of elements is possible, from which our ordinary notions of space and time, along with those of separately existent material particles, are abstracted as forms derived from the deeper order. These ordinary notions in fact appear in what is called explicate or unfolded order, which is a special and distinct form contained within the general totality of all the implicate orders. According to Bohm, both relativity and quantum theories agree in that they both imply the need to look on the world as an undivided whole, in which all parts of the universe, including the observer and the observer's instruments, merge and unite in one totality. In this totality, the atomistic form of insight is a simplification and an abstraction, valid only in some limited context. Can this alternative paradigm teach us something new about how we should design our SETI strategies?

Bohm observed that Indo-European languages are inadequate to describe the properties of this wholeness and to enable our cognitive maps to understand the universe from the wholeness point of view. In his writings, he presented several examples of how these language limitations restrict our perception of quantum reality. He proposed an alternative and hypothetical language—the rheomode—that is strongly verb based.

We can find examples of this type of cognitive map within the human species. For instance, the Blackfoot culture's language is organized in such a way as to perceive constant flux and change, as required by the rheomode proposal. Rather than space and time being separate, the Blackfoot cognitive map presents a space-time in which the choice of tenses may depend on special distances as needed to explain quantum realities (Peat 2002).

Bohm also emphasized that the act of reason is essentially a kind of perception through the mind, similar in certain ways to artistic perception, and not merely the associative repetition of reasons that are already known. For example, in Lomberg's painting "The Great Chain of Being" different scales of reality are shown: those of atoms, protein chains, phage viruses, diatoms, insects, humans, the Sun, stars, the Milky Way, and clusters of galaxies (e.g., Lemarchand and Lomberg 2009, 396). There is no doubt that science cuts reality into pieces and considers each scale separately. Reality has no divisions, is a "wholeness." This particular image shows a perception of reality encompassing everything from atomic nuclei to clusters of galaxies in one scene. Humans exist between the macrocosm and the microcosm. Science reveals the details; art unites the whole. Seeing the wholeness in such a complex multileveled reality is in many ways also an experience of transcendence.

Unfortunately, there are very few systematic studies that attempt to perceive the properties of nature, aesthetics, and ethics under these alternative cognitive maps. These epistemological approaches might open new windows into our understanding not only about universal patterns in physics and the perceived reality of our universe but also about questions related to aesthetic perceptions (Bohm and Peat 1987) and spiritual conceptions (Bohm 1985).

It seems that the implicate order paradigm and the perception of wholeness include cognitive maps that might be more universal than those used by our mainstream physics, which organizes the universe and life in a fragmented way. Any serious attempt to establish contact with other galactic civilizations must explore these alternative epistemologies to infer new contact strategies.

Lomberg (1974) explored the application of aesthetic and musical cognitive universals for the design of interstellar messages. Later on, some of these novel conceptualizations were applied as selection criteria in the "pictures of Earth" included in the Voyager I and II interstellar records (Lomberg 1978). At that time, it was obvious that the concept of "picture" was by no means universal. Even on Earth, members of several cultures that do not use photographs have had to be educated on the concept before they could see pictures as Westerners do. The selection process had to take into account that images should contain as much information as possible, but should also be very easy to comprehend. For this reason, the first sequences had very little information, primarily to help the recipients understand how to see pictures. Using engravings provided a way of comparing images with an object that can be touched.

In a local context, Lomberg and Hora (1997) have explored the use of aesthetic principles in the design of warning markers for long-term nuclear waste storage sites intended to prevent inadvertent human intrusion in the distant future. Some of the principles elucidated in the design of messages for extraterrestrials were incorporated here, such as the need to begin from first principles in attempting to describe anything. To use a symbol such as the radiation trefoil, the message must include a pictograph explaining the meaning of it. The proposed design employed a redundant system of large and small markers, inscribed with markings in written languages, symbols, and pictographs, warning people against digging into the nuclear repository. The entire site might be marked by arranging large mounds into symbolic shapes, using materials with radar brightness different than the surrounding desert floor. The site might thereby be made visible to aircraft or spacecraft observations.

Proposal for Sounds on Voyager Record
by Jon Lomberg, May 1977

It has been proposed that part of the record to be placed upon the Voyager spacecraft be a montage of various Earth sounds. The following considerations may prove useful in designing this part of the record.

If we can meet the other demands involved in selecting material (political, Earthside demands), the Voyager record should be a message as easy as possible for recipient races to read. At least, the information contained on the record should be presented in such a way as to maximize the likelihood of recipients understanding the contents of the record.

The record as conceived already has two parts: musical and nonmusical. We would prefer that recipients recognize the difference between the two parts, and that they listen to the music in a different way than they listen to the sounds. We want them to attempt to identify the nature of' the sound; (what they are sounds of) and to analyze the structure and, if possible, enjoy the music (rather than asking themselves what the fugue is the sound of). We want them to group all the musical pieces together as examples of one phenomenon and to realize that the other parts of the message are different phenomena. Therefore, we should keep the number of parts of the message as small as possible—if possible keeping it just two parts.

I believe that all nonmusical parts of the message could be unified into a single montage, arranged logically and hierarchically, and presented in a way that would make each part of the montage more explicable and clearly indicate a two-part message, and perhaps even something of the nature of the second part (the music).

We would like to include in the montage the sound of human speech (the UN delegates greeting the recipients), the sound of animals, some sounds of our culture and technology, and some natural sounds of our planet. We would like to link the human sounds together in such a way as to say these sounds (speech, music, technological sounds) are our sounds.

This could be done by having some background sound that was played whenever human sounds were played. This sound should be distinctly human, yet not so obtrusive or confusing that it would interfere with other sound being played over it. I propose that this sound should be the sound of a human heart beating. This should open the record, and consistently be identified with human sounds throughout the sound montage. Thus, throughout the catalogue of greeting in various Earth languages, the heart would beat in the background. But it would not sound during the songs of whales or the sound of rain. Recipients should notice that we are grouping the sounds in two distinct classes—those identified with the heartbeat and those not. If the distinguishing sound was the first sound heard on the record, they might guess that this sound was in some way the signature of the makers of the record.

There are two advantages for using a heartbeat as the identifying sound. First, it avoids any regional or cultural squabble as to what is a distinctive human sound. Second, the sound itself is simple enough so that it would not be confusing as a background noise (as speech would be), and it could easily be removed electronically from the record by recipients if they wanted to study the sounds on the collage in more detail without the distraction of a background sound.

I propose the montage begin by establishing some characteristic human sounds:

1. Heartbeat (10 sec.)

2. Breathing (over heartbeat) (5 sec.)

3. Greetings from humans (over heartbeat) (about 7 min.)

4. Heartbeat (fadeout)

This establishes some sounds associated with us. I hope that some of the UN greetings will be recorded so that an intake of breath before syllables could be heard. This would link breathing with speech, and perhaps give a clue as to the respiratory nature of speech, and link the sounds of speech with the heartbeat.

From here, we move into a whole range of sounds. I think all the sounds (except for a few important ones we wish to call attention to) should be of exactly the same length (five seconds) and separated by a pause or characteristic sound (to punctuate and separate them).

How should the sounds be ordered? One way that would be very apparent would be to order them tonally or by frequency—putting similar sounds together. This would be a logical arrangement, but it would tell recipients nothing about the individual nature of the sounds. Some of the sounds will probably be familiar or deducible to recipients. The natural sounds of water (in thunder, or rivers) should be present in other planets with atmospheres and weather systems. There will be liquid water in many other places. There will be rocks in some of the places. Therefore the sound of water flowing over rocks will not be as alien as the sounds of speech. So we should arrange the message in such a way that the sounds they might identify tell them something about the sounds they can't identify. I propose that the sounds be ordered to reflect in a general way the evolution of life on the planet. After our opening, which identifies us, the sequence of sounds should move through natural sounds (mostly sounds), sounds of nonhuman life, sounds of human life, and sounds of society. Darwinian evolution may not operate on all worlds, but there is some chance that it does, and surely an arrangement that is ordered according to some possibly common principle has a better chance of being decoded than a random ordering.

The montage could continue this way:

5. Wind (gentle)

6. Wind (stormy)

7. Roll of thunder

8. Clap of thunder (5 sec. each)

9. Rain

10. Trickling brook

11. Running river

12. Waterfalls

13. Ocean surf (fadeout) (10 sec.)

This section should be the most easily identifiable to recipients. It describes part of the water cycle on our planet. This may tell them that we are a water-based life form. Also, if they recognize thunder, its particular sound may tell them something of the nature of our atmosphere (its density, composition, etc.) The sound of ocean should be twice as long as the other sounds, both to indicate its importance in the water cycle and to connect it with the next section—the sounds of life.

14. Fishy, swimming sound (life in water)

15. Frog jumping and splashing

16. Crickets chirping

17. Bees buzzing

18. Snake hissing

19. Birdsong (simple)

20. Birdsong (complex)

21. Horse clopping down road (slowly)

22. Horse galloping and snorting

23. Elephant trumpeting

24. Lion roaring

25. Dog barking

26. Whale or dolphin song (3 min.)

A sudden disproportional amount of time given to an animal sound may indicate to recipients that this is a special animal, though not as special as the sending species, since it is included with other animal sounds.

27. Monkey sounds

28. Chimpanzee sounds

We now have come through the evolutionary ladder to humans. This shows our connection with the rest of life on this planet, but to mark our sounds off, we bring up the heartbeat as a background to the sounds that follow.

29. Heartbeat (after 5 sec. by itself, it fades down and continues as background)

30. Slap and infant crying

31. Baby noises (gurgles and coos)

32. Baby vocalization

33. Baby's ma-ma

34. Children talking

35. Children singing

36. Recapitulation of first five hellos from UN greetings

By this, what we have done is say, "These are the sounds of life (coming from water); these are the sounds of our species; and that's what the first thing you heard [the UN greetings] was all about—it was some message in the sound of our species."

Now we continue with other sounds of our species, moving from body sounds, to mechanical and technological sounds.

37. People at party

38. Crowd at concert or lecture

39. Crowd at sporting event

40. Music and applause (last five sec. of singing from music section—such as the Mozart *Alleluiah*) (10 sec. plus 5 sec. applause)

This can give a very important clue as to the nature of the music part of the record. By previewing a snatch of the music that comes later, in the context of other human sounds (the heartbeat is running through all of this), we show that the music that occupies the second part of the record is a human sound. Perhaps very astute recipients will deduce the social nature of concerts. The fact that this sound is ten seconds long also indicates we want them to notice it especially as a clue.

41. Church bell

42. Hammer chipping flint

43. Hammer on nails

44. Jackhammer

45. Sawing wood

46. Power saw

Sequences such as 42–44 or 45–46 show a pretechnological sound with its technological descendent.

47. Bellows and fire

48. Hammer and forge

49. Blast furnace

50. Metal press or punch (heavy industry)

51. Crackling fire

52. Small internal combustion engine

53. Dynamo or generator

Sounds 47–50 show various levels of fabrication processes; sounds 51–53 sounds of energy usage and production. The remainder of the montage might show the development of forms of transportation.

54. Footsteps (slow)

55. Footsteps (running)

56. Horse clopping (same as item 21, but with addition of harness sounds and sounds of creaking wheels)

57. Car engine turning over

58. Car screeching down road

59. Small airplane engine

60. Jet engine

61. Saturn V liftoff

62. Heartbeat (fadeout)

This montage does not even come close to filling even twenty-three minutes (the time I have suggested is sufficient for a detailed montage). Thus, there is room for plenty of sounds that I have not thought of putting in (bacon sizzling, laughing, Ping-Pong, etc.) Additional sounds could be included in appropriate spots on the hierarchy. The structure of this montage connects sounds in a way that might inform recipients of their nature. On a more poetic level, the sound of the systole and the diastole, not specific to race or country, but characteristic of all our species, assures that a human heart will beat in space forever.

Note: These notes were written by Jon Lomberg, in April 1977.

4.0. The Aesthetic Dimension within the Interstellar Species Communication Haystack

In order to maximize the probability of the discovery of an extraterrestrial civilization, we have studied how to minimize the number of unknown dimensions usually employed by SETI scientists (Lemarchand and Lomberg 1996; Lemarchand 2002). We have explored the advantages of using several radial and bilateral symmetries and other aesthetic principles, such as the GS, to restrict some free variables of the Cosmic Haystack. We concluded that symmetry, as well as other aesthetic considerations, can contribute to the solution of general and specific SETI problems, including choices among the various frequencies, directions, and times that define particular search strategies. Figures 24.1 and 24.2 show how to apply some of these ideas.

The epistemological restrictions on the concept of the "universality of laws of nature" are basic assumptions that lie behind the concept of spatial-temporal symmetries and the hypothesis that universal laws have the property of isotropy and homogeneity in space and time. Our human cognitive maps use assumptions of "symmetries" to define the cognitive universals of the laws of physics.

There is also a strong link between aesthetic principles and the perception of scientific truth. For example, symmetry is the necessary condition to consider a law of physics of universal character. It is interesting that several prominent scientists consider aesthetics and symmetry as criteria of truth to be employed when choosing between one theory and another. Paul Dirac, more than any other modern physicist, became preoccupied with the concept of mathematical beauty as an intrinsic feature of nature and as a methodological guide for its scientific investigation. He originated the epigraphs: "A physical law must possess mathematical beauty" and "A theory with mathematical beauty is more likely correct than an ugly one that fits some experimental data." Similar statements are found in Einstein, Weyl, and Chandrasekhar.

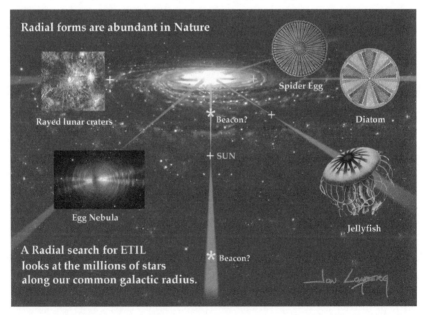

Figure 24.1. The ubiquity of radial symmetry in nature might be used as a cognitive universal to design transmitting and receiving strategies among galactic civilizations.© Jon Lomberg.

The Golden Section cognitive universal

φ and Fibonacci numbers

If $u_3 = u_2 + u_1$, and $u_1 = 1$ and $u_2 = 1$, then we derive the following series:

1, 1, 2, 3, 5, 8, 13, 21, 34, 55, 89, 144, 233, etc.

This is called the Fibonacci series.

The Golden Ratio, φ may be expressed as

$$\phi \rightarrow u_{n+1}/u_n \text{ as } n \rightarrow \infty$$

For u_{30}/u_{29}, $\phi = 1.6180339887498744831$

In *ABCD*, in which *AB:BC* = φ:1; through *E*, the golden cut of *AB*, draw *EF* perpendicular to *AB* cutting off from the rectangle the square *AEFD*. Then the remaining rectangle *EBCF* is a golden rectangle. If from this square *EBGH* is lopped off, the

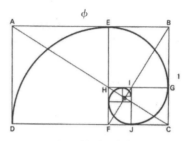

Logarithmic spiral

remaining figure *HGCF* is also a golden rectangle. We may suppose this process to be repeated indefinitely until the limiting rectangle *O*, indistinguishable from a point, is reached.

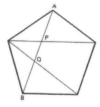

Nautilus pompilius   **Galazy M74 in Pisces**

Golden Cuboid

$$\frac{1}{\phi} = 1.618$$

$$\phi = .618$$
$$V = \phi \times 1 \times \frac{1}{\phi} = 1$$

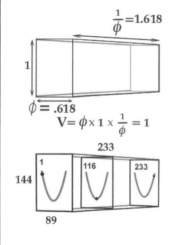

The point of intersection P of two diagonals of a pentagon divides each in the golden ratio *phi*. P divides AQ and AB internally and QB externally in this ratio.

Figure 24.2. The ubiquity of the Golden Mean, a proportion that connects the parts with the whole, is present in nature, art, pure mathematics to advanced computed algorithms, and might be used as a cognitive universal to encode interstellar messages. © Jon Lomberg.

A systematic epistemological analysis of beauty as a criterion of truth was conducted by McAllister (1996).

Lemarchand (2002) has given an example of how to combine Bohm's paradigm with the GS aesthetic principle. He proposed transmitting three-dimensional holograms using the Golden Section, which is a unique mathematical proportion between the parts and the "whole," as the key to unfolding the message. To do so, he described the construction of a Golden Cuboid (GC), in which there are sides of three different lengths, and these lengths are related to one another through the GS.

Any electromagnetic artificial interstellar message (e.g., a short movie) is translated into a time order, which is carried by the radio wave. Points that are near each other in the visual image are not necessarily near in the order of the radio signal. Thus, the radio wave carries the movie according to an implicate order. The function of using hypothetical aesthetic cognitive universals is, then, to explicate this order, that is, to unfold it in the form of a new sequence of visual images.

Using elementary algebra, it is easy to show that the dimensions of a cuboid (rectangular parallelepiped) of unit volume can have edges with lengths (a, b, c) related to the GS, if we let the shortest side (a) be Φ = 0.61803 . . . units long, the next longest side (b) is 1 unit long and the longest side (c) is the reciprocal of Φ (or Φ^{-1} = 1.61803 . . .) units long. When we calculate the volume of such a GC by multiplying the three sides ($\Phi \times 1 \times \Phi^{-1}$), we note that the volume is 1.

Lemarchand (2002) proposed to use the first thirteen terms of the Fibonacci Series as a calling signal that could be unmistakably recognized as artificial and that would also show a subtle implicate aesthetic order. A message organized in a temporal sequence of two level signals (on-off, 0-1, etc.) with 2,986,128 different time intervals can be regularly repeated. This is a representation of a three-dimensional GC message constructed as the product of three special Fibonacci numbers: $89 \times 144 \times 233 = 2,986,128$. The latter product will also place a limit on the equivalent number of bits for each GC message unit. The number 144 is the only square number of the Fibonacci series; moreover, it lies between two prime numbers that are also contiguous members of that series (89 and 233). Due to the particular character of the number 144 in the series, this number may be taken as a second-order basic unit of measurement. Dividing each of the three numbers in the sequence by 144 results in the following ratios: a = 89/144 $\approx \Phi$; b = 144/144 = 1; and c = 233/144 $\approx \Phi^{-1}$. The result is a three-dimensional aesthetic module, in which it is possible to insert representations of solid objects or a sequence of two-dimensional images to generate a short movie (see Fig. 24.2).

We may also encode more complex information reflecting some interesting aesthetic and mathematical properties of this GC such as: (1) the ratios between the areas of the faces follow the relation $\Phi:1:\Phi^{-1}$; (2) the total surface area of the cuboid is $2(\Phi + 1 + \Phi^{-1}) = 4\Phi$; (3) four of the six faces of the GC are also GS rectangles; (4) the ratio of the area of the sphere circumscribing the cuboid to that of the cuboid is $\Pi:\Phi$, an interesting relation between two incommensurable numbers; and finally, (5) the GC also shows a self-similar or fractal pattern: If two cuboids of square cross section ($\Phi^{-1} \times \Phi^{-1}$) are cut from the GC, the edge lengths of the remaining cuboid are in the same ratio as those of the original GC: $1:\Phi^{-1}:\Phi^{-2} = \Phi:1:\Phi^{-1}$. This results in a new GC, Φ^{-3} times the original size, providing a property that might be useful for encoding a higher level of information. In all cases, the content or meaning that is enfolded and carried is primarily an order and a measure, which might be unfolded by an aesthetic cognitive universal.

On Earth, several terrestrial cultures have long traditional schemes that connect aesthetics and symmetrical patterns (also present in nature) with transcendental values. Their human cognitive maps generate a perception of "Sacred Geometry" in several natural forms, ranging from crystals, sunflowers, and Nautilus pompilius seashells to galactic spirals. Those patterns are also found in many artistic works worldwide. There is also the belief that certain forms found everywhere in nature are useful for evoking a feeling of transcendence in art and religion. Examples include Hindu and Tibetan Mandalas, European church windows, and GSs. For the philosopher A. N. Whitehead and his theological followers, the purpose of the cosmos consists of its aim toward the intensification of beauty (Haught 2001). As in science, aesthetic principles are also thought to provide a connection with immanent and transcendental truth.

5.0. The Technological Adolescent Age Stage and the Long-Term Evolution of Ethical Values

The *Principle of Mediocrity* proposes that our planetary system, life on Earth, and our technological civilization are about "average" in the universe, and that life and intelligence will develop by the same rules of natural selection wherever the proper surroundings and the needed time are given. In other words, anything particular to us is probably average in comparison to others. From a Lakatosian point of view, the Principle of Mediocrity and the assumption of Universality of the Laws of Nature are within the hard core hypotheses behind any search for life in universe (from bacteria to intelligent beings).

Biological and ecological studies have shown that in terms of geological times different species on Earth emerged, developed, and became extinct with similar evolutionary patterns. Human species may not be an exception. The *Homo sapiens* has broken the "ecological law" that establishes that big, predatory animals are rare. Two crucial innovations in particular have enabled our species to alter the planet to suit ourselves and thus permit unparalleled expansion: speech (which implies instant transmission of an open-ended range of conscious thoughts) and agriculture (which causes the world to produce more human food than unaided nature would do). However, there is no evidence that natural selection has equipped us with long-term sense of self-preservation.

In this context, we have shown (Lemarchand 2000, 2006, 2010) that humans are facing a new type of macro-transition: the technological one. Lemarchand (2000) defined the Technological Adolescent Age (TAA) as the stage in which an intelligent species has the capability to become extinct due to: (1) self-destructive technological behavior (e.g., global war, terrorism, etc.); (2) environmental degradation of the home planet (e.g., global warming, overpopulation, increasing rates of species extinction, etc.); or simply by (3) the misdistribution of physical, educational, and economic resources (difference between the degree of development among developed and developing societies). The last factor might cause the collapse of the civilization due to the tensions generated by the inequities among different factions of the global society.

Since the invention of mass destruction weapons (e.g. nuclear, chemical, biological, etc.) we have for the first time—in the whole evolutionary history—the possibility of becoming extinct as a terrestrial species. The cultural differences are not a real issue when the whole species is facing the possibility of self-annihilation. From small towns of hunters in the Savanna to the most exclusive neighborhoods in the metropolis, everybody is in danger. The application of S&T discoveries to the military sector generated an exponential growth in the weapon's lethality (see Fig. 24.3). During the last one hundred years the new weapons increased in lethality by a factor of sixty million, while during the previous 2,300 years this number was only twenty.

Using data from 1800 to 1930, Richardson (1945) discovered that the distributions of wars over time follow a power-law. Levy (1983) introduced a definition of intensity of a war, I, as the ratio of battle deaths to the population at the time of the war. In our previous works (Lemarchand 2004, 2006) we have represented the distribution of the number of battles N_B against the intensity I. When we consider these distributions of deadly quarrels using normalized values of technologies, we found a correlation between the coefficient of lethality and the slope of the distribution that follows very well the behavior of a self-organized criticality (SOC) system. Our results show a very similar slope to the ones found by Clauset, Young, and Skrede Gleditsch (2007) in their analyses of the frequency of severe terrorist events worldwide between 1968 and 2006. Both studies cover the dynamics of the deaths—generated by violence among humans—from a few persons to tens of millions of people.

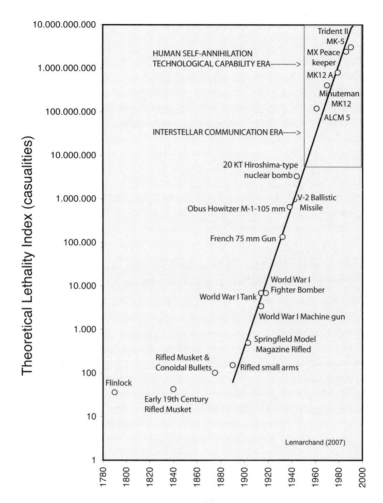

Figure 24.3. Historical evolution of weapons' lethality. Here we show the evolution of the Theoretical Lethality Index between 1780 and 2000. The vertical axis represents the theoretical maximum number of deaths that each weapon can generates per hour. The TLI increased in a factor of 60,000,000 during the last 100 years or at a rate of 600,000 per year. After Lemarchand (2010).

In a frequency versus intensity log-log graph the slopes found in both studies are close to −1.4. In one way, this shows that the dynamics on interhuman violence is governed by the same type of processes observed in a great variety of other complex systems that have the SOC property (Jensen

1998). A rough analysis shows that there is a small probability of having a violent event at which the complete human population will self-annihilate sometime between the next decades to the next hundreds years.

Using a large amount of empirical data across the last five centuries, Lemarchand (2010) showed the most dangerous combination of elements, which could start the end of our technological civilization within the next few decades. The elements that favor a human extinction event are the following:

1. Opportunity: Humankind is facing a demographic transition at global level, similar to the one that appeared with the invention of agriculture. This fact generates an out of equilibrium situation for the long-term global demographic dynamic that might amplify any perturbation to the system.

2. Probability: The analysis of the long-term interhuman violence—using five centuries of war data—shows a self-organizing criticality (SOC) behavior, which can be used to determine the probability of self-destruction by a global war within a window of few hundred years.

3. Technology: Since the invention of nuclear weapons, for the first time in the history of our species we have the technological capability to self-annihilate the whole species and most life on Earth.

4. Funding: Since World War II humankind has been increasing its military expenses at a global level. Nowadays, more than $1,400 billion is expended each year.

In this way, we have the opportunity, the probability, the technology, and the funding, all the necessary ingredients to become extinct. This is the reason why we are living in the most dangerous period of our history: the technological adolescent age. In this context, it is clear that it is impossible to have superior science and technology and inferior morals. In the long run this combination is dynamically unstable and we can guarantee self-annihilation within the lifetimes of any advanced societies (10^2 to 10^6 years).

At some point, in order to avoid self-destruction, all intelligent species in the universe must produce an ethical breakthrough among the members of their societies in order to achieve harmony in their planetary environment. Otherwise, the probability of global extinction will be very high, and consequently their societal life expectancy will be very short. Applying the Principle of Mediocrity, if we are average, the probability of detecting any evidence of extraterrestrial technological civilizations will be totally negligible.

For these reasons, our main thesis is that all civilizations should evolve ethically at the same time as they evolve technologically. When these civilizations reach their technological adolescent stage, they must perform an ethical societal mutation or become extinct. Any other possibility is very unlikely in the long run.

What kind of ethical principles should guide this transformation or social mutation? We consider that Kantian ethics provides some good elements to start the discussion. Kant's outstanding contribution to moral philosophy was to develop with great complexity the thesis that moral judgments are expressions of practical, as distinct from theoretical, reason. For Kant, practical reason or the rational will does not derive its principles of actions by examples from the senses or from theoretical reason; it somehow finds its principles within its own rational nature. Kant argued that willing is truly autonomous if, but only if, the principles that we will are capable of being made universal laws.

Such principles give rise to categorical imperatives, or duties binding unconditionally, as distinct from hypothetical imperatives, or commands of reason binding in certain conditions that we desire for certain ends. Kant seemed to hold that universalizability is both necessary and sufficient for moral rightness. Kant arrived at the ideal of "the kingdom of ends in themselves," or of people respecting each other's universalizing wills. This has been an enormously influential idea, and its most distinguished recent exponent has been John Rawls (1980).

Some useful ideas in the direction of the evolution of societal ethical stages—applied to the study of several terrestrial cultures—were developed originally by Piaget (1971) and extended by Kohlberg (1973). In his pioneer works, Kohlberg established a correspondence between Piaget's cognitive evolutionary stages and his moral judgment stages. According to his view, the final ethical evolutionary stage is based on "universal principles."

From the early works of Hamilton (1972) to Novak (2006) and West and Gardner (2010) there has been an increasing amount of evidence that altruism is a characteristic that is found in a great variety of species on Earth. The history of life on our planet shows that cooperation is needed for evolution to construct new levels of organization. Genome, cells, multicellular organisms, social insects, and human society are all based in cooperation. The fundamental principles of evolution are mutation and natural selection. But evolution is constructive because of cooperation. New levels of organization evolve when the competing units on the lower level begin to evolve (Novak 2006). Perhaps the most remarkable aspect of evolution is its ability to generate cooperation in a competing world. There is no doubt that we are facing a phase transition period in the history of our species. We must learn to cooperate together increasing the levels of solidarity and altruism if we want to maximize the number of our descendants or at least to guarantee the continuity of our genetic pool for the next millenniums.

At a certain point, all the advanced technological civilizations will need to reach a synergetic harmony among their species' individual members, their groups, and their relation with their own habitat. Probably, very advanced civilizations (measured in terms of technological capabilities) would extend this praxis to all living beings, including their hypothetical galactic neighbors. Their own evolutionary history will teach them the Kantian principle of respecting each other's universalizing wills. Being aware that each planetary evolutionary path is unique, these advanced civilizations will have a noninterference policy with the evolutionary process of underdeveloped societies. These species will not want to place potentially destructive knowledge at the disposal of any "ethically underdeveloped" society. Such knowledge could be a threat to the emerging society's survival. Any civilization needs time to work out adequate moral restraints on its own behavior.

These advanced civilizations could use a different approach to call our attention. Instead of sending hundreds of terabits of scientific and technical knowledge, they could send us manifestations of their artistic production. The manifestation of human symbolic thinking started with our first artistic expressions. It would be natural to think that it would be much easier to "contemplate" an extraterrestrial piece of art than to "interpret" the correct application of an extravagant technology. An extraterrestrial symphony, an abstract image, or a new aesthetic manifestation might help us to expand our symbolic capacities to new, unexpected frontiers and eventually diminish the incommensurability of cognitive maps between intelligent species in the galaxy.

All these different approaches, from the physical and natural sciences to the social, cognitive, artistic and philosophical disciplines, represent only tiny sparks in the dark. In fifty years of astronomical searches for extraterrestrial artificial signals, we have only explored a small fraction of the Cosmic Haystack (0.00000000000001%) with null results (Lemarchand 1998). We now need the strongest interaction among physical, natural, social, and humanist scholars to search together for new "universal cognitive maps" that might enhance our strategies for detecting the first evidence that we are not alone in the universe.

Acknowledgments

We want to thank the editors of *Leonardo* for their authorization to reproduce here the material published in an early version of this chapter. We are thankful for the valuable comments and suggestions made by Douglas Vakoch.

Note

This chapter is an expanded version of an article that appeared as G. A. Lemarchand and J. Lomberg. 2009. Universal Cognitive Maps and the Search for Intelligent Life in the Universe, *Leonardo* 42(5): 396–402.

Works Cited

Bohm, D. 1980. *Wholeness and the implicate order*. London: Routledge.

———. 1985. Fragmentation and wholeness in religion and science. *Zygon* 20 (2): 125–33.

———, and F. D. Peat. 1987. *Science, order, and creativity*. New York: Bantam Books.

Clauset, A., M. Young, and K. Skrede Gleditsch. 2007, On the frequency of severe terrorist events. *Journal of Conflict Resolution* 51: 58–87.

Dyson, F. J. 1979. Time without end: Physics and biology in an open universe. *Reviews of Modern Physics* 51 (3): 447–60.

Hamilton, W. D. 1972. Altruism and it related phenomena, mainly in social insects. *Annual Review of Ecology and Systematics* 3: 193–232.

Haught, J. F. 2001. Theology after contact: Religion and extraterrestrial intelligent life. *Annals NY Academy of Sciences* 950: 296–308.

Hoerner, S. von, 1974. Universal music? *Psychology of Music* 2 (2): 18–28.

Jameson, K., and J. Lomberg, 2010. Color processing universals and the construction of deep time messages. In *Proceedings of the Society for Psychological Anthropology* (in press).

Jensen, H. J. 1998. *Self-organized criticality*, Cambridge: Cambridge University Press.

Kohlberg, L. 1973. The claim to moral adequacy of the highest stage of moral judgment, *Journal of Philosophy* 70: 630–45.

Kuhn, T. 1962. *The structure of scientific revolutions*. Chicago: University of Chicago Press.

Lemarchand, G. A. 1992. *El Llamado de las Estrellas*. Buenos Aires: Lugar Científico.

———. 1998. Is there intelligent life out there? *Scientific American Quarterly* 9 (4): 96–104.

———. 2000. Speculations on the first contact: Encyclopedia Galactica or the music of the spheres?" In *When SETI succeeds: The impact of high information contact*, ed. Allen Tough, 153–64. Bellevue, WA: Foundation for the Future. www.futurefoundation.org/documents/hum_pro_wrk1.pdf.

———. 2002. Counting on beauty: The role of aesthetic, ethical, and physical universal principles for interstellar communication In *Between worlds:*

The art and science of interstellar message composition, ed. D. Vakoch. Cambridge: MIT Press. http://arxiv.org/abs/0807.4518.

―――. 2004. The technological adolescent age transition: A boundary to estimate the last factor of the Drake Equation. In *IAU Symposium 213: Bioastronomy 2002, Life among the stars*, ed. R. P. Norris and F. H. Stootman, 460–66. San Francisco: A.S.P.

―――. 2006. The life-time of technological civilizations. In *The future of life and the future of our civilization*, ed. V. Burdyuzha, 457–67. Dordrecht: Springer.

―――. 2010. The lifetime of technological civilizations and their impact on the search strategies. In *Bioastronomy 2007: Molecules, microbes, and extraterrestrial life*, ed. Dan Werthimer, Karen Meech, Janet Siefert, and Michael Mumma. APS Conference Series 420. San Francisco: Astronomical Society of the Pacific.

―――, and J. Lomberg. 1996. Are there any universal principles in science and aesthetics that could help us to set the unknown parameters for interstellar communication? Presented at the International Astronomical Union Colloquium 161, Capri, Italy, 1–5 July, 1996, 4–12. <www.jonlomberg.com/articles-Capri_paper.html>.

―――. 2009. Universal cognitive maps and the search for intelligent life in the universe. *Leonardo* 42 (5): 396–402.

Levy, J. S. 1983, *War in the modern great powers*. Lexington: University of Kentucky Press.

Lomberg, J. 1974. *Some thoughts on art, extraterrestrials, and the nature of beauty*. Mimeo publication by the National Air and Space Museum. Washington, DC: Smithsonian Institution.

―――. 1978. Pictures of Earth. In C. Sagan, F. D. Drake, A. Druyan, T. Ferris, J. Lomberg, and L. Salzman-Sagan, *Murmurs of Earth*, 71–121. New York: Random House.

―――, and S. C. Hora. 1997. Very long term communication intelligence: The case of markers for nuclear waste sites, *Technological Forecasting and Social Change* 56 (2): 171–88.

McAllister, J. W. 1996. *Beauty and revolution in science*. Ithaca: Cornell University Press.

Nowak, M. A. 2006. Five rules for the evolution of cooperation. *Science* 314: 1560–63.

Peat, F. D. 2002. *Blackfoot physics: A journey into the Native American universe*. Grand Rapids: Phanes Press.

Piaget, J. 1971. *El Criterio Moral en el Niño*. Barcelona: Fontanella.

Rawls, J. 1980. Kantian constructivism in moral theory. *The Journal of Philosophy* 77: 515–72.

Richardson, L. F. 1945. Distribution of wars in time. *Nature* 155 (3942): 610.

Shklovskii, I. S., and C. Sagan. 1966. *Intelligent life in the universe*. San Francisco: Holden Day.

Tsevetkov, V. D. 1997. *Heart, the Golden Section, and symmetry*. Puschino: ONTI, PNZ RAU (in Russian).

Vakoch, D. A. 1998. Signs of life beyond Earth: A semiotic analysis of interstellar messages. *Leonardo* 31 (4): 313–19.

West, S. A., and A. Gardner. 2010. Altruism, spite, and greenbeards (review). *Science* 327: 1341–44.

The Chemiosmosis Message

A Simple and Information-Rich Communication in the Search for Extraterrestrial Intelligence

Charles S. Cockell

1.0. Introduction

A large number of attempts have been made to construct interstellar messages (Drake 1992), which are either sent into space by some electromagnetic means, such as the Arecibo message transmitted toward globular cluster M13 in 1974 using the Arecibo radio telescope, or they are physically inscripted on a media of some type, for example the records (gold-plated copper disks) on board the Voyager 1 and 2 spacecraft.

Apart from the technical question of how to construct messages, an important question is what content the message should have (Vakoch 2008), and the relative content of art and science (Vakoch 2004). For example, for the Voyager 1 and 2 records, considerable effort was made to collect music, examples of written work, and photographs that represented many scientific and cultural aspects of our world (Sagan 1978). The construction of the Voyager records rested on the luxury of having a large amount of space compared to most messages beamed into space by electromagnetic means, where attempts are made to make the messages short and decipherable.

Regardless of the technical scope for a large information content, the minimization of the size of the message and the maximization of its information content will enhance the chances that the message will be deciphered and understood, and that it can convey useful information. Therefore, a challenge in interstellar message construction is to attempt to find short, information-rich messages that can be transmitted into space, whether by electromagnetic or physical means.

2.0. Characteristics of a Message

A message transmitted to an extraterrestrial intelligence or inscribed on a spacecraft should have a number of characteristics. They might include: (1) a scientific content to demonstrate that our civilization accepts the primacy of science—showing that we can engage in objective communication; (2) a technical insight of importance that demonstrates a willingness to communicate information; and (3) it should be simple, maximizing the chances that it will be deciphered and then understood (Sukhotin 1973). Simplicity will also reduce the chances that the message is degraded during transit, such that at the point of reception it is undecipherable.

3.0. The Chemiosmotic Message

One approach to interstellar message construction is to seek out a biological process that is common to all life on the Earth and that lends itself to message encryption. The process should be one that says something fundamental about the nature of terrestrial life and its function. Chemiosmosis is the biochemical process at the heart of energy capture in life on Earth (Mitchell 1961), and it evolved very early in the history of life (Lane 2010). The process works by the transfer of electrons through a cell membrane. This electron transfer is used to pump protons from the inside of a membrane to the outside. The protons subsequently move back through the membrane to generate energy-rich molecules (in terrestrial life, adenosine tri-phosphate ATP) (Figure 25.1). The remarkable beauty of the system is its ubiquity in life on Earth and its use of two subatomic particles to drive energy acquisition as the fundamental basis of operation. Many molecules are required to operate the system, including cytochromes, etc., and other electron transfer proteins, but these are terrestrial-specific innovations that are not important to describe the basic physical process at work. The total system is complex, even in pictorial form (Figure 25.1).

A simple message can be constructed based on the chemiosmotic system (Figure 25.2), which reduces the process to its simplest elements—transfer of electrons through a membrane, exchange of protons across a membrane, which leads to biomass (and obviously in the case of the transmitting entity, intelligence).

The effectiveness of the message lies in its simplicity—the depiction of only a membrane, an electron, and a proton—which lends itself to decryption by the receiving entity, but it contains a large information content about the biochemistry of life on Earth and the nature of our society For example, things that it reveals include: (1) that we have unraveled this system and must have mastered biochemical science; (2) that we consider this system

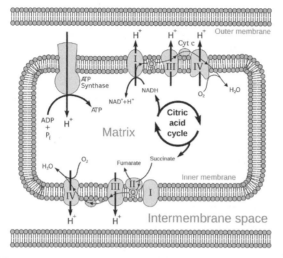

Figure 25.1. The chemiosmotic process within a mitochondrion, the energy-producing organelle of eukaryotes.

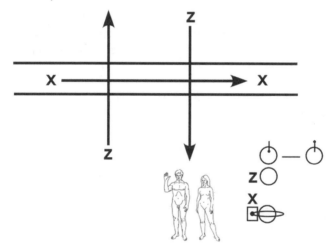

Figure 25.2. Depiction of chemiosmosis message. Message shows electron transfer through a membrane (two parallel lines) resulting in proton exchange as the basis of life. The outcome of this exchange is biology, depicted as humans (after the Pioneer 10 and 11 plaques). Electrons are depicted as in orbit around the center of an atom (bottom right) and marked with the symbol "X." Protons are depicted as the hydrogen atom without electron for simplicity (symbol "Z"). The hydrogen atom is shown above in its two states with a line depicting transfer from one state to the other. The hydrogen atom provides scale for the atom beneath it to show that X depicts a subatomic particle, not a planetary system.

important enough to transmit and must therefore cherish scientific knowledge; and (3) that we have mastered particle physics and have an understanding of the nature of matter.

If the intelligence recognizes the system, they will know that they share a common biochemical basis with us; if they do not, then they can attempt to create such a process in a laboratory—a technical insight or gift from our civilization.

Other molecules have received attention in designing messages. Although the structure of DNA was transmitted in the Arecibo message of 1974, unless detailed chemical information is provided on such a molecule it is not clear how the molecule would work, or whether it is in fact related to biochemistry at all. The details of the macromolecular structures involved in chemiosmosis are not required to depict the basic biochemical and physical core of the system.

4.0. The Value of Messages

An obvious question is whether short messages such as the one proposed here are of any value outside SETI. As with the Arecibo message, creating extraterrestrial missives, even if they are never received, makes us think about how to transmit scientific information across language and cultural barriers on the Earth, and how to create depictions of scientific processes and concepts that can be understood internationally. Many concepts in science are complex. The chemiosmotic process is no exception. Diagrams assist in explanations, particularly for metabolic pathways and other biochemical pathways, which are difficult to describe in words. However, even diagrams can be highly complex, as Figure 25.1 attests. The construction of extraterrestrial messages forces us to strip away information to its minimum and essential components, and thus to construct diagrams that depict the essentials of the process that we seek to transmit. In the case of the message proposed here, Figure 25.2, or variants of it, might be used to explain the basics of chemiosmotic energy conservation in pictorial format in textbooks and other printed and electronic materials used in schools.

Works Cited

Drake, F. 1992. *Is anyone out there?: The scientific search for extraterrestrial intelligence.* New York: Delacorte.

Lane, N., J. Allen, and W. Martin. 2010. How did LUCA make a living? Chemiosmosis in the origin of life. *Bioessays* 32: 271–80.

Mitchell, P. 1961. Coupling of phosphorylation to electron and hydrogen transfer by a chemi-osmotic type of mechanism. *Nature* 191: 144–48.

Sagan, C., ed. 1978. *Murmurs of Earth.* New York: Ballantine.

Sukhotin, B. 1973. Methods of deciphering of a message from extraterrestrial intelligences. *Acta Astronautica* 18: 441–50.

Vakoch, D. 2004. The art and science of interstellar message composition. *Leonardo* 37: 33.

———. 2008. Representing culture in interstellar messages. *Acta Astronautica* 63: 657–64.

A Proposal for an Interstellar Rosetta Stone

Stéphane Dumas

1.0. Introduction

Interstellar radio messages are probably our best way to communicate with an alien civilization. Since the delay of communication over long distances is considerable, the content of the broadcast is very important. The possibility for contact may occur once, and it should not be missed.

The current chapter reports on a proposal for an "Interstellar Rosetta Stone," which should be seen as an introduction to the whole message. Such an introduction is intended to teach the recipient how to read the rest of the communication.

The first part of the message is concerned with mathematics. The following sections are related to elementary notions of physics, chemistry, and biology. The objective is to communicate enough information so the receiver will understand the concepts, providing a common reference for future discussions.

The choice of science as the language of the preamble chapter is dictated by the fact that any alien civilization receiving it must have built some sort of device capable of detecting such transmissions. Possession of this device implies knowledge of engineering and therefore of mathematics and physics. It is also expected that the message itself will be analyzed, upon reception, by a group of people with the appropriate knowledge.

The process of writing a message to an extraterrestrial civilization is similar to the process of analyzing a message received from the same. The tools that we are going to use—information theory and statistics—should be the same for designing and decoding messages.

While mathematics and physics serve to introduce reference points for communication, they are not adequate to communicate social and human concepts, that is, to talk about our civilization. This will be the subject of further chapters of the message.

The message should be written using a two-dimensional structure (i.e., images) rather than a one-dimensional structure (such as Morse code). The second dimension offers redundancy of information, which is more resistant to noise than a linear message.

2.0. Starting with Science

The first section of the preamble introduces numbers. Numbers are easy to teach and do not involve many abstract notions. Learning to count is the first step in learning mathematics. Figure 26.1 provides an example of what could be transmitted to teach how to count. The numbers from zero to nine are shown using two different approaches: (1) counting dots and (2) binary representations. The choice of base-ten numerals is arbitrary. There is no reason to use or not use this base system. It is for pure simplicity for us.

The characters, or glyphs, are from a message transmitted in 1999 from Evpatoria in the Ukraine. Each character (a matrix of five by seven pixels) of this alphabet is different from other characters by at least seven bits out of the total of thirty-five bits for any orientation of the characters. Each character is unique and cannot be mistaken for others. This avoids characters similar to one another, which is common in natural languages, as exemplified by the letters p and q and the numbers 6 and 9.

Mathematical operators (e.g., +, −, *, /, and =) are then introduced via the same kind of notation. The idea is to write several expressions using the same operator, while keeping most of the symbols unchanged (Table 26.1).

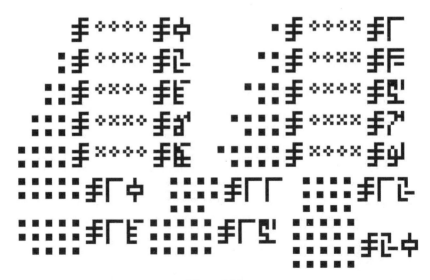

Figure 26.1.

Until this point in the tutorial, the numbers used to teach mathematics have been integers.

Division is used to introduce the floating point (Table 26.2). Floating point notation is very useful for representing very small and very large numbers in a compact way, as when we give numerical measures of the mass of the proton or the mass of the Sun. Division can also be used to introduce the concept of periodicity (e.g., 1/3 = 0.33333...) which is useful when describing groups of objects.

The notion of raising a number to a power is also important to teach early in the message. The typical way of writing a power uses exponential notation (e.g., $2^3 = 8$). However, it is preferable to keep all symbols on the same line, written in the same size. This will help with decoding. Therefore, instead of using exponential notation, we borrow from computer languages the "^" symbol (e.g., 2^3 = 8). Table 26.3 shows some examples of this.

Table 26.1. Learning How to Use Mathematical Operators.

0+1 = 1	1+1 = 2	2+1 = 3
0+2 = 2	1+2 = 3	2+2 = 4
0+3 = 3	1+3 = 4	2+3 = 5
0−1 = −1	1−1 = 0	2−1 = 1
0−2 = −2	1−2 = −1	2−2 = 0
0−3 = −3	1−3 = −2	2−3 = −1

Table 26.2. Division and Floating Point.

1/1 = 1	
1/2 = 0.5	1/3 = 0.333etc.
1/4 = 0.25	1/9 = 0.111etc.
1/5 = 0.2	2/3 = 0.666etc.

Table 26.3. Exponents and Powers of 10.

1^1 = 1	2^1 = 2	3^1 = 3
1^2 = 1	2^2 = 4	3^2 = 9
1^3 = 1	2^3 = 8	3^3 = 27
10^1 = 10	10^-1 = 0.1	
10^2 = 100	10^-2 = 0.01	1.23 * 10^2 = 123
10^3 = 1000	10^-3 = 0.001	123 * 10^-2 = 1.23

Notions of physics can be used to introduce length, time, and mass. In the 1999 Evpatoria message, the hydrogen atom was used to illustrate those notions. The hydrogen atom was schematized using the Rutherford Model. While the use of drawings may or may not be useful for interstellar communication, depending on ability of extraterrestrial recipients to interpret images, the glyphs in the message represent the masses of protons, neutrons, and electrons. The glyphs representing numbers are easily read if the mathematics section is understood. From those values, and more precisely, the ratio of those values, one can discover the mass and charge of the three particles. Whatever the unit used for the mass, the proton is 1,836 times heavier than the electron. The neutron has a neutral charge (= 0). The electron's charge is the inverse of the proton's charge. Those values are universal throughout the universe. Once the proton, neutron, and electron have been defined, their masses can be introduced.

Hydrogen can be also used to introduce the notion of length. Figure 26.2 shows the energy levels of the hydrogen atom. Once again, the drawing may be difficult to interpret by an alien reader, but the values are not.

With the discovery of exoplanets and new orbital observatories (i.e., Kepler and Corot), it is now possible to observe other solar systems. It is quite possible that an extraterrestrial civilization will have similar technologies and therefore can observe our own solar system. Figure 26.3 shows a diagram

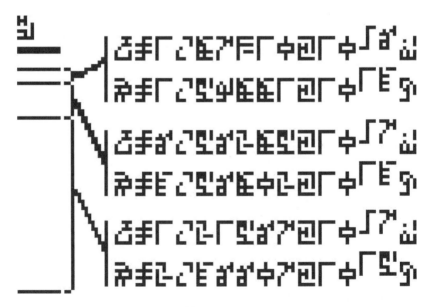

Figure 26.2.

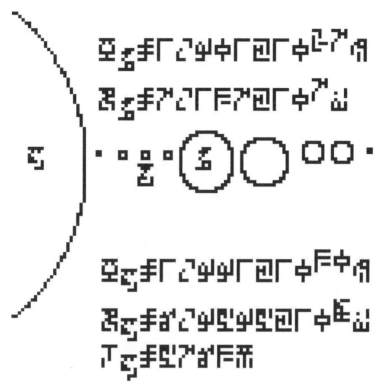

Figure 26.3.

of our solar system, with masses for Jupiter and the Sun indicated. Another way to illustrate our solar system without the use of drawings would be to list all planets with their masses and periods of revolution around the Sun.

Using the hydrogen atom as a starting point, other atoms may be introduced (Table 26.4). From there, chemistry and biology can also be used to convey additional information in the message.

Groups of atoms can be shows as molecules (Table 26.5). The previous notation must be respected. A number preceding a symbol indicates the quantity of that symbol (e.g., 2H = H H).

Table 26.4. More of the Periodic Table.

C : = 6*P + 6*N	O : = 8*P + 8*N
N : = 7*P + 7*N	Si : = 14*P + 14*N

Table 26.5. Defining Molecules

water : = 2*H+O
methan : = C + 4*H
carbonate : = C + 3*O

Chemical reactions can be used to introduce a new operator =>, meaning "the product of" or "yields." For example, 2*N + 6*H => 2*(N+3*H).

This scientific preamble serves only to establish a common ground with the extraterrestrial civilization. There is nothing new that an extraterrestrial civilization should not already know. The rest of the message would require a more complex language. From the point of view of the alien recipient, it will be also more interesting to learn something about human civilization. Consequently, any further communications will be centered, not on technology, but on social interactions. This level of communication requires a real language, or at least a more elaborated one.

3.0. Examples of Using Group Theory and Logic to Introduce New Concepts

Some fields of mathematics can help us explain subjects outside the realm of mathematics. Using logical propositions (Table 26.6) and some concepts from group theory (Table 26.7), it is possible to explain a little about ourselves.

Given all the previous suppositions then the following could be relatively easy to understand that:

Human: = Man or Woman
Seven billion humans are included on Earth

Table 26.6. Some Logical Propositions.

T : = 1 = 1		
F : = 1 = 0		
T or T = T	F or F = F	T or F = F
T and T = T	F and F = F	T and F = F

Table 26.7. Some Group Theory Expressions.

A : = {a,b,c}	B : = {d,e,f}	A union B = {a,b,c,d,e,f}
a is included in A	d is included in B	A inter B = {}
d in not included in A	a is not included in B	

It is also possible using this technique to talk about our biology. An example would be human procreation, where "M" stands for Man and "W" for Woman:

$$M+W => M+W+W \text{ or } M+W+M$$

Similarly, we could communicate about our longevity:

$$M + 80 \text{ years} => \{\}$$
$$W + 80 \text{ years} => \{\}$$

Conceptually, the "+" should add only the same type of objects. Here we used it to illustrate that humans have a life span of eighty years.

4.0. Why Not Use Human Languages?

Human languages are culturally dependant, not precise, and difficult to use. Human languages are too imprecise and ambiguous. For example, the sentence, "What disturbed John was being disregarded by everyone" has two different meanings, depending on the intonation (Chomsky 1966). Interstellar communication requires an artificial language to avoid potential misunderstandings and to be independent of social contexts. The process of building an artificial language should start with the already defined concepts from the scientific part of the message, and then expand beyond this. The structure of the language (e.g., the grammar, word sequence, etc.) can easily be found using statistical tools (Elliott 2001; Friedman 1992):

Given a bit stream, the coincidence test (kappa test) and entropy level can detect the size of the group required to represent a character. These tests will detect the 8-bit characters of an ASCII message.

Zipf's Law stipulates that the frequency of any word is inversely proportional to its rank. This a good method to produce a first grammar.

Markov's chain can be used to reconstruct the grammar of any language as it finds relationships between words.

The real focus should be on teaching vocabulary. This is the really difficult part of the whole message. This second chapter of the message should build on the science and introduce new words.

5.0. Examples of Artificial Languages

Several artificial languages exist already that could be used to create something suited for interstellar communication.

1. Lambda Calculus (Turner 2006): Created by Alonzo Church in 1930, it relies heavily on the use of mathematical functions. It is considered to be the first computer language. From this a pseudo-computer language can be created to describe processes and concepts.

2. Latino Sine Flexione (Peano 1903): This artificial language was proposed by Giovanni Peano in 1903. It is similar to Latin but without the heavy grammar, inflexion, double roots, and so forth. This would be close to a real human language but still dependent on social context.

3. Lincos (Freudenthal 1960): Lincos was created by Hans Freudenthal in 1960 as a language to be used for communication with extraterrestrial intelligence. Lincos primitives must be defined prior to being used. It is not a self-taught communication system, and it bears the problem of being a metalanguage. Preceding the Lincos text with a preamble could provide a solution.

Table 26.8 illustrates an example of Lincos in which Freudenthal used a dialogue to convey information. In summary, Ha asks Hb what is x such that 10x = 101. Hb answers 101/10, and Ha says this is good. Ha and Hb can be interpreted as humans A and B; inq stands for inquit (say) and ben represents bene (good). The dialog form is another example of structure to present the reader with information without entering into overly complex descriptions of concepts.

Table 26.8. Example of Lincos

Ha inq Hb ?x 10x = 101
Hb inq Ha 101/10
Ha inq Hb ben

6.0. Conclusion

In order to be able to communicate with an extraterrestrial civilization, a synthetic language must be created. The exercise of creating this artificial language may also indicate what to look for in a message from another civilization.

Once the basics of the artificial language have been understood, with the help of a basic dictionary, a more complex message could follow. During the course of the last chapter of the message, new words (and concepts) would be introduced via a combination of previous words. The dictionary would expand and a true conversation would begin. The different subjects to be discussed in the third part should be carefully addressed by a multidisciplinary group.

Works Cited

Chomsky, N. 1965. *Cartesian linguistics.* New York: Harper and Row.

Dresher, M. 1981. *The mathematics of games of strategy.* Toronto: Dover Publications.

Elliott, J., E. Atwell, and W. Whyte. 2001. *First stage identification of syntactic elements in an extraterrestrial signal.* 52nd International Astronautical Congress, Toulouse, France.

Freudenthal, H. 1960. *Lincos: Design of a language for cosmic intercourse, part 1.* Amsterdam: North-Holland.

Friedman, W. F. 1992. *Military cryptanalysis part III.* Walnut Creek, CA: Aegean Park Press.

McCowan, B., S. F. Hanser, and L. Doyle, 1999. Quantitative tools for comparing animal communication systems: Information theory applied to bottlenose dolphin whistle repertoire, *Animal Behaviour* 57: 409–19.

Peano, G. 1903. De Latino sine flexione. *La Revue de Mathematiques*, t. 8.

Revesz, G. E. 1983. *Introduction to formal languages.* Toronto: Dover Publications.

Stolayar, A. A. 1970. *Introduction of elementary mathematical logic.* Toronto: Dover Publications.

Turner, D. 2006. Church's thesis and functional programming. In *Church's thesis after 70 years*, ed. A.Olszewski, 518–39. Berlin: Logos Verlag.

Processes in *Lingua Cosmica*

Alexander Ollongren and Douglas A. Vakoch

1.0. Introduction and Basics

The present chapter is concerned with the nontrivial matter of incorporating descriptions of cooperating sequential processes without reference to time in a linguistic system for interstellar message construction. The linguistic system used in the present chapter is the new *lingua cosmica* (LINCOS) proposed and discussed in some detail in a number of papers by the first author (Ollongren 1999, 2001, 2003, 2004, 2010a, 2011a, 2011b; Ollongren and Vakoch 2003, 2011). LINCOS is based on constructive logic and its terms (logical forms) are represented as *types*.

Processes can be represented in LINCOS by a map in the form of an *inductive type*

$$\text{Proc} := \text{seq} : \text{Proc} \rightarrow \text{Proc} \rightarrow \text{Proc} \mid \text{arb} : \text{Proc} \rightarrow \text{Proc} \rightarrow \text{Proc}.$$

This injective, total map represents either a sequential process, or a process of arbitration. The *selectors* seq and arb identify the terms representing the two cases. A process p of type Proc, written p : Proc, occurring in a sequence is supposed to "do something," carry out an action (elementary or compound), and then is either followed by a process because

(seq p) : Proc $\rightarrow$ Proc, *or* it terminates.
If q is next after p in a sequence, (seq p q) : Proc.

Termination occurs also when q is the elementary process **stop**, as there is no next process in that case. Using these ingredients, sequential processes of any length (in terms of the number of steps taken), and also recurrent processes, can be represented in LINCOS. There is no reference to time. Elementary processes (actions, indivisible single steps) can be given fixed names, for instance **start** : Proc, **go** : Proc, and as mentioned **stop** : Proc. A terminating sequence of processes **go**, p_1, p_2, p_3, **stop** might look like this:

p_1 = (seq **start go**); p_2 = (seq **go** p_1); p_3 = (seq p_1 p_2);
stop = (seq p_2 p_3); (seq p_3 **stop**) undefined.

Here, p_1, p_2 and p_3 can but need not be elementary processes. Useful elementary processes are **wait** : Proc and **exit** : Proc. The "waiting for" situation can be characterized by the fact that for some process p, (seq **wait** p) : Proc is an idling state, meaning that the sequential successor of p is not automatically invoked. At the same time an arbiter (identified by arb) can be informed of that fact by the process (arb (seq **wait** p)) : Proc → Proc expecting some other process to succeed p. The arbiter can instruct the idling process to continue eventually and in that case (arb (seq **exit** p)) is the sequential successor of p. In practise, however, heavier machinery for arbitration may be needed, as the example of the seven dining philosophers farther on in this paper shows.

2.0. Representation of Processes

In the context of the present chapter it is relevant that processes are supposed to carry out actions, but there is no need to consider the internal constituents of them: these can be considered to be hidden. Therefore, processes can be abstracted and represented by variables. However, processes might have to communicate with the environment they reside in, or with other processes (for instance because cooperation is called for). So we need ways and means to achieve that. We shall use methods for registering states of multiple processes acting simultaneously: in the simplest case each of them executing ("doing something"), idling, or terminating. In this chapter we use *state vectors* for this purpose. In addition we arrange for processes to communicate with the environment via *channels*. Letting these objects be processes as well, there is no need for leaving the processes' realm as defined above.

In order to model cooperating sequential processes, configuration information is usually needed. That kind of information can be registered also in the form of a state vector. This suggests a special interpretation: a state vector, also a process, carries information, accessible to arbiter processes. The same is the case for channels. Below we work out in somewhat detail a specific case of cooperating sequential processes.

3.0. Seven Dining Philosophers

As a representative example we consider here a simple situation of limited resources and resource-demands, where arbitration is called for to ensure a fair kind of distribution between the process actions. The way arbitration can be modeled in LINCOS for this purpose is described. The following situation, rather famous in informatics, is provided with a model.

A round table is available for seven philosophers to dine. On the table seven plates are placed, and seven forks, each plate has one fork to the right and one fork to the left of it. A philosopher sitting at the table needs both forks to be able to dine. A philosopher is either thinking (process t) or dining (process d), not both at the same time. Dining and thinking is done in a finite fixed number of steps. Evidently only three philosophers can dine simultaneously (they need six forks). Let us name the philosophers 1, 2,...7 and let the initial situation be that philosophers 1, 3, and 5 are dining. Then there are initially three dining processes d1, d3, d5 and four thinking processes t2, t4, t6, and t7. Note that these seven processes run simultaneously, in parallel. We write for the collection of processes

$$d1 \parallel t2 \parallel d3 \parallel t4 \parallel d5 \parallel t6 \parallel t7 : \text{Proc.}$$

This initial situation can be represented by a *state vector of processes* ordered in some way, *e.g.* (d1 t2 d3 t4 d5 t6 t7) : Proc, as mentioned itself a process. A state vector contains in general information on "a state of affairs," but carries out no actions. So a state vector is a *silent process*. After a while (a finite number of discrete steps) a new configuration and state vector of processes will occur: some philosophers will leave the table, others will stop thinking and wish to dine. We are interested in a sequence of state vectors representing a fair regime in which all philosophers spend equal numbers of steps thinking and equal numbers of steps dining. We choose the following regime, a sequence of state vectors, that is, silent processes:

$$(d1 \ t2 \ d3 \ t4 \ d5 \ t6 \ t7)$$
$$(t1 \ d2 \ t3 \ d4 \ t5 \ d6 \ t7)$$
$$(t1 \ t2 \ d3 \ t4 \ d5 \ t6 \ d7)$$
$$(d1 \ t2 \ t3 \ d4 \ t5 \ d6 \ t7)$$
$$(t1 \ d2 \ t3 \ t4 \ d5 \ t6 \ d7)$$
$$(d1 \ t2 \ d3 \ t4 \ t5 \ d6 \ t7)$$
$$(t1 \ d2 \ t3 \ d4 \ t5 \ t6 \ d7)$$

This sequence can be considered to be a *program*. Note that a program like this does not exist as a type within the apparatus of LINCOS—in fact, one needs another level for reasoning about programs like this one. We will not go into this matter here, but the design of the *lingua cosmica* contains facilities for that purpose (Ollongren 2010b).

4.0. The Arbiter

In the definition of processes we have introduced the alternative

$$\text{arb} : \text{Proc} \rightarrow \text{Proc} \rightarrow \text{Proc, of arity 2.}$$

We assume now that the first argument of the selector arb is a silent process, in fact a state process vector and that arb maps process vectors on process vectors. So the arbiter can be used to change a set of processes running in parallel to another set of processes running in parallel (both are multiple processes). This objective is realized by associating to any process a *channel*, mapping the process to its state vector

$$\text{channel} : \text{Proc} \rightarrow \text{Proc.}$$

supplemented by a *return* process, mapping a state vector to the corresponding process

$$\text{return} : \text{Proc} \rightarrow \text{Proc.}$$

Let p : Proc be a process and (p) : Proc its vector. Then

$$(\text{channel p}) = (p) \text{ and } (\text{return (p)}) = p,$$

so (return (channel p)) = p. Next we show how a process p is "switched" to another process q using the arbiter. Let (arb (channel p)) = (q). Then

$$(\text{return (arb (channel p))}) = q.$$

Another way of achieving these results is by making use of elementary actions such as **wait** : Proc and **exit** : Proc. In more simple examples, for example, in modeling interrupts, these can be used fruitfully—but this will not be detailed here.

The first step of the pattern (program) for the case of the dining philosophers mentioned above, is then realized as follows using the first two "statements" of the program:

(channel d1 || t2 || d3 || t4 || d5 || t6 || t7) = (d1 t2 d3 t4 d5 t6 t7).
(arb (d1 t2 d3 t4 d5 t6 t7)) = (t1 d2 t3 d4 t5 d6 t7).
(return (t1 d2 t3 d4 t5 d6 t7)) = t1 || d2 || t3 || d4 || t5 || d6 || t7 : Proc.

416

5.0. Conclusion

In the example of the seven dining philosophers the specific choice of state process vectors and channels shows how an arbiter can be used to ensure a fair distribution of process demands given a limited number of resources. In this particular case the arbiter together with a program arrange an acceptable series of process actions satisfying the constraints and meeting the fairness criterion.

In the basic setting (the initial environment), two types of processes are introduced: sequential processes and processes of arbitration (associated with seq and arb), but the processes themselves are not detailed in any way—apart from the requirement that state vectors are associated with them. In addition there is no reference to time or the number of steps executing processes require.

Whether or not a program of arbitration specifies an acceptable series of actions for a given purpose is a matter for discussion at a higher level, in fact a *meta level*, where methods from mathematics or logic can be used freely. One enters then the domain of multilevel astrolinguistics (Ollongren 2010b). In the case described above, statistical analysis shows that the chosen distribution program together with a suitable rule for sequencing yields fairness.

Works Cited

Ollongren, A. 1999. *Large-size message construction for ETI: Typing static relations*, Paper presented at the 50th International Astronautical Congress, Amsterdam, Netherlands.

————. 2001. *Large-size message construction for ETI: An experiment in CETI*, Paper presented at the 52nd International Astronautical Congress, Toulouse, France.

————. 2003. *Large-size message construction for ETI: Inductive self-interpretation*, Paper presented at the 54th International Astronautical Congress, Bremen, Germany.

————. 2004. *Large-size message construction for ETI: Non-deterministic typing and symbolic computation in Lincos*, Paper presented at the 55th International Astronautical Congress, Vancouver, BC, Canada.

————. 2010a. On the signature of Lincos. *Acta Astronautica* 67: 1440–42.

————. 2010b. Representing sequential processes, Chapter 5.1. In *Astrolinguistics: Logic design of a system for interstellar communication, LINCOS*, Alexander Ollongren (publication pending).

————. 2011a. Large-size message construction for ETI: Aristotelian syllogisms. *Acta Astronautica* 68: 549–53.

————. 2011b. Large-size message construction for ETI: Recursivity in Lingua Cosmica, *Acta Astronautica* 68: 544–48.

————, and D. A. Vakoch. 2003. Large-size message construction for ETI: Self-interpretation in LINCOS. In *Bioastronomy 2002: Life among the stars*, ed. Ray Norris and Frank Stootman, 499–504. San Francisco: Astronomical Society of the Pacific.

————, and D. A. Vakoch. 2011. Large-size message construction for ETI: Typing logic contents. *Acta Astronautica* 68: 535–38.

Testing SETI Message Designs

Michael W. Busch and Rachel M. Reddick

1.0. Introduction

The search for extraterrestrial intelligence consists of two parts: conventional receive-only "Passive SETI" and "Active SETI," where signals are transmitted with the goal of an extraterrestrial observing eventually detecting them. Passive SETI is now highly developed, and our ability to detect and identify artificial radio beacons continues to improve (e.g., Siemion et al. 2008). Active SETI has been comparatively neglected, in both theory and implementation.

While the mechanics of transmitting radio signals across interstellar distances are well understood (e.g., NAIC 1975; Zaitsev 2006), there has been little effort spent on ensuring that a transmitted message will be understandable to an alien listener. Possible counterexamples include the Arecibo Message (NAIC 1975) and the Cosmic Call Message transmitted from Evpatoria (Zaitsev 2006), but neither was blind tested for decipherability.

Regardless of whether Active SETI is desirable, designing a message that is deliberately easy to interpret with a minimum of additional information is an interesting exercise. To this end, one of us (Busch) developed a coding scheme and a possible message and provided the other (Reddick) with the encoded data, in a blind test of the effort required for decryption.

2.0. The Test Message

The coder based the encryption scheme on the general purpose binary languages proposed by several authors (e.g., Freudenthal 1960; McConnell 2001), to avoid the potential for bias inherent in pictorial representations (Vakoch 2000). Blocks of code (in this case, eight-quad words) represent numbers, mathematical operators/verbs, variables/nouns, or delimiters, and are assembled into a series of statements. The message totaled 113,960 quads, but the content of the message was on average repeated three times, so that the true length is ~75 kilobits.

We assumed that for an interstellar beacon, a watcher would be able to identify the signal as artificial due to its low bandwidth, frequency modulation, and periodic Doppler shifts due to the Earth's rotation and motion around the Sun. Such a signal would be detected by almost all SETI data analysis programs (e.g., SETI@home, Anderson et al. 2008), provided it was strong enough to be detected and at an appropriate carrier frequency. In addition, the coder assumed the watchers would be able to locate the Sun as the likely source of the message and had at least an equivalent knowledge of mathematics, astronomy, and physics.

The watcher was provided with a version of the message that was missing a randomly selected amount of material from the beginning (10–20% of the total) and ~2 percent of later quads, to represent the initial detection of the beacon and intermittent instrument downtime. We did compromise the blind nature of this test in one way. The watcher was forewarned that the coding scheme was not one of those used in previous Active SETI messages, which have relied on images with a prime number of pixels in each row or column (NAIC 1975; Zaitsev 2006).

3.0. Initial Decryption: Pattern Recognition and Mathematics

To decrypt the message, the watcher used mostly pencil and paper for analysis and search-and-replace to replace deciphered blocks of the message. The lack of a need for high-power computing reflects the relatively small amount of data in the message. The watcher first recognized that the two most common strings in the message are delimiters separating individual pieces of code from each other.

After decoding the delimiters, the watcher observed statements such as:

$$(\; 10000000 \;\; 01000001 \;\; 10000000 \;)$$
$$(\; 10000003 \;\; 01000001 \;\; 10000003 \;)$$
$$(\; 10022133 \;\; 01000001 \;\; 10022133 \;)$$
$$(\; 10000000 \;\; 01000001 \;\; 20000000 \;)$$
$$(\; 10000001 \;\; 01000100 \;\; 20000001 \;)$$
$$(\; 10000001 \;\; 01000100 \;\; 10000000 \;)$$
$$(\; 10031242 \;\; 01000100 \;\; 10031243 \;)$$

and recognized "01000001" = "=" and "01000100" = "≠." She then determined the notation for integers ("1XXXXXXX" for positive, "2XXXXXXX" for negative). Additional statements illustrate variables ("30XXXXXX"), assigning values to them.

The next set of code the watcher decrypted consisted of fundamental arithmetic, followed by the notation for floating point numbers, and with these a representation of the first "nouns": "31000001" = e and "31000002" = π.

To provide additional perspectives on the decryption, we provided the message independently to six undergraduate students, who each spent no more than an hour attempting to decode it, again working without any pattern recognition software. Four of the students correctly identified the delimiters, with the other two also identifying "=," "≠," and the notation for integers.

4.0. Defining Physical Constants and "Nouns"

The nuclear fine structure constant, the gravitational coupling constant, the proton-to-electron mass ratio, and the neutron-to-electron mass ratio are dimensionless numbers that should be universally recognized, given that the watcher is proficient in physics. To define a system of units, the message contains a series of formulae relating the various Planck units to each other and to these numbers. The watcher found this transition from mathematics to physics to be the most difficult portion of decryption.

After recognizing the units, the watcher found the remaining portions of the message readily understandable. The proton, neutron, and electron are defined as nouns equal to statements providing their masses, charges, and, for the proton, charge radius. A partial chart of the nuclides defines the stable and most common isotopes of twelve elements, providing their masses and number of protons, neutrons, and electrons. For hydrogen, it also includes the radius, nuclear and electric binding energies, and a selection of spectral lines. Given the nouns for the atomic species, chemical definitions follow naturally. In a slight subversion of the intended decryption, the watcher recognized the chemical formulae for molecular hydrogen, water, and oxygen, and used them to more rapidly decrypt this section of the message.

At this point, almost all of the symbols and more than 75 percent of the content of the message had been decrypted. While we have described the decryption in a linear fashion, the message is not structured this way. Individual blocks of code (series of statements) are separated from each other and repeated a varying number of times. For example, the blocks defining =, ≠, and + are given five times, while the chart of the nuclides and the chemical formulae are given only twice. Overall, the message is three times longer than the individual blocks of code. This redundancy serves the obvious purpose of allowing the message to be decrypted regardless of when the watcher started observing.

The next-to-last block of code defines a new noun "31130000," which describes something with a mass of 1.989×10^{30} kg, a temperature of 5778 K, a radius of 6.955×10^8 m, and a power of 3.839×10^{26} W. The block also describes the mass in terms of fractions of hydrogen; deuterium; the isotopes of helium, carbon, nitrogen, oxygen, and neon; and amounts of iron, sulfur, silicon, and magnesium. Together, this composition information defines the

Sun, which would presumably be the subject of scrutiny when it is in the same direction as the source of the message. The watcher did know the composition of the Sun beforehand, which slightly compromises the results.

The Sun block also refers to the eight subsections of the last block, which provide masses, radii, temperatures, and a set of distances and times that follow Kepler's third law. These blocks define the planets. One block contains two sets of composition information: one typical of rocky terrestrial planets and one a list of gases dominated by oxygen. In addition to defining the noun for "atmosphere," this indicates certain chemical disequilibria on the object. Its orbit also matches the Doppler behavior of the beacon. So the message defines the Earth and provides a minimalist description of terrestrial life. The watcher recognized all of this and accurately decoded it after a total of approximately twelve hours of work.

5.0. Comparison to Other Message Designs

As a comparison, the watcher was also provided with one of the Cosmic Call messages (Zaitsev 2006). Deciphering this message was trivial, since it is a pictorial message based on a grid and the watcher merely had to adjust the width of the display window of her text editor. As mentioned above, the coder used a general purpose language rather than a pictorial representation to avoid such biases. There is a more important comparison between the two message designs, however.

The information contained in our test message is roughly equivalent to that contained in twelve images of the 2002 Cosmic Call message. These images total 193 kilobits, or ~2.5x the length of our message. In addition to avoiding a possible human bias toward interpreting images, constructed languages convey information more densely than purely pictorial messages.

6.0. Conclusions

In a sharply limited number of bits and assuming only a common knowledge of radio and stellar astronomy and the physics and mathematics required to build a radio telescope, we can establish a common vocabulary and describe the solar system in considerable detail. Tentatively, those who receive it can reliably decrypt such a message—with the caveat that we have been using human astronomers as proxies for extraterrestrial intelligence. Finally, a general purpose constructed language is significantly more information-dense than a series of prime number by prime number images, an advantage for beacons to be detected over large volumes.

We ask two additional questions. How much data are required for a blind decryption of more complicated ideas? And, given that there are no technical or apparent theoretical limitations on communicating intelligibly with unknown extraterrestrial watchers, should Active SETI be developed on a large scale?

Acknowledgments

We thank C. J. White and B. R. Lawrence of Caltech and four anonymous contributors for their efforts in decrypting the test message.

Works Cited

Anderson, D. P., J. Cobb, E. Korpela, M. Lebofsky, and D. Werthimer. 2002. SETI@home: an experiment in public-resource computing. *ACM Communications* 45: 56–61.

Freudental, H. 1960. *Lincos: Design of a language for cosmic intercourse,* Amsterdam: North Holland Publishing.

McConnell, B. S. 2001. *Beyond contact: a guide to SETI and communicating with alien civilizations.* Sebastopol, CA: O'Reilly Media.

NAIC—The staff of the National Astronomy and Ionosphere Center. 1975. The Arecibo message of November 1974. *Icarus* 26: 462–66.

Siemion, A., J. Von Korff, P. McMahon, E. Korpela, D. Werthimer, D. Anderson, G. Bower, J. Cobb, G. Foster, M. Lebofsky, J. van Leeuwen, and M. Wagner. 2009. New SETI sky surveys for radio pulses. http://arxiv.org/abs/0811.3046.

Vakoch, D. A. 2000. The conventionality of pictorial representation in interstellar messages. *Acta Astrononautica* 46: 733–36.

Zaitsev, A. 2006. Messaging to extra-terrestrial intelligence. arXiv:physics/0610031v1.

The DISC Quotient

A Post-Detection Strategy

John R. Elliott and Stephen Baxter

1.0. Introduction

In this chapter we present a strategy to follow the receipt of a complex and potentially decipherable signal from extraterrestrial intelligence (ETI). This includes a signal-processing algorithmic procedure that contains analysis stages based on the signal's data quantity, information-theoretic and linguistic structure, and affinity to known human languages. In addition, we propose a numerical scoring system based on the procedure's algorithmic steps to characterize the significance of the signal and its subsequent analysis for the purpose of public communication. This "DISC Quotient" scale is modeled on the example of the "Rio Scale" for characterizing the discovery of an ETI.

2.0. DISC: Decipherment Impact of a Signal's Content

Suppose SETI succeeds.

It is an indisputable fact that positive identification of a signal from an extraterrestrial source will have a profound effect on the human race. And, because of this, we now have initial strategies in place to cater for such a "contact" situation. We have methods for calculating the significance and impact of announcing a signal and the risk factors for replying to such a signal (the Rio Scale, the San Marino Scale, and the First SETI Protocol); much has also been discussed about how we manage a post-detection announcement situation.

But what next? In the event that we can prove that we have detected an extraterrestrial technology and that the signal displays intelligent-seeming structured content, we will be in a much more complex situation. No longer do issues of dissemination merely focus on announcing facts surrounding the existence of a technological "beacon"; we now find ourselves facing the complexities involved in understanding and glimpsing the intellect of the author—while the world's fears and expectations would demand immediacy of information.

The public demand to know can be summarized by a simple question: "What does the message say?" This in turn can be broken down to subquestions. "Is this a message? How large is it? How complex is it? What level of cognition produced it? Can we ever translate it—and if so, how long would it take?" To put the challenge of decipherment into context, we still have many scripts from our own antiquity that remain undeciphered, despite many serious attempts, over hundreds of years (Pope 1999).

With these issues in mind, we look at the immense difficulties involved in trying to decipher the content of a signal and in communicating information on progress, while that decipherment is underway. In developing a strategy for message detection and decipherment, comparators from existing protocols for "catastrophic" and globally significant events that have high societal impact are presented as supporting rationales. Nevertheless, a post-detection scenario has very particular challenges that form its core metrics, dictating logical stages and subsequent information flow, which are significantly affected by the unknown aspects of its structure and content. To assist our capabilities in tackling such a complex task, prior research conducted on identifying structural "universals" and decipherment strategies (Elliott 2000, 2002, 2007), based on aspects of these computational phenomena identifiable in the constructs of language, provide essential insights into the difficulty factors each phase is likely to present.

Building on a "position" paper rationale (Elliott 2008), we propose an algorithmic rationale based on previous research into signal decipherment techniques as an initial methodology for attempting to unlock the content of an extraterrestrial signal.

In terms of communication, the "DISC Quotient," in which a numerical significance factor is assigned to each of four signal analysis algorithmic steps, is a numerical method to characterize the cultural significance of the receipt of a complex and potentially decipherable signal from ETI. DISC Quotient scores would be published and updated as an analysis was underway. The purpose is to facilitate the public communication of work on any such claimed signal as such work proceeds, to provide a predictor of the work's likely outcome, and to assist in its discussion and interpretation. Analogies from human civilization are given. The most significant outcome would be a large, information-rich, and fully translatable signal from another culture: an encyclopedia, perhaps, or a Bible.

At present, the DISC Quotient is envisaged as summarized into a four-axis diagram, to be populated in the course of the signal analysis. Ultimately the four factors may be combined into a single numerical scale analogous to the Rio Scale (Almár 2001), with appropriate weightings based on example cases.

This technical analysis may of course be just the first stage of the wider scholarly study of the message, as we addressed the ultimate question of the meaning and significance of a message from another culture.

3.0. Previous Work

There are no existing algorithmic methods or numerical communication scales that address the specific issue of decipherment. As previously discussed in Elliott (2008), previous work has focused on the societal impact of receiving a message as a general issue (SETI 2008a, 2008b). We have scales to assess the risk associated with message transmission (Marino Scale), and how seriously to regard claims that we have detected an ETI (Rio Scale). Issues tackled in these papers do have some relevance and therefore aspects are included here for discussion.

The Post-Detection SETI Protocol (Billingham 1996) is in a sense a misnomer as it focuses on transmissions from Earth rather than protocols regarding how we manage a positive detection. It presents a set of principles concerning the sending of communications to extraterrestrial intelligence based on SETI being successful in detecting an extraterrestrial civilization, and raises the question of whether and how humanity should attempt to communicate with the other civilization.

In essence, the San Marino scale (Almár and Shuch 2005) is a method of quantifying the potential impact of active SETI (transmitting signals into space from Earth). It is an ordinal scale between one and ten, used to quantify the potential exposure of employing electromagnetic communications technology to announce Earth's presence to our cosmic companions, or replying to a successful SETI detection.

The Rio Scale (Almár and Tarter 2000; Almár 2001) is an attempt to quantify the importance of a candidate SETI signal. Again, it is an ordinal scale but in this case between zero and ten, used to quantify the impact of any public announcement regarding evidence of extraterrestrial intelligence.

It should be noted that the Rio Scale is a tool for dynamic, rather than static, analysis. Throughout the life of any unfolding SETI event, as research is conducted and verification measures pursued, new information is constantly being made available that will impact our perceptions as to the significance and credibility of the claimed detection. Thus, the categorization of social impact from any post-detection decipherment of its content is likely to ebb and flow from an initial high value on the Rio Scale, rather than yielding a monotonically rising or static set of values. In fact, the aim of such a strategy is to identify potential peaks and probable periods of complexity that will result in "quiet" periods and devise methods that maintain information dissemination across an otherwise variable set of complex tasks.

This ebb and flow of significance as seen by the public is likely to continue during decipherment exercises, reflecting a complex and changing perception of a complex and dynamic process. As an example, initial confirmation of intelligent content is likely to have high public impact, as will any latter stages confirming semantic content; other stages, which look at cognition, may have less immediate but still far-reaching impact. For example, expectation and possible anxiety could be heightened if it is not possible to disprove a cognitive content to the message—or, later, anticipation may be reduced if an inability to assign semantic "values" indicates that a full translation will be much delayed if possible at all, even if evidence of cognition is present. The latter of these is likely to present a worst case scenario, as "hobbyists" may well attempt their own analysis, which could be disruptive and increase anxiety levels as pessimistic (even apocalyptic) scenarios are debated in the media.

Thus, DISC Quotient scores would be updated repeatedly during the course of a decipherment process, in an attempt to guide and inform the public reaction.

4.0. DISC Processing Stages and Numerical Factors

Elliott's post-detection decipherment strategy (2008) can be summarized in four stages as set out below. These four stages are steps toward the ultimate goal of translation, which is a full semantic assignment of message content, perhaps with probabilistic prioritization of alternative interpretations.

For the purposes of the DISC Quotient numerical factors δ, ω, σ, α have been assigned to each processing stage, each ranging in value from 0 to 10, 0 meaning least significance and 10 meaning maximal: DISC = (δ, ω, σ, α). These parameters are defined below.

1. *Characterization as a Signal and First Estimate of Size:* δ. Given that a signal has been characterized as message-like (as opposed, for example, to image-like), a first indication of its significance is its size, perhaps in the first instance of the length of a no-repeating bit stream, and later based on counts of internal components. δ is a measure of data quantity: "How big is the message?" The scores are assigned by comparison to human analogues, as defined below. Of course a comparatively short work may have a disproportionate impact; the Koran is the length of a short novel, yet has shaped human history. See examples presented in Table 29.1.

There is a minimum meaningful size in this context. It is estimated that a signal of complexity equivalent to human speech would require twenty

Table 29.1. Characterization as a Signal and First Estimate of Size: δ.

δ	Message size (bits)	Comparison
0	<$10\exp5$	e-mail; short message; Phaistos Disc; page out of an encyclopedia
1	~$10\exp5$	1 image low-resolution
2	~$10\exp6$	Koran; Sagan's novel *Contact* (Sagan 1985)
3	~$10\exp7$	whale song [30 minutes average duration]; Bible; collected works of Shakespeare
4	~$10\exp8$	average language corpus; large dictionary
5	~$10\exp9$	*Encyclopaedia Britannica*; fictional Vegan "Message" in Sagan's *Contact*; Renaissance; symphony in high-fidelity sound
6	~$10\exp10$	human genome; estimated memory capacity of a human being's functional memory
7	~$10\exp11$	library floor of academic journals on shelves
8	~$10\exp12$	one-hour Hollywood movie
9	~$10\exp13$	all the data from Microsoft's WorldWide Telescope
10	~$10\exp14$ or greater	Library of Congress; estimated text content of Internet; all human knowledge, or greater; ~$10\exp32$ bits to store the "pattern" of a human being (with ~$10\exp27$ atoms per human body)

Illustrates a range of information examples ranging in [δ] data quantity: "how big is the message?" The scores are assigned by comparison to human analogues.

thousand words (~$10\exp6$ bits) to enable a full semantic analysis (of any comparable complex/information rich system).

2. *Information-Theoretic Analysis:* ω is a measure of data quality, and a first indication of complexity and likely cognition. In early stages this can be measured using the Shannon entropy order analysis of information theory, in which any given signal can be broken down into a distribution of entropic values, a measure of internal structure and correlations, with human languages reaching a typical maximum order of 9 (see Table 29.2). Maximum values for different species correlate with encephalization quotients. (Note that the "chimpanzee" assessment given here is extrapolated from encephalization quotient rather than measured directly from speech analysis.)

3. *Linguistic Analysis and Cognitive Assessment:* σ. For a language-like signal, σ is a further measure of data quality, complexity and likely cognition based

Table 29.2. Information-Theoretic Analysis: ω.

ω	
0	White noise
1	Pulsar-like repeating pattern
2	Ant chemical signaling
3	Chimpanzees
4	Bee "waggle dance"
5	Whale "song"
6	Dolphins
7	Proto-language: petroglyphs/pictographs
8	Music
9	Human—Modern languages
10	Exceeds complexity of human speech

This table ranks examples against.[ω] data quality, and a first indication of complexity and likely cognition. Complexity exceeding human speech can be also characterized using these techniques.

on linguistic (rather than information-theoretic) analysis. Numerical scores can be assigned analogously to ω using more direct comparisons, perhaps subjectively assigned, to human speech and other data sets.

4. *Semantic Analysis:* α. α is an assessment of the likely outcome of the semantic analysis of a complex signal, with values ranging from 0 implying no affinity to any human language to 10 implying very human-like language. The various outcomes have analogies to previous experience in the analysis of human languages.

At these "higher" levels of analysis, resources such as the "Human Chorus Corpus" (Elliott 2003) and the "Affinity Matrix" (Elliott 2007) will be instrumental in assigning values for impact assessment.

The Affinity Matrix measures characteristics, which comprise the behavioral "footprint" of a given script; this will then provide a set of comparable measures by which its affinity can be ultimately measured against another linguistic "system." The Human Language Chorus Corpus will be a central resource from which all known language constructs are ultimately modeled and contrasted. Not to constrain ourselves to human communication, research into other "intelligent" communicators, such as dolphins and humpback whales, is ongoing. Currently, these include knowledge of their surface and internal communication, which show similar structural signatures to our own. As this research progresses, these resources will be enriched to provide an ever-increasing ability to accurately measure all elements of an unknown (ETI) signal.

As indicated in Table 29.3, α is a key indicator of the time likely to be required to complete any translation, if possible at all; the time of translation is likely to be a key determinant of cultural impact (Baxter 2008).

Table 29.3. Linguistic Analysis and Cognitive Assessment: σ.

α		Analogy	Likely translation time
0	No affinity	Unintelligible or impervious format	Infinite; likely to be impossible
1	Complex signal; low but non-zero affinity	DNA/Chemical-type constructs forming communication carrier	Likely to be impossible, without crib or context
2	Partial affinity: Nonlinear complex (possible sensory synaesthesia) system with embedded (ellipsis or undetectable layers) content prevalent with no primer	Sound (speech) communicated as spectrum of colors, some of which beyond our capabilities	Likely to be impossible, without crib or context
3	Music/Mathematical-based communication with no primer	Lincos-type constructs	Syntactic—rapid; Semantic—protracted period extends depending on system's internally stated relationships
4	Primitive language (akin to petroglyphs) with some affinity; no crib	Cave paintings: ~50 symbols to depict all concepts	Rapid or Generations, depending on symbol universality
5	Mature language with some affinity; no observation possible to aid translation	Computer/programming-type language construct; embedded functions (objects), loops, etc.	Generations?
6	Mature language with no affinity; some observation possible to aid translation	Cognitive and contextual (anaphoric) constructs beyond our capabilities; Apes trying to understand our language	Years +?
7	Nontransparent content; hidden/embedded information	Ideograms; cuneiform	Decades?
8	Transparent content	Linear B	Years?
9	Crib present	Sagan *Contact* message	Rapid
10	Like known human language	Martian English	Very rapid

Presents examples of (α)—an assessment of the likely outcome of the semantic analysis of a complex signal, with values ranging from 0 implying no affinity to any human language to 10 implying very human-like language. The various outcomes have analogies to previous experience in the analysis of human languages.

Table 29.4. DISC Quotient Applied to a Fictional Example.

	δ	ω,	σ	α
Stage 2: type of message: 261 primes (10exp5 bits?)	0	1	1	0/10
Stage 2: type of message: Hitler images (10exp10 bits)	6	1	1	0/10
Stage 2: type of message: manual (10exp9 bits)	5	10	10	7
Stage 3: structure analysis (before crib found)	5	10	10	7
Stage 3: structure analysis (after crib found)	5	10	10	9
Stage 4: rapid translation proceeds	5	10	10	9

DISC quotient analysis to the fictional case-study scenario of a receipt of a complex message from ETI in Carl Sagan's novel *Contact* (Sagan 1985); in the "α" column, two values are recorded. The first value is for language like structural affinity, at this stage; the second is for information identifiable from non language like structure, such as images that convey meaning and therefore communication.

5.0. A Case Study: DISC Quotient Applied to a Fictional Example

In this section we apply the DISC quotient analysis to the fictional scenario of a receipt of a complex message from ETI in Carl Sagan's novel *Contact* (Sagan 1985). This follows similar trial applications of the Rio Scale (Shostak and Almár 2002).

In Sagan's novel the analysis begins as a prime-number sequence is detected from Vega (Rio scale 9). Then the signal is progressively analyzed and deeper levels found: first, 261 prime numbers, then "echoed" TV images of Hitler, then a construction manual to build a wormhole "Machine." Eventually, a primer is found for the manual: see Table 29.4 for DISC measures applied to this scenario. In addition, Table 29.5 presents some additional examples of varying size and complexity, as comparators.

Table 29.5. DISC Quotient Applied to Comparator Scenarios.

	δ	ω	σ	α
Bible	3	9	9	10
Whale song	3	4	3	0
Sagan Vega message	5	10	10	9
Encyclopedia Galactica (translatable)	10	10	10	9
Encyclopedia Galactica (untranslatable)	10	10	10	0

DISC Quotient Applied to additional message scenarios for measurement comparison.

6.0. Summary Representations

Regarding the communication significance to be derived from a DISC analysis we have experimented with the summary diagram representation shown in Table 29.6.

The examples indicate that cases of particular types—such as a complex and translatable message versus a complex but intractable message—yield particular at-a-glance patterns, which might aid communication. A possible further development of the Quotient is to produce a weighted total, a single number to indicate the significance of a message and an analysis event, after the manner of the Rio Scale.

These initial examples serve to demonstrate the dynamic flexibility of the defining metrics, which comprise the DISC Quotient. Each of its numeral factors is derived from a rigorous analysis of linguistic and complex communicative phenomena, which form algorithms and models representing

Table 29.6. DISC Quotient Summary Representation Grid.

				ω			
				10 *A,B*			
				9 C			
				4 *D*			
σ	10 *A,B*	9 C	3 *D*	0 B, D	3 C, D	10 *A,B*	δ
				B			
				5			
				9 A			
				10 *C*			
				α			

Presents the proposed summary representation for communication significance derived from a DISC analysis.

Key: A = translatable Encyclopedia Galactica; B = nontranslatable Encyclopedia Galactica; C = Bible in close-to-human language; D = Whale song our knowledge of known constructs and "universals." This suite of analytical resources ranges from initial "physical" analysis, which incorporates pattern matching and entropic assessment of the structure, to detecting cognitive and semantic constructs through syntactic relationships. Furthermore, the factors are computationally tractable: they are measures designed to provide us with scientific and objective assessments of the "signal," rather than just best guesses and personal opinion.

7.0. Dissemination Mechanisms and Rationales

Contact with an ET civilization will, of course, present unique issues and impacts on the individual and each sociocultural group, as well as humanity in general. It is therefore essential that the delivery mechanism for informing the public is such that it does not provoke (or optimizes reduction of) a sense of overwhelming fear or panic. Strategies currently in place, which address comparable issues; include the disseminating information for a potential pandemic, global disaster, and terrorism alerts. On a grid of qualitative categories of risk (OECD 2001), depicting scope and intensity measures, the endurable-global intersection, where pandemics lie, confirms the appropriateness of such strategies.

Another factor in the equation of communicating in such a considered manner is the contribution played by the media. Although much of the media's reporting is both factual and responsible, there are elements that will speculate to a level that may create false and negative impacts, if an information vacuum occurs, especially during the initial phases, where much of the core information will be discovered. Additional to this will also be the vast communication networks on the Internet, where there are no safeguards for maintaining factual information. Nevertheless, responsible media coverage can be considered a positive force in such high-impact situations, where at their respective "local" levels they are a trusted source of information (Burkeholder-Allen 2000). It is therefore paramount that "official" channels are as effective and considered as possible, to facilitate clear, nonsensationalistic, and accurate information in a timely and trusted manner.

As an effective conduit for this mechanism, it is also submitted that "paper" people (people who convey cooperation and can represent all peoples of the world) may be needed as a team of spokespersons, rather than faceless statements from experts, to avoid negative effects, such as a heightened perception of threat, belittlement, and hopelessness, seen in focus groups interviewed on strategies for communicating a potential pandemic (Jones et al. 2007). However, experts not representing or attached to any government, are less likely to rouse these negative issues (seen in these trials), which should allow SETI scientists to fully participate in phases of dissemination where appropriate. Other mechanisms that will support positive dissemination of information are attention to accurate translations with regard to cultural constraints (Duggan and Banwell 2004) and associated regional communicators (CELI 2008).

Immediate impact on discovery of a signal (message) is likely to result in: fundamental change in accepted knowledge (wisdom), behavior, and attitude; individual and sociocultural interaction with new information (truths); awareness and implications of the source.

Key Factors influencing information dissemination are therefore:

- Reinforcement of and/or new adjustment to information and assumptions;
- Perception of relevance and impact on human civilization;
- Conjecture beyond physical, geographical, and decipherable facts;
- Religious/societal impact and reaction;
- Recognition of need for new knowledge;
- Willingness to accept and change as a result of contact;
- Need/willingness to "participate" (be involved) in process.

In the initial phase (announcement of signal detected), the location and therefore perceived immediacy of any impact for a potential encounter will most certainly facilitate the opportunity to convey the vast distance and therefore time separation involved. However, in the subsequent decipherment phase, the DISC Quotient's decipherment dissemination rationales can only convey optimal effective and timely information, if they are supported by such responsible media conduits and appreciation of the factors mentioned. But if such conduits are available the DISC Quotient is a means to enable a detailed and scientifically well founded communication of the decipherment process to a public audience, and thus to reinforce trust.

8.0. Conclusions

As stated previously in (Elliott 2008): "To prevent hobbyists and speculation filling an information 'vacuum,' wherever possible, it is paramount that we provide timely and accurate information that conveys our best interpretation of any received signal."

It may be that the ingenuity of an advanced ETI will enable a message to be made accessible to us even without any prior contact, by the provision of a "primer," "crib," or "Rosetta stone," or even by the provision of an

AI-embedded signal with some self-translating capabilities. However, most of the discussion in this chapter centers on the assumption we do not have a primer (crib) to assist decipherment; if so, then the latter stages of analysis will have high overheads in time cost. It will be at these stages that more complex societal impacts are likely to occur; because of this, these areas will need to be subsequently looked at in more detail, as a precursor to any final agreed strategy.

Future work will also need to be done on refining the DISC Quotient scoring system for information dissemination. This should then serve to reduce potential negative societal impact over any protracted period required for decipherment.

If a semantic "wall" does occur, that is, if full semantic assignment and ultimate translation prove impossible, we will at least know, with a high degree of certainty, that we have received an extraterrestrial message and that the required components are evidence of intelligent communication; it will then be in our collective gift to decide whether we respond with a message of our own, including all devices to assist decipherment, to promote further and decipherable contact. Nevertheless, positive "forces" of global unity can be seen as probable counters to some of the likely negative impacts of anxiety, amateur intervention, and media hype.

In conclusion, it is submitted that the development, refinement, and implementation of the DISC Quotient's measures, associated mechanisms, and rationales from scientific observations of language and communication, as an integral strategy during the complex processes involved in post-detection and decipherment, is essential if we wish to minimize cultural disruption and optimize dissemination of necessary information in the event of the receipt of a complex signal from ETI. Recent global events have shown us that a lack of confidence, trust, or negative rumors are enough to send a stock market into meltdown or cause extreme reactions globally, when in reality the facts would hardly cause a ripple.

Works Cited

Almár, I. 2001. *How the Rio Scale should be improved.* Paper IAA-01-IAA.9.2.03 presented at the 52nd International Astronautical Congress, Toulouse, France.

Almár, I., and J. Tarter. 2000. *The discovery of ETI as a high-consequence, low-probability event.* Paper IAA-00-IAA.9.2.01 presented at the 51st International Astronautical Congress, Rio de Janeiro, Brazil.

Almár, I., and H. P. Shuch. 2005. *The San Marino Scale: A new analytical tool for assessing transmission risk.* Paper IAC-05-A4.1.03 presented at the 56th International Astronautical Congress, Fukuoka, Japan.

Baxter, S. 2008. *Renaissance v. revelation: The timescale of ETI signal interpretation*. Paper IAA-S8-1005 presented at the First IAA Symposium on "Searching for Life Signatures," Paris, France.

Billingham, J. 1996. Post Detection SETI Protocol (2008) [online] http://www.coseti.org/setiprot.htm.

Burkholder-Allen, K. 2000. *Media relations and the role of the public information officer*. NDMS Training and Education Online Program.

CELI. 2008. *Information dissemination strategy*. Workshop report. Yerevan, Armenia: Centre of European Law and Integration.

Duggan, F., and L. Banwell. 2004. Constructing a model of effective information dissemination in a crisis. *Research* 9 (3) paper 178, http://InformationR.net/ir/9-3/paper178.html.

Elliott, J. 2002. *The filtration of inter-galactic objets trouvés and the identification of the lingua ex machina hierarchy*. Paper IAA-02-IAA.9.2.10 presented at the 53rd International Astronautical Congress, Houston, TX, USA.

———. 2003. *A human language corpus for interstellar message construction*. Paper IAA-03-IAA.9.1.04 presented at the 54th International Astronautical Congress, Bremen, Germany.

———. 2007. A post-detection decipherment matrix. *Acta Astronautica*.

———. 2008. *A post-detection decipherment strategy*. Paper IAC-08-A4.2.5 presented at the International Astronautical Congress, Glasgow, UK.

Jones, S. C. et al. 2007. *Developing proactive communication strategies for a potential pandemic*. Paper presented at the International Non-profit and Social Marketing Conference, Brisbane.

OECD. 2001. *Communicate risk*. Reykjavik: Public Management Service, Public Management Committee, PUMA/MPM.

Pope, M. 1999. *The story of decipherment*. London: Thames and Hudson.

Sagan, C. 1985. *Contact*. New York: Doubleday Books.

SETI. 2008a. *Declaration of principles concerning activities following the detection of extraterrestrial intelligence*. http://www.seti.org/Page.aspx?pid=680.

SETI. 2008b. *Cultural aspects of SETI*. http://www.seti.org/Page.aspx?pid=682.

Shostak, S., and I. Almár. 2002. *The Rio Scale applied to fictional SETI detections*. Paper IAA-02-IAA.9.1.06 presented at the 53rd International Astronautical Congress, Houston, TX, USA.

On the Universality
of Human Mathematics

Carl L. DeVito

1.0. Introduction

It has often been stated that mathematics would serve as a universal language, one suitable for communication between totally alien societies. Our purpose here is to examine that statement in detail. We shall see that while mathematics is often motivated by scientific applications, it is equally likely to arise from internal sources, sources that have nothing to do with the world of science. Nevertheless, we argue that human mathematics can be understood by any race that has a science, and can be an effective means of mutual communication. There are a number of "philosophies" of mathematics (Pinter 1971) but, in this connection, the views of only two of these need concern us: Extreme Platonists and Strict Formalists. The difference between them is apparent in how they answer the following question: Are the natural numbers, 1, 2, 3, 4,... merely creations of the human mind or do they exist independently of us? The Platonic view is that these objects, and indeed all mathematical objects, really exist, perhaps in some hyperworld. In this view the mathematician is rather like the scientist. He, or she, discovers objects that are "out there." So if an alien intelligence exists then they, too, could discover the same mathematical objects that we have found, for instance, real and complex numbers, functions, topological spaces, etc.

The strict formalist, however, has a very different view. To her, or him, mathematics is a kind of game played by specific rules. Somewhat like chess. An unsolved problem gives a kind of goal, and solving such a problem constitutes a "win." So, to a strict formalist, while five fingers, five cars, and five dollars certainly exist, the number 5 does not. It is a creation of the human mind and an alien, however intelligent, might have no knowledge of 5 or of any other human mathematical object.

My own position is a strange combination of the two. I think the natural numbers do exist independently of us. The rest of mathematics, however, might not exist anywhere but in our minds. But since, as we shall see, all of mathematics can be based on the notion of natural number, all of our

439

mathematics could, in principle, be communicated to any intelligent alien who understands these numbers, certainly to any race whose members can count. It will become apparent, however, that the world of mathematics is not the world of physical reality. It is an artificial world, a world of abstractions and idealizations that human mathematicians have created over many centuries. It may be more reflective of our minds than we realize and may say more about human nature than it does about the real world. Still, one must not forget that human mathematics has an uncanny habit of becoming useful either in explaining some aspect of reality or modeling that reality (Dantzig 2007). So as strange as it might appear to an alien he, or she, or it will be able to appreciate its value.

2.0. Geometric Problems

Humankind has been aware, for many millennia, of the fact that space has properties and that these properties could be usefully exploited. The annual flooding of the Nile forced the ancient Egyptians to find ways to correctly reset the boundaries between adjacent farms, and this involved some geometric insight. As early as 1700 BC, the peoples of Mesopotamia knew about the Pythagorean Theorem, a rather sophisticated piece of information for such an early civilization (Edwards 1984). In the nineteenth century, when it was believed that the moon was inhabited, the mathematician Gauss suggested we use this theorem as a basis of a message to our lunar neighbors. It was the ancient Greeks, however, who first systematically studied space and gave us an organized geometry. This has come down to us in the books of Euclid written about 300 BC. His approach was axiomatic (Pinter 1971). From a few simple assumptions, he deduced all that was then known. This provided us with an ideal, a canonical model, to which all succeeding generations of mathematicians aspired, and there were attempts to base all of mathematics on geometry; numbers, for example, were to be thought of as geometric ratios. Geometry is, of course, a major branch of modern mathematics with sometimes striking developments occurring centuries apart. In the 1600s it was combined with algebra, by Descartes and Fermat (Dantzig 2007), to give us analytic geometry, a subject that enabled us to bring our considerable geometric insight to bear on problems of algebra, and to use algebraic techniques to solve problems of geometry. With the advent of calculus in the seventeenth century mathematicians applied these ideas to the study of curves and surfaces, giving us differential geometry. The subject blossomed in the nineteenth century with the surface theory of Gauss, the brilliant insights of Riemann (Dantzig 2007), and the tensor calculus of Ricci and Levi-Civita (Adler, Bazin, and Schiffer 1965). This was the mathematics drawn upon by Einstein to formulate his far-reaching general theory of relativity. It is no diminution of his genius to point out that the mathematics was there, fully developed by those whose interests were purely mathematical, for him to use. This happens far more often than is generally recognized; mathematics is developed for purely mathematical reasons long before it finds application to some area of science. This is an important point. It shows that human reasoning is an effective tool for understanding physical reality. As Galileo insisted, we can understand the world and unravel the mysteries that surround us (Hawking 1988). This understanding does not come only from mathematics, of course, but from the work and insights of physicists, chemists, geologists, etc. Communication with an alien race should be possible, at least up to a point, if that race has a science; because science is a study of physical reality and we share the same physical reality (DeVito and Oehrle 1990). Geometry is by no means a dead subject. There were remarkable developments in the twentieth century and research continues today.

Geometric problems led to other branches of mathematics even in ancient times. Due to an unfortunate series of incidents, Dido, a Phoenician princess, needed to relocate to Africa. But those already there were reluctant to sell her land. Finally, after the exchange of a great deal of money, it was agreed that she could have all the land she could enclose in an ox hide. A bad deal it would seem. But Dido was clever and resourceful. She cut the ox hide into thin strips, tied the strips together to make a rope and demanded all of the land she could encompass with the rope. Here we have a problem, what shape encloses, in a fixed perimeter, the most area. To see that there really is a problem here, note that with 100 ft., say, of rope you can encompass a rectangle with two sides 49 ft. and two sides 1 ft. each, giving you an area of 49 ft^2. But with the same 100 ft. rope you could also enclose a square having each side 25 ft., giving you an area of 625 ft^2., considerably more land. There is a branch of mathematics that deals with problems of this kind (van Brunt). It is called the calculus of variations. The subject was developed in the eighteenth century by Euler, Lagrange, and the Bernoullis and now has numerous applications to physics, such as Fermat's principle of optics, Hamilton's work in mechanics, etc. The answer to Dido's problem turns out to be a circle. This curve encloses the maximum area for a given perimeter. According to legend, Dido used her rope to enclose a large circle, claimed the land within it, and founded the city of Carthage upon it.

3.0. Numerical Problems

Humans had a sense of numerical relationships long before numbers themselves were discovered. If the elders of a prehistoric clan wanted to know if they had enough spears to equip a hunting party all they had to do was have each man pick up a spear. If all the men were armed and there were still spears left on the ground, they had plenty of weapons; there were more spears than men since the men were in one-to-one correspondence with a sub-collection of the spears. On the other hand, if all the spears were taken and some men were left emptyhanded then they clearly had an equipment shortage. There were more men than spears because here the spears were in one-to-one correspondence with a sub-collection of the men. The final case, of course, is when each man is armed and there are no spears left over. In this case the two collections, men and spears, are equinumerous, because there is a one-to-one correspondence between them. This idea was used by ancient peoples all over the world to keep track of their herds or even their armies. They set up one-to-one correspondences between the collection they wanted to keep track of and notches on a stick or pebbles in a pile. No counting is involved here. Counting is a rather sophisticated process, and to begin doing it we must first have a standard set of models for our numbers. The wings of a bird gave a natural model for the number two, the fingers on a hand gives a model for the number five, and other models suggest themselves gradually giving rise to the concepts of two, five, etc. Here, however, the numbers are seen as cardinals, the number of elements of a given collection. To begin counting we need to discover the ordinal aspect of numbers; the fact that they form an ordered sequence with each number occupying a specific place. When we count a collection of objects we mentally label them as first, second, third, fourth, and, let's say, fifth, and we conclude that there are five objects in our collection. We pass from ordinal number to cardinal number with such ease that we are rarely aware of it. I strongly suspect that the ideas outlined here arose from human observation of the day-night cycle and the realization that that cycle could be usefully exploited. In keeping track of a herd you could lead the animals one by one past a carver, a person who would record the passing by carving a notch on a stick. The order here is unimportant. You choose the animals randomly. But in keeping track of the days needed for a particular journey, this is what was important for early man, not distance, which he couldn't measure anyway; you had to record the days as they came, in their natural sequence. This is what may have led to the idea of ordinal number and its relation to cardinal. It wasn't easy, and it took time for the idea to crystallize, but eventually humans realized that ordinal numbers were important because they gave the only practical way of finding the number that really interested them, the cardinal number of a collection. The natural numbers may exist, waiting to be discovered by us or by some alien race. But we were led to this discovery by the presence on our planet of a day-night cycle. This astronomical property of a planet may be the trigger needed to lead the inhabitants of that planet to this important discovery.

4.0. Infinity

The term *infinity* is the source of much confusion and general misunderstanding. In elementary calculus it is used as a shorthand, a "way of talking." The numbers ½, 2/3, 3/4, 4/5...increase, as is obvious from their decimal representations 0.5, 0.666..., 0.75, 0.8...but they never exceed the number 1. We say they have an upper bound. A sequence of numbers that increases without an upper bound, like the sequence 2, 4, 8, 16, 32, 64..., is said to "tend to infinity." A sequence that tends to infinity is simply a sequence that increases without an upper bound. There is a geometric interpretation of this (Conway 1978). In function theory the complex plane is mapped stereographically onto a sphere, called the Riemann sphere. You set the sphere on the plane with its' "south pole" at the origin and, given a point in the plane, you draw a line from the "north pole" of the sphere to the given point. The point where this line intersects the sphere is taken as the image on the sphere of the given point. All points of the plane are, in this way, put in one-to-one correspondence with points on the sphere, but the north pole of the sphere corresponds to no point in the plane. This is called the "point at infinity" and a sequence of points in the plane whose distance from the origin increase without bound is said to converge to infinity; the corresponding points on the sphere tend toward the north pole.

In classical mathematics a distinction was made between a collection that was actually infinite, in which infinitely many objects are thought of as existing simultaneously, and a collection that was potentially infinite (Pinter 1971). The collection of all natural numbers, for example, was potentially infinite. No matter how many of these numbers we write down, we are aware that there are always more. It was Georg Cantor (1845–1918) who, while working on a problem in the technical area of trigonometric series, was led to consider collections that were actually infinite. This was very disturbing to the mathematicians of his time and led to much criticism. The idea of a "set" was making its way into mathematics; it is a remarkably useful idea and leads to an underlying unity that is elegant and insightful. Cantor's work cast some doubt on the wisdom of relying too heavily on this concept. Today, mathematicians distinguish two types of set theory: axiomatic and naïve. In the latter a set is any well-defined collection of objects. Well-defined means that it must be clear just what objects are in the set, just what its members or "elements" are, and just what objects are not in the set. To Cantor the collection of *all* natural numbers, 1, 2, 3, 4..., is a perfectly good set even though it contains infinitely many elements.

A good deal of Cantor's early work consisted of setting up one-to-one correspondences between various sets, something that, as we have seen, humans have been doing for millennia. But applying this idea to infinite sets led to some counterintuitive, and even disturbing, problems. The first surprise was already noted by Galileo (Dantzig 2007). In the course of his "Dialogs Concerning the New Sciences," the question arises as to whether there are more squares, 1,4,9,16,25,…, or more natural numbers, 1,2,3,4,5,… . One of the characters notes that every square is a natural number, meaning that the set of squares is a subset of the set of natural numbers, but there are many numbers that are not squares, for example, 2,3,5,6,… . Thus, the squares are a proper subset of the natural numbers and hence contain fewer members. A second character, however, points out that the two sets can be placed in one-to-one correspondence: 1 corresponds to 1, 2 corresponds to 4, 3 corresponds to 9, etc. Galileo's conclusion is that all one can say is that both sets are infinite; neither is larger than the other. Cantor would say that the two sets are equinumerous. This is, in fact, a characteristic property of infinite sets: A set is infinite if, and only if, it can be placed in one-to-one correspondence with one of its proper subsets (Dantzig 2007; Hrbacek and Jech 1984).

The second surprise is, perhaps, even more striking. Cantor tried to set up a one-to-one correspondence between the set of natural numbers and the set of real numbers, the so-called continuum. He found that no such correspondence exists. There are more real numbers than there are natural numbers even though the two sets are both infinite! So there are "degrees" or "orders" of infinity. Some sets are infinite, but some sets are even "more infinite." There is, in fact, no limit to the infinities. Given any set, one can show that the collection of all of its subsets (sets whose elements are also in the original set) is larger than the given set. If you start with an infinite set and collect all of its subsets into a new set, the new set is more infinite than the one you started with (Enderton 1977).

The results of set theory and our ideas on geometry, outlined above, combine to show forcefully that the world of mathematics is not the same as the world of physical reality. Using a technical result due to Hausdorff, Stefan Banach and Alfred Tarski showed that one can cut a sphere the size of a pea into a finite number of pieces, reassemble the pieces and obtain a sphere the size of the Sun (Wapner 2005)!

5.0. Putting It All Together

It has long been the goal of mathematicians to unify the entire subject; to deduce all of mathematics from a small number of basic axioms. We have mentioned that attempts were made to base all of mathematics on geometry, but these never got very far. The work of Karl Weierstrass, Richard Dedekind, and others in the nineteenth century, showed that all of mathematics could be based on the notion of natural number (Dantzig 2007). Since, in my view, these numbers will be known to any race that has the radio telescope, I think that all of human mathematics can be communicated to such a race. This does not mean that they will have arrived at our mathematics themselves. They may think very differently and go off in directions we cannot anticipate. Just as alien chemists, using the same chemical elements we know here, might go off into aspects of chemistry our chemists would never think of; not because they aren't smart enough, but because they are human and hence have the same limitations as all other humans, limitations the aliens might not have. But whatever they do will be understandable to earth chemists, and aliens and humans should be able to understand each others' mathematics insofar as these are based on natural number.

Cantor's work, because of its elegance and inherent beauty, began to be accepted and incorporated into mathematics when, ironically, certain disturbing paradoxes arose in connection with it. Some of these were logical, some were semantic. The latter led mathematicians to study formal languages. Some of the early computer languages were based on this work and Hans Freudenthal drew on these ideas when he developed Lincos, a language specifically designed for cosmic communication.

Let's sample two of these paradoxes here. Perhaps the best known of the logical paradoxes is that due to Bertrand Russell (Enderton 1977). We can present this in ordinary language, avoiding technical jargon. There is a town so small that it has only one barber, and he shaves those men, and only those men, who do not shave themselves. Now, who shaves the barber? If he does not shave himself then he must be shaved by the barber. But he is the barber. So if he does not shave himself, then he does shave himself—a clear contradiction. Well, maybe he does shave himself. But then, as a man who shaves himself, he is not shaved by the barber—another contradiction. At this time, and even now, people were working with sets of sets, like a league, which is a set of teams each of which is a set of people. So Russell considered the set of those sets that were not members of themselves (the set of all books is not a book, but the set of ideas is an idea). This set leads to the same contradictions that the barber led us to.

An interesting semantic paradox is the one due to Berry (Enderton 1977). It is a deep and useful fact that any non-empty set of natural numbers has a smallest member. The smallest odd prime is 3, the smallest perfect number is 6, and the smallest seven digit number is 1,000,000. Now suppose we choose a dictionary and consider all numbers that can be described in twenty-five or fewer words from that dictionary. There are only finitely many words in the dictionary and only finitely many ways of combining these words into sentences of twenty-five words or less, so the set of numbers that cannot be described in this way is non-empty. So it has a smallest member, say, N.

Then: "N is the smallest number that cannot be described in twenty-five words or less from our dictionary." But the sentence in quotes describes this number in just eighteen words!

Some wanted to reject the whole of set theory because of these paradoxes, but many others felt that that was going too far. In 1900, an international congress of mathematicians was held at the University of Paris. Here one of the greatest mathematicians then active, a young German, gave a list of problems to challenge the mathematicians of the coming century. His very first problem involved Cantor's work, thereby lending his support to Cantor and to the importance of set theory. The young German was, of course, David Hilbert, and the problem he posed was this: We have seen that the set of real numbers, the continuum, is larger than the set of natural numbers. Is there an infinite set that is between these two, that is, a set that is larger than the natural numbers but smaller than the continuum? It was conjectured that this could not be and this conjecture was shown to be consistent with the other axioms of set theory (see below) by Kurt Gödel in the 1930s. To the surprise of all, in 1963, Paul Cohen showed that the negation of this conjecture was also consistent with the axioms of set theory. Here is a statement, "There is no set strictly larger than the natural numbers but strictly smaller than the continuum," that cannot be proved and cannot be disproved! Gödel had shown that, if mathematics is consistent, then statements of this type, not provable and not disprovable, had to exist, but no one even imagined that the continuum problem was such a statement (Enderton 1977).

The paradoxes of set theory led mathematicians to try to formulate this subject in terms of a collection of axioms that, it was hoped, would avoid these paradoxes and allow us to keep the useful aspects of that subject. Perhaps the best known are those of Zermelo and Fraenkel; let's call this ZF. But one more axiom had to be added to this collection, the so-called axiom of choice; let's call that AC (Hrbacek and Jech 1984). When first stated, this axiom seems harmless, even obvious. But it has far-reaching consequences (Hrbacek and Jech 1984). The result of Banach and Tarski, stated above, relies on it as do many other results of analysis, topology, and even modern algebra. Let me give a whimsical illustration of the AC. Imagine a forest in which infinitely squirrels live. Each squirrel has its own tree and hidden in that tree has a hoard containing infinitely many acorns. Does it not follow that there are more acorns than squirrels? To pin this down we would have to set up a one-to-one correspondence between the squirrels and some subset of the acorns. This would be easy if we could assign to each squirrel one of the acorns in its hoard. But how would you select that acorn? What criterion would you use? You can't do it one at a time because there are infinitely many squirrels and you'd never finish. The only way to do this is to give some rule whereby each squirrel can select an acorn in its hoard.

But it is not at all clear how to do this—can you interview infinitely many squirrels and find out which acorns they would select? The AC says that for any collection of non-empty sets there is a rule whereby you can assign to each set one of its members. The axiom gives no clue as to how one can do this. How would you select from each galaxy a particular star in that galaxy? Even an astronomer might have trouble coming up with a rule for doing this and the set of galaxies is finite.

It is now known that the AC is independent of the other axioms of ZF and that if ZF is consistent, then adding this axiom does not produce any inconsistency (Enderton 1977). I should mention that there is a small set of axioms, the so-called Peano Postulates (Hrbacek and Jech 1984), from which all of the properties of the natural numbers can be deduced. There is no need to go all the way back to set theory in our communication of human mathematics to an alien race. Perhaps it is the human fascination with the infinite that attracted us to the direction pioneered by Cantor and maybe it is that same fascination that keeps us engaged in set theory and the foundations of mathematics.

6.0. Conclusion

I have taken the natural numbers as the set 1, 2, 3..., which is the same as the set of positive integers (in some books the natural numbers start with 0). I think these numbers exist and are, or could be, known to any intelligent race. The rest of mathematics is, to my mind, a human creation. As Kronecker might have said it, "God made the positive integers, all the rest is man's work" (Dantzig 2007). Mathematics is as much a part of us as is our Music and our Art. Aliens may not share any of it, but they should be able to understand it because it can all be based on natural numbers. To an alien, human mathematics might appear curious and exotic. It may give them insights into physical reality that they could never come to on their own, just as their mathematics might give us such insights. Here is one aspect of humanity that can be communicated to any society whose members can count. What deductions they will make about human nature from our mathematics is impossible to predict. Our extensive study of space, our geometry, may stem from our reliance on the sense of sight. Our study of the mathematics of motion, calculus, may stem from our having played the role of prey and that of predator. When something is chasing you, you have no time to think about how to run or climb. These things must be internalized. Similarly, you cannot think about how to aim your spear at a fleeing animal; you must "know" what to do. This might be why many find calculus very natural and comfortable. But even here our sense of sight enters in. In calculus we seek to approximate a given curve at each point by a line. The derivative at that point gives us the slope of that line. Would an alien race with a different evolution think this is natural? Should we choose to discuss set theory, the human preoccupation with the infinite will soon become apparent. Would an alien race share this with us also, or would they find it strange, even weird? The mathematics of an alien race may, in a similar way, tell us a great deal about that race. Surely we would like to learn all we can about any alien race we contact. The realities of this kind of interaction, however, severely limit what we can hope to learn, at least in the near term. The main advantage of mathematics over other subjects is that, in the context of alien-human communication, it has a good chance of being understood.

Works Cited

Adler, R., M. Bazin, and M. Schiffer. 1965. *Introduction to general relativity*. New York: McGraw-Hill.

Conway, J. B. 1978. *Functions of one complex variable*. New York: Springer-Verlag.

Dantzig, T. 2007. *Number: The language of science*. New York: Penguin Group.

DeVito, C., and R. Oehrle. 1990. A language based on the fundamental facts of science. *Journal of the British Interplanetary Society* 43: 561–68.

Edwards, H. M. 1984. *Galois theory*. New York: Springer-Verlag.

Enderton, H. B. 1977. *Elements of set theory*. New York: Academic Press.

Freudenthal, H. 1960. *Lincos: Design of a language for cosmic intercourse*. Amsterdam: North Holland Press.

Hawking, S. W. 1988. *A brief history of time*. New York: Bantam.

Hrbacek, K., and T. Jech. 1984. *Introduction to set theory*. New York: Marcel Dekker.

Pinter, C. C. 1971. *Set theory*. London: Addison-Wesley.

Van Brunt, B. 2004. *The calculus of variations*. New York: Springer-Verlag.

Wapner, L. M. 2005. *The pea and the sun*. Wellesley, MA: A. K. Peters.

Cognitive Foundations of Interstellar Communication

David Dunér

1.0. The Problem of Interstellar Communication

This is a message, a message addressed to you constructed in a code-system called English marked with Roman letters in ink on sheets of cellulose, or as liquid crystals on the screen. It is here and now, perceived by the senses, interpreted by a being with a brain, body, and history, living in the world.

By using this code-system I evidently hope to make myself understood, to awaken in the mind of the receiver similar thoughts and ideas that I have when I formulate this message. Ultimately—above and beyond the mere question of language skills—this hope of mine stems from the fundamental fact that we share the same human cognitive abilities that are a function of our common evolutionary history here on Earth, the planet Tellus. But if we extend this communicational situation beyond Earth, however, the question naturally arises: *How could communication be possible between intelligent beings of different environments that differ physically, biologically, and culturally, and have developed through separate evolutionary lines?* This is the problem of interstellar communication. It is right, as Michael Arbib wrote, that "we are in fact at this very time receiving messages from intelligent civilizations, messages transmitted hundreds or even thousands of years ago" (Arbib 1979, 25; cf. Finney and Bentley 1998). Arbib had the writings of Newton, Euclid, and others in mind. Even though these messages do not contain all facts needed for the right interpretation of them, they are in fact possible to understand. The explanation for this, I would say, is that we all share the same history, the same evolutionary and cognitive setup. And this we cannot count on in respect to extraterrestrial cognition. The problem of interstellar communication is far more complicated. These circumstances—that we will have no kinship, that we will not share similar bodies or cultures, or even similar physical realities—will have far-reaching consequences for how we will be able to construct and interpret messages from distant civilizations.

2.0. Recasting the Objective

The usual strategy to overcome the problem of interstellar communication has been to try to construct a message that is: *A universal symbolic information transfer that is independent of context, time, and human nature.* This can be called "the universal-transcendental interstellar message objective." My point here is that this strategy and its requirements are not reconciled with what we presently know about cognition, communication, and evolution. Particularly, it presupposes the universality it aims at, and thereby ignores the facticity of evolution, and the situatedness and embodiment of symbolization. It ignores the context, that the living organisms—and consequently their cognition and communication—are planet-bound, tied to and constrained by certain physical conditions. It leaves out time and history—the evolution, the phylogenetic, ontogenetic, and cultural-historical time—in which the organisms are evolving. And finally it ignores the nature of the communicators—that they have bodies and brains evolved in interaction with their environment.

For these reasons, I propose a different strategy that in my view does justice to the known facts about cognition, communication, and evolution. In this chapter I put forward instead an "embodied-situated interstellar message objective"; that is, a search for: *A local, embodied, situated, and concrete nonsymbolic interaction.* I will later on in this chapter explain more in detail why it is problematic—if you would like to be understood in the cosmos—to exclude the context, the situation, space, time and human nature, and also why symbols (or conventional signs) and information transfers are probably less effective ways of starting a communicative interaction. But first I will say something on the often unproblematized assumption of universality.

3.0. Assets for Solving the Problem

There are three different fields of research that I think are inevitable for any future interstellar message construction: cognitive science, semiotics, and history. Firstly, regarding cognitive science, I build on the basic observation that our cognitive and communicational skills are embodied, situated in, and adapted to our terrestrial environment. In short, cognitive science studies how the external world is represented, how we use cognitive tools for our thinking—such as language, image schemas, mental maps, metaphors, and categories—but also how we use and interpret, for example, drawings and images to enhance communication. It is about perception, attention, memory, learning, consciousness, reasoning, and other things that we include in what is called "thinking." Elsewhere, I have discussed that cognitive science can play a vital role in our studies of extraterrestrial intelligence, and reveal new perspectives on human encounters with the unknown. The field of "astrocognition" can be defined as the study of human cognitive processes in extraterrestrial environments (Dunér 2011).

Secondly, the time-honored insights of classical hermeneutics, which theorize on the many aspects of symbol formation, the transfer of meaning, and the decoding of messages, cannot be neglected in the quest for interstellar symbolization. What research fields are actually dealing with communication and meaning? It is the humanities, the arts, anthropology, history, and other studies of the human being as creators of and searchers for meaning. Inasmuch as interstellar communication is thought to be an exercise in coding and decoding signs, the relevance of these insights, especially in semiotics, should be obvious. Douglas Vakoch is one of the few who have observed the potentiality in an exosemiotic perspective (Vakoch 1998). Another is Göran Sonesson, who has delivered a sharp semiotic analysis of the problem (Sonesson 2007). It could be argued that the problem of intersteller communication is not just a problem within natural science primarily, but a true humanistic problem in its true sense; a human problem. It is we humans who will send and receive, code and decode the messages.

Finally, I focus on history, understood, on the most basic level, as the interaction of organisms with their environment over time. Our cognition and communication are a result of time, of history; both evolutionary history and sociocultural history. Communication is not something pre-given, but rather evolves in interplay with the environment; a process during millions of years. The latter factors of Drake's equation are in fact social, historical problems, and are probably the most difficult to answer for the moment: f_c—the fraction of intelligent sites that develop a technological communicating civilization; and L—the longevity of a communicative civilization (Sagan 1973; McConnell 2001; Tarter 2001; Shostak 2009). We know just one such civilization and we have not seen the end of it. We need to place interstellar communication in time, the fourth dimension.

These fields, I suggest, contain vital tools for any future interstellar message construction. In fact, without firm insights into these fields I think the problem of interstellar communication will be hard to solve. These, together with science, mathematics, and technology, will give us the indispensable, collective human mental powers for deciphering interstellar messages. Before I go on to the cognitive foundations of communication, I have to say something about the often unproblematized assumption of universality that has commonly been taken for granted as a necessary requirement in interstellar message construction.

4.0. The Universality Principle Reconsidered

Behind the universal-transcendental objective is a standard assumption that there are some universal facts and laws, especially of a scientific and mathematical nature, that are the same throughout the universe (DeVito and Oehrle 1990; Vakoch 2009a). From our knowledge about the universe we infer that these laws we have found are valid for the whole universe, by a way of analogical reasoning that what is true here might be true there. It could also be said to assume a reality that transcends the physical reality of here and now, reminding us of a Platonic world of ideas that exists independently of the world of matter in which man lives and functions. The question here is not if there are universal facts that are the same for all observers—it might be true—but rather if we can know that we have come to the indisputable final, universal facts (the epistemological problem), and whether we actually can use our assumed universal propositions of reality for our purposes and as a foundation for an interstellar message that aims to be possible to decode and be understood by an extraterrestrial intelligence (the semiotic problem). The problem concerns, as it did for George Sefler and Nicholas Rescher, the postulating of our models of reality as universally commensurable (Sefler 1982; Rescher 1985; Vakoch 1998). There are, however, critics, such as the physicist Andrei Linde, of the idea that the universe and the laws of physics are everywhere the same (Michaud 2007). Our traditional "universal" laws of physics might be just local bylaws with limited range.

Against the assumption that our present-day understanding of universality could be a starting point is, firstly, a historical-epistemological argument. Can we know that the universal a posteriori statements we come to by inductive reasoning are true everywhere and whenever? Throughout the history of science, man has time after time been surprised over reality. That which we once thought was a universal fact turned out to be false. A classical example is how the Newtonian gravitational theory was once thought to be universal but later on found to be just a special case of Einstein's relativity theory. Many such examples can be found in the history of science. The whole history of science could be said to be a history of surprises that violate our first assumptions of a universal and uniform world. Reality is more complex and diversified than we originally imagined. It is more reasonable—in respect to extraterrestrial intelligence—to expect diversity than uniformity in understanding and representing the world. Diversity is what we should expect, not least concerning different ways of communicating ideas in the universe.

Secondly, there are cognitive arguments against postulating human science and mathematics as a foundation for universality. Our universe is seen from our point of view, from our planet, through our species-unique senses and cognition. It is an anthropocentric and terrestrial view. Our understanding of the universe depends on our terrestrial brain. Science and mathematics are products of the evolution of human species-unique cognitive skills, of human embodiment and sensorimotor interaction in the gravitational field specific to our Earth, and a particular cultural evolution. Basic mathematics, which usually is said to be universal, rests, like language, on human ways of experiencing the world as consisting of distinct discrete objects in a Euclidean three-dimensional space, and is a function of a mental adaptation of the organism to its environment. More advanced mathematics was developed very late in the human history, and can be said to be a cultural product to a much greater extent than language. Language and symbolic reasoning can be found in all cultures, but only a few have independently developed more complex and advanced mathematics (Tomasello 2005). Not everyone within a culture has knowledge about complex mathematics, and this knowledge has to be transferred through conscious instruction to new generations.

There are reasons to believe that an extraterrestrial intelligence would have some understanding of something that we call mathematics. I do not mean just in order to be able to develop advanced technology, but in a deeper cognitive sense. As organisms they adapt to and have to orient themselves in their specific spatial environment. Seeing the environment as discrete instances for detection, and being able to recognize spatial relations, have certainly had survival benefits. Mathematics has no different status than other skills of the human mind. It is also based on cognitive abilities evolved through the history of our species (Lakoff and Núñez 2000). Carl DeVito has recently proposed natural numbers (1, 2, 3, 4…) to be something that can be understood by an extraterrestrial intelligence. The rest of mathematics can be said to be a human creation (DeVito 2011). This can be workable if we assume that we share the same cognitive categorization of reality in similar cluster-statistics. Humans categorize things in discrete instances, doing some sort of cluster-statistics of continuity (Rosch 1975; Rosch 1978; Taylor 2003). For example, adult listeners of a particular language classify sounds as one phoneme or another, and show no sensitivity to intermediate sounds (Kuhl 2004). In a similar way we categorize the continuum in natural numbers. For an alien, real numbers might be more "natural," or mathematics based on true continua.

A convergent psychic uniformity of all intelligent species is often taken for granted. Some commentators have challenged this (McNeill 1973; Baird 1987; Westin 1987; Finney and Bentley 1998). In addition to this critique, we cannot from a cognitive standpoint assume that the extraterrestrials will have the same ability to reason as humans. Cognition is not universal; neither is it transcendental, belonging to a Platonic world of ideas, independent of the thinking subject; cognition is embodied in an subject, and is situated in a particular environment.

5.0. Cognitive Foundations of Communication

In order to discuss interstellar communication in a more fundamental sense we have to ask ourselves: Why do we have communication in the first place? What is it and how has it evolved? From the research within the fields mentioned above, we can find some fundamental characteristics that seem to underlie all communication as we know it. I do not aim to list all sufficient or necessary qualities that characterize communication, just some that I think are particularly relevant in the quest for a possible interstellar communication, and which will force us to reconsider the universal-transcendental message objective. These are the cognitive foundations of interstellar communication: evolution, embodiment, situatedness, and symbolization.

5.1. Evolution: Communication Has an Evolutionary History

Communication has an evolutionary origin (Christiansen and Kirby 1997; Deacon 1997; Tomasello 2008). Human communication, whether it is of lingual, symbolic, or bodily expressions, depends of how our brains work and are constructed, and how humans interact with their physical, biological, social and cultural environment. The human mind has to a large extent evolved as an adaptation to certain problems that our ancestors have faced during the evolutionary development of our species (Gärdenfors 2006). That is, the human brain is adapted to, firstly, the physical and biological environment of the Earth—to understand and interpret, interact and deal with, and orient itself in the Earth's physical and biological environment, in relation to its specific conditions, such as planetary orbit, gravitation, light conditions, atmosphere, radiation, temperature, chemistry, geology, ecology, fauna, and flora. Secondly, the human brain is also adapted to the mind and culture of of the tellurian species *Homo sapiens*, to understand and interact with other beings of our species, to understand human feelings, thoughts, motives, etc., in a psychological and social interplay that forms our human culture. Culture can here be defined as the existence of intraspecies group differences in behavioral patterns and repertoires, which are not directly determined by ecological circumstances, and which are learned and transmitted across generations (Sinha 2009; Tomasello 1999).

Tens of millions of years of social evolution have adapted our species to be highly sensitive to human social signals. Language has been around for almost two million years. Our language, with its phonemes and symbolization, has evolved due to its enhancement of communication between humans. Yes, it is true, our language can be used for describing the world around us, and to transfer information between ourselves. But perhaps more important for its appearance is its use for social interplay: to express feelings, to gossip, for socializing and creating bonds, etc. As Robin Dunbar and others have suggested, language has interpersonal functions, and emerged for the improvement of social bonding, to hold together and sustain large groups, and to incite desired behavior in other members of the group (Dunbar 1996). It evolved because of social reasons, for social grooming, making social contracts, to coordinate action, and pass on knowledge. Alison Wray has discussed protolanguage as a holistic system for social interaction, and proposed that we started with holistic utterances that later were segmented (Wray 1998). Like primates' use of vocal and gestural signals, human language has also specific interactional goals. The signals are intended to have an effect upon the world of the sound maker (and hearer), by inciting a reaction in the hearer. In Wray's understanding, signals are holistic; they have no component parts that could be recombined in order to create new messages. The early hominids might have communicated in a similar way, but perhaps using a larger set of utterances that later on gave rise to a segmented and combinatoric language.

So, language is a social device for expressing feelings, but also to disguise them. As Terrence Deacon clarifies, a species' communication has evolved not just to solicit another to assume the same state, but to a large extent for manipulative purposes (Deacon 1997). Language is not just for communicating honest statements, and due to this deceitful use of the communicating practices, the communication must, then, be seen in the specific social and evolutionary context in which it is produced. Communication should in that sense be seen in a social context, rather than as an objective, decontextualized information transfer. Communicating with the aliens will actually be a socializing practice.

Language had evolutionary benefits that extended the range of our capabilities, and had also an impact on the evolution of the brain itself. The gradual increase in communication led to new possibilities for the mind (Workman and Reader 2004). The symbolic artifacts that are grounded in particular structures of human cognition and communication have been invented and modified over historical time by members of a particular group of intelligent organisms. All the symbols and constructions of a given language are not invented at one and the same time. They evolve and change over historical time as they are used and adapted to changing circumstances. Accordingly, communication is a biocultural hybrid, a changing product of the genetic-cultural co-evolution. There is, consequently, a very distinct possibility that the respective evolutions of the cognitive and communicational skills of terrestrials and extraterrestrials have taken very different and separate paths. If their civilization is much older than ours, which is very likely, not just their technical skills, as many have assumed, might be at a more advanced stage; even their evolutionary process will have been working during a longer period of time and through many more generations, and so also, because of biocultural changes, their way of communication. It might have evolved into something far beyond the capacity and efficiency of our present earthly way of communication.

5.2. Embodiment: Communication Is Embodied

We think with the body. The mind is not detached from the body. A key concept in contemporary cognitive science is "embodiment" (Varela, Thompson, and Rosch 1991; Lakoff and Johnson 1999; Krois et al. 2007). According to the theory of embodied cognition, intelligence, both natural and artificial, depends upon interaction with the environment. Computation alone is not sufficient for explaining intelligence. It is also necessary to take into account sensorimotor interaction with the world. Our cognition is embodied and dependent on our bodily activity. We do not think with our brains alone, but to be able to think we need also the rest of our bodies. This embodiment gives rise to species-unique forms of cognitive representation. For example, our human body schema—the bodily organization of space, the orientation of front-back, in-out, up-down, etc.—is dependent on how our bodies are and function in the environment. Different sensory equipments would change the perceptions, but also the cognition, and in the end culture and social structure. Thus, bodies of other kinds and evolutionary backgrounds would have other minds and ways of thinking.

Obviously, in regard to interstellar communication the bodily constructions of the aliens and ourselves would probably be very different due to our isolated evolutions. Our different brains and bodies, and because of that different cognition and body schemas, will decrease the possibility of finding a common ground for mutual understanding and communication. The sender and receiver are embodied, and their mental understanding cannot be separated from their bodies. We cannot assume that the extraterrestrials would have senses like ours—that would be anthropocentric. But we could assume that they are in someway or another embodied, and that they interact with their environment; that they possess some sort of spatial awareness and feeling for directionality, up and down, in and out, etc., according to the physical structure of their celestial body. Pictorial representations are difficult to use due to the cognitive differences we will have in interpreting visual sensations. A picture is nothing objective; it is changed by the perceiver's sensory apparatus, but also not least by her understanding and former experiences. John Michael Krois suggests, however, that chiral organization of the physical world is universal, and that this would make it possible to communicate with extraterrestrials by pictures (Krois 2010). Recognizing an image is a kind of sensorimotor activity, not just a purely mental process, according to Krois. Pictures can in that case be easier to understand than symbolic messages, because they are more basic, and are based on sensorimotoric activities.

5.3. Situatedness: Communication Is Situated

The theory of "situated cognition" claims that our cognitive processes are not just inside our brains; we also use our environment for thinking (Clark 1997; Brinck 2007). The environment has an active role in driving cognitive processes, or, as Andy Clark and David Chalmers called it, the "extended mind" (Clark and Chalmers 1998). The brain does not only need the body but also the surrounding world in order to function efficiently. Thus, cognition emerges in the interaction between the brain, the body, and the world. There exists no sharp line between the brain and the world. We cannot be isolated observers. In other words, cognitive activity cannot be separated from the situations in which it occurs. To this we can add what is called "distributed cognition": that we are using our environment and objects for enhancing thinking; that we place our ideas and memories outside us, in things—in books, computers, and other external objects (Giere and Moffatt 2003). Interstellar messages are in fact distributed thoughts outside our brains.

It is thus necessary to take into account the intelligent organisms' sensorimotor interaction with the world. Cognition emerges from a history of actions in the world that are performed by an organism. Communication is therefore a situated practice. It is constrained by its surroundings, and is adapted to specific circumstances. This means that we cannot exclude the situation where the message is performed, and the physical, biological, and sociocultural context of the communicators. We are planet-bound creatures. We have to bear in mind that our spatial understanding is a result of our evolution and is adapted to our needs in this particular terrestrial environment, and is very well adapted to our needs for activities in the local environment. Our innate spatial understanding is of a Euclidean space, which is something different from an Einsteinean universe or the ten dimensions of string theory. So, in conclusion, where we are in time and space is totally fundamental for cognition.

5.4. Symbolization: Communication as a Symbolization of Thought

Intelligence could be seen as an evolved mental gymnastics required to survive and reproduce within its specific environment. This includes the capability of representing activities and being able to make inner models of reality. If the extraterrestrials are intelligent, they probably have some kind of symbolization abilities and abstract thinking detached from the environment, with which they can reason about things not existent; things that are not right in front of them, facing their senses, in a specific moment in time. A very effective tool for symbolizing thought is our communicational devices. According to John Taylor, language can be understood as a set of resources that are available to the language user for the symbolization of thought, and for the communication of these symbolizations (Taylor 2002). Language frees us from the here and now, that is true, but anyhow it rests on cognitive abilites that are a result of biocultural evolution here on Earth.

Cognitive linguistics aims to situate language within more general cognitive capacities. Contrary to Arbib, I believe that linguistic theory can help us in constructing and decoding interstellar messages—in fact, I would say, it is very much needed. A lot has happened in this field since Arbib wrote his article thirty years ago. Current cognitive linguistics can show how language is a result of an evolutionary cognitive process. If we believe that cognition and communication are not something pre-given, these perspectives can situate interstellar communication in an evolutionary context. Communication is actually an extension of prelinguistic cognitive capacities.

Spatial experience is fundamental for cognition, as mentioned above, which also leads to the fact that many abstract concepts relate to bodily experiences. Our cognitive capacities, especially concerning concept formation, can be explained as a kind of metaphorical extension of spatial reasoning (Lakoff and Johnson 1980; Gärdenfors 2008). Abstract concepts relate to concrete, basic human experiences. Light-darkness, up-down, and other physical experiences give rise to metaphors such as "knowledge is light," "ignorance is darkness," "good is up," and "bad is down," etc. Based on observations of the extraterrestrials' astrophysical environment, we might be able to make qualified guesses about their metaphors.

According to the linguists Nicholas Evans and Stephen C. Levinson, humans have a communication system that varies at all levels: phonetic, phonological, morphological, syntactic, and semantic (Evans and Levinson 2009). They put forward strong arguments from a cognitive perspective against a Chomskean universal grammar. Languages on Earth are much more diverse in structure than we expect. It is certainly even less likely to assume a universal uniformity of communication that would be valid for all communicating creatures out there in space. We are misled by anthropocentrism, and start with our preconceived understanding of human communication, which we superimpose on presumed intelligent beings that we have no knowledge of whatsoever. There are reasons to expect diversity and plasticity of cognition and communication in the universe. The limits of what is possible in the universe are wider than what our earthbound minds can imagine.

Many of our present-day attempts at interstellar message construction have much in common with the search for a universal language in the seventeenth century, performed by, among others, George Dalgarno and John Wilkins, who had a rather simplistic conception of language as a mirror of reality and thought, believing that it was possible to find a structural connection between symbol, concept, and things (Eco 1997). The interstellar communication problem is very much a semiotic problem: how meaning can be transferred and interpreted. An exosemiotic analysis is needed. The problem with symbols is that they are conventional, or arbitrary, as Ferdinand de Saussure called them (Saussure 1916). They are detached representations and, as such, dependent on culture and human interaction. The sign (the expression) and the signified (the content) have no intrinsic connection. The symbol refers to the inner world, not to something in the outer environment, in contrast to the signal, which refers to something in the latter. We may figure out the reference of the signal, but will probably have severe problems understanding extraterrestrial symbols. It is not impossible to imagine that the aliens would have certain knowledge about their environment that in its content is similar to our own knowledge of mathematics, physics, and chemistry. But their expression of it, as Göran Sonesson has clearly pointed out, would most likely be very different from ours (Sonesson 2007). In fact, most attempts at interstellar message constructions violate this basic semiotic understanding of signs that distinguishes between expression and content. In a very convincing way, Sonesson shows the inevitable role semiotics must play in message constructions. If we have different ways of expressing mathematics, and then send human mathematical messages, how could the aliens—who might have exactly the same mathematical understanding—know what we are referring to? How we, and the aliens, transfer meaning in different ways, I would say, is the result of dissimilar evolution, bodily and cognitive construction, and sociocultural history. The symbols in use depend on how our brains work, what our bodies are like and interact with in their environment, how our sensations are processed, and not least the history of our culture.

The first problem that arises in a situation of interstellar communication is realizing that it really is a message at all, as Sonesson has pointed out (Sonesson 2007). Some regularity and order, finding a repetition in the pattern, is not enough. We have to understand that someone has an intention with it that we should understand as a message. Next comes the problem of deciphering what the message means. Cultural semiotics, developed by, among others, Yuri Lotman, studies sign systems and the correlations between different systems (Lotman 1990). In order to understand a message the receiver must be able to fill in the gaps between the receiver's perception of the message and the sender's intention with it. The problem is that the creator of the message and the receiver of it are situated in different and specific cultural and social contexts. Relating to interstellar communication, this gap will be huge, with totally different ecological and cultural contexts. As Vakoch clearly states: "In the absence of knowledge of physical and cultural clues, communication between two species can be almost impossible" (Vakoch 1999, 26). Designing a language for cosmic intercourse, such as Hans Freudenthal's *lingua cosmica*, will probably be in vain (Freudenthal 1960). The famous Pioneer plaque now traversing deep space is also too firmly restricted by human culture and cognition, and will most likely be incomprehensible for an extraterrestrial

(Crane 1995). The aliens and ourselves live in different cognitive or, if you wish, semiotic worlds with ways of thinking and signification that are not in agreement with each other. This is another way of expressing the "incommensurability problem" (Vakoch 1999). Nicholas Rescher's legitimate critique of an assumed universality of science can, I would say, be based on cognitive and semiotic foundations (Rescher 1985).

5.5. Conclusion

So we can conclude that: *Communication is based on cognitive abilities embodied in the organism that has developed through an evolutionary and sociocultural process in interaction with its specific environment.* This is the case for human cognition and communication according to recent research in cognitive science and cognitive linguistics.

Our communication is adapted to an earthly environment and for communication with our co-species. Our communication and symbolization have evolved through an evolutionary and cultural-historical process here on Earth, and are thereby constrained by our human bodies, terrestrial environment, and the sociocultural characteristics of our species. So our human communication is in fact maladapted to interstellar communication. This understanding of human cognition might be crucial for future interstellar communication and should be taken into account in order to be able to transfer messages to other minds in the universe.

6.0. Research Initiatives and Solutions

How can we go from this conclusion to more plausible ways than the universal-transcendental objective of interstellar communication? To begin with, we need as much observational data as possible about the lifeworld of the extraterrestrials. To enhance our chances of establishing communication with an extraterrestrial intelligence, we should look for exoplanets with very similar physical characteristics as ours, but even if we find such a planet, this will not prevent the intelligent life forms on that planet from having taken a very different evolutionary path. We know that evolution to a great extent is not only an adaptation to specific environments, but also a result of mutations and accidental occurrences. But if the aliens have adapted to similar physical forces and conditions such as gravitation, light, planetary dynamics, chemistry, etc., this will enhance the possibility of having a mutual understanding of the physical environment. We also have to, of course, focus on what we at this moment can study: our own understanding of our communicational abilities and constraints. We need more knowledge about how our human cognitive and communicational abilities emerged phylogenetically and ontogenetically.

6.1. Phylogenesis and Ontogenesis of Interstellar Communication

We have earlier concluded that communication is an evolving phenomenon. It might be the case that the phylogenesis and ontogenesis of communication and symbolization can guide us. I propose that, based on the knowledge of how complex communication skills have evolved since the early hominoids, but also how present-day humans in their early ontogenesis acquire a language and first learn symbols, we might have clues to interstellar communication. In the phylogenesis of communication we see the transition from animal communication to human language, from an interpersonal function to an ideational function, from signals to symbols. In the ontogenesis of communication we find how infants acquire a complex language with culture-specific symbols. Infants use computational strategies in order to detect statistical and prosodic patterns in language input, which leads to their discovery of phonemes and words of a specific language (Kuhl 2004). With a set of initial perceptual abilities that are necessary for language acquisition, they approach language and rapidly learn from exposure to language, combining pattern detection and computational abilities with social skills.

In other words, we must find methods and strategies of learning how to communicate with the extraterrestrials. We must find a way to learn their symbols. In one way or another we must try to tune into their accumulated communicational tools and symbols of their culture. This would probably demand a continuing flow of social interaction. If we receive something that must be interpreted as an artificial message, terrestrial scientists will be occupied for centuries in deciphering its meaning. But, of course, the immense distances in space will perhaps forever restrict us from more lively conversations. However, we can still listen to their transmissions.

Instead of directly trying to accomplish an information transfer by means of a symbolic abstract language with a content that we think consists of universal scientific facts, we could initiate interaction on the lower cognitive levels, and from there go on to more complex communication systems, starting with concrete messages in signals, indexical and iconic signs in interstellar cognitive semiotics. So rather than focus on shared knowledge, we could focus on shared experiences that are fundamental for a successful mutual understanding. We should search for the basic cognitive processes underlying communication that we have in common.

I suggest that we skip the search for a vehicle for information transfer and symbols, and in fact an abstract message construction altogether. In order to increase our chances of being understood, we should not send an abstract symbolic message based on presupposed universal scientific or mathematical facts. Instead, we should be very concrete and interact with them. The message must force the receiver to interpret it in just one way—it must be direct, immediately understandable, nonambiguous, and tied to the situation and locality.

6.2. Interstellar Intersubjectivity

What we eagerly strive to attain in our efforts toward interstellar communication is ideally to establish a way of transferring information, to get knowledge about their world, nature, and culture, and say something to them about us in return. Before we can reach so far in communicational interaction we first have to determine if they are intelligent at all, that they are intentional, self-conscious beings that show attention to us. And we have to show that we are something alive, something intelligent and self-conscious. What is needed is intersubjectivity.

Intersubjectivity, the sharing and representing of others' mentality, is an important part of our inner worlds (Thompson 2001; Zlatev et al. 2008). Empathy, the representing of other human beings' emotions, motives, intentions, and desires, bodily expressions of emotions, beliefs, and knowledge, are impossible without a rich inner world. Cooperation about detached goals requires advanced coordination of the inner worlds of the individuals. Future encounters with aliens will face severe problems concerning intersubjectivity, in coordinating our inner worlds, feeling empathy, etc., due to our totally different biological and cultural attributes. A human and an extraterrestrial will probably even have trouble perceiving the same target, in aligning their attention, adjusting their actions, and imitating each other. Because of our divergent evolutions, empathy and intersubjectivity toward extraterrestrials would probably be even more problematic than in the case of interspecies communication on Earth.

What is in fact needed for all successful communication is intersubjectivity: that is, shared devices for sharing and manipulating attention. In human conversation, for example, we constantly monitor each other's attentional status. There are strong arguments, according to the psychologist Michael Tomasello, that an infant can only understand a symbolic convention if it understands its communicating partner as an intentional agent with whom one may share attention toward something. A linguistic symbol can in that case be said to be a marker for an intersubjective and shared understanding of a situation. The linguistic symbol is also perspectival in the sense that it embodies one way of many other available ways a situation may be construed (Tomasello 2005). Symbols do not represent the world directly, but are rather used to induce the receiver to construe certain perceptual and conceptual situations. To reach understanding in interstellar communication we need to establish an intersubjectivity that could lead to the possibility of entering the others' inner thoughts and views of reality. It is crucial to find out whether the others are, like ourselves, intentional agents, so that we in that case could relate to their world, and have perspectives on our worlds that can be followed, directed, and shared.

In order to reach an interstellar intersubjectivity I suggest, then, that we try to establish joint attention; that we develop a mutual referential behavior, directed gaze, or mutual gaze. One option is to try to tune in our spatial organizations, and together observe things observable to both terrestrials and extraterrestrials. We can use certain astronomical landmarks in their very neighborhood, to which we can direct our joint attention—for example with reference to known pulsars in the neighborhood, as the Pioneer plaque represented the sun's relationship to fourteen known pulsars. Or the Andromeda Galaxy, our nearest galaxy, which, as Carl Sagan said, would be the only object that both we and the recipients could see firsthand. The best way to find out if they understand natural numbers would be to count concrete objects in our shared physical reality, together with them. Everett M. Hafner proposed transmissions simulating astronomical objects, for example, the fluctuation of the sun's cycle back and forth between the stars (Hafner 1969; Vakoch 1998). The sounds of geological activity, such as volcanoes, earthquakes, thunder, and ocean waves, included in the Voyager recording (Vakoch 2009b), are something we both might experience if we both hear in the same frequency range. By using such indexical references toward some concrete phenomena in the physical environment, we do not need to presuppose a universal science that we should have in common, and do not have to point to our models of the phenomena. Instead, we firmly connect our interaction in the physical reality. If we succeed in this, we will have taken a first step toward an interstellar intersubjectivity.

From joint attention, or perception, we might go on to indexical reference. This can be explained as an outgrowth of the repeated experience of pairing stimuli or events. And then to imitation—to mimic their actions, and attempt to reproduce the other agents' intended actions in the world; and then further to iconic signs. The icon is a sign that has some nonarbitrary similarity to the signified, in contrast to the symbol's arbitrary relation. Vakoch has observed that icons, such as transmissions simulating natural phenomena, will be superior to symbols of our models of the phenomena (Vakoch 1998).

Research in the phylogenesis and ontogenesis of language is rapidly making progress and expanding. Here, I do not intend to postulate a particular notion of a fixed track toward higher order communication. What I want to put on the agenda are the prelinguistic cognitive capacities that we use for communication. Merlin Donald has proposed an evolutionary scale from perception, signs, sign systems to embodied signs (Donald 1991; Donald 2001; Sonesson 2009). The first stage in his system is a transition from episodic to mimetic culture, from the culture of apes and australopithecines to *Homo erectus.* The second transition goes further on to *Homo sapiens*, from mimetic to mythical culture. The recent and largely nonbiological transition led to a third stage of cognitive evolution, including external symbolic storage and theoretic culture.

In the second stage, mentioned above, we use iconic signs, including mime, imitation, and gestures. In a similiar way, we can engage in interstellar mimetics, long before we have any clue of what the message means; that is, to mimic their actions and attempt to reproduce the other agents' intended actions in the world. Imitations and modeling actions is a way of doing something together, including action, object, and result. If we get a message we could send it back. If we start to try to communicate we could send an iconic message, preferably representing something well known in their neighborhood, maybe the emission spectrum of their sun. If they answer with the spectrum of our sun, then we have begun aligning our thoughts. With icons to mimes we achieve something together. Doing something like this, we are going through an imitative learning process—not unlike the infant's strategy to learn symbols—to comprehend the symbolic conventions of their species and culture that finally can result in a mutual understanding. Thus, we go from bodily acts to imitative acts, and then further to representative imitation, and symbolic representation.

To conclude, instead of constructing a universal and abstract message, we should be local and concrete. Instead of using arbitrary symbols and trying to transfer information, we should try to interact and establish joint attention. Instead of leaving out the context, space, and time, we should tie the message firmly to the situation and the environment. So what I am proposing is a shift from the message itself to what precedes it, logically as well as temporally, phylogenetically as well as ontogentically: the mental activities, the interaction, the embodied and situated cognitive processes of the minds of the universe. That is the message.

Acknowledgments

The author is grateful for valuable suggestions and comments on the manuscript made by Professor Göran Sonesson, research leader of the Centre for Cognitive Semiotics, Lund University; and the philosopher of religion Per Lind, Centre for Theology and Religious Studies, Lund University.

Works Cited

Arbib, M. A. 1979. Minds and millenia: The psychology of interstellar communication. *Cosmic Search* 1 (3): 21–25.

Baird, J. C. 1987. *The inner limits of outer space.* Hanover, NH: University Press of New England.

Brinck, I. 2007. Situated cognition, dynamic systems, and art: On artistic creativity and aesthetic experience. *Janus Head.* Special issue: The situated body. Guest ed. S. Gallagher, 9 (2): 407–31.

Christiansen, M. H., and S. Kirby, eds. 1997. *Language evolution.* Oxford: Oxford University Press.

Clark, A. 1997. *Being there.* Cambridge: MIT Press.

———, and D. Chalmers. 1998. The extended mind. *Analysis* 58 (1): 7–19.

Crane, T. 2003. *The mechanical mind: A philosophical introduction to minds, machines, and mental representation.* 2nd ed. London: Routledge.

Deacon, T. 1997. *The symbolic species: The co-evolution of language and the brain.* New York: Norton.

DeVito, C. L. 2011. On the universality of human mathematics. In *Communication with extraterrestrial intelligence,* ed. D. A. Vakoch. Albany: State University of New York Press.

———, and R. T. Oehrle. 1990. A language based on the fundamental facts of science. *Journal of the British Interplanetary Society* 43: 561–68.

Donald, M. 1991. *Origins of the modern mind: Three stages in the evolution of culture and cognition.* Cambridge: Harvard University Press.

———. 2001. *A mind so rare: The evolution of human consciousness.* New York: Norton.

Dunbar, R. 1996. *Grooming, gossip, and the evolution of language.* London: Faber.

Dunér, D. 2011. Astrocognition: Prolegomena to a future cognitive history of exploration. In *Humans in outer space—interdisciplinary perspectives,* ed. U. Landfester, N.-L. Remuss, K.-U. Schrogl, and J.-C. Worms. Vienna: Springer.

Eco, U. 1997. *The search for the perfect language.* New ed. Oxford: Blackwell.

Evans, N., and S. C. Levinson. 2009. Language diversity and its importance for cognitive science. *Behavioral and Brain Sciences* 32: 429–92.

Finney, B., and J. Bentley. 1998. A tale of two analogues: Learning at a distance from the Ancient Greeks and Maya and the problem of deciphering extraterrestrial radio transmissions. *Acta Astronautica* 42 (10–12): 691–96.

Freudenthal, H. 1960. *Lincos: Design of a language for cosmic intercourse.* Amsterdam: North-Holland.

Giere, R. N., and B. Moffatt. 2003. Distributed cognition: where the cognitive and the social merge. *Social Studies of Science* 33 (2): 1–10.

Gärdenfors, P. 2006. *How Homo became Sapiens: On the evolution of thinking.* Oxford: Oxford University Press.

———. 2008. Concept learning. In *A smorgasbord of cognitive science,* ed. P. Gärdenfors, and A. Wallin, 165–81. Nora: Nya Doxa.

Hafner, E. M. 1969. Techniques of interstellar communication. In *Exobiology: The search for extraterrestrial life,* ed. M. M. Freundlich, and B. M. Wagner, 37–62. Washington, DC: American Astronautical Society.

Krois, J. M. In press. Enactivism and embodiment in picture acts, or: the chirality of pictures. *Actus et Imago: Berliner Schriften für Bildakt-Forschung*. 1. Berlin: Akademie Verlag.

——, M. Rosengren, A. Steidele, and D. Westerkamp, eds. 2007. *Embodiment in cognition and culture*. Amsterdam: Benjamins.

Kuhl, P. K. 2004. Early language acquisition: Cracking the speech code. *Nature Reviews Neuroscience* 5: 831–43.

Lakoff, G., and M. Johnson. 1980. *Metaphors we live by*. Chicago: University of Chicago Press.

——, and M. Johnson. 1999. *Philosophy in the flesh: The embodied mind and its challenge to Western thought*. New York: Basic Books.

Lakoff, G., and R. E. Núñez. 2000. *Where mathematics comes from: How the embodied mind brings mathematics into being*. New York: Basic Books.

Lotman, Y. 1990. *Universe of the mind: A semiotic theory of culture*. London: Tauris.

McConnell, B. 2001. *Beyond contact: A guide to SETI and communicating with alien civilizations*. Sebastopol, CA: O'Reilly.

McNeill, W. H. 1973. In *Communication with extraterrestrial intelligence (CETI)*, ed. C. Sagan, 342–46. Cambridge: MIT Press.

Michaud, M. A. G. 2007. *Contact with alien civilizations: Our hopes and fears about encountering extraterrestrials*. New York: Springer.

Rescher, N. 1985. Extraterrestrial science. In *Extraterrestrials: Science and alien intelligence*, ed. E. Regis Jr., 83–116. Cambridge: Cambridge University Press.

Rosch, E. 1975. Cognitive representations of semantic categories. *Journal of Experimental Psychology: General* 104: 192–233.

——. 1978. Principles of categorization. In *Cognition and categorization*, ed. E. Rosch, and B. B. Lloyd, 27–48. Hillsdale, NJ: Erlbaum.

Sagan, C., ed. 1973. *Communication with extraterrestrial intelligence (CETI)*. Cambridge: MIT Press.

Saussure, F. de. 1916. *Cours de linguistique générale*. Lausanne: Payot.

Sefler, G. F. 1982. Alternative linguistic frameworks: Communications with extraterrestrial beings. In *Philosophers look at science fiction*, ed. N. D. Smith, 67–74. Chicago: Nelson-Hall.

Shostak, S. 2009. The value of "L" and the cosmic bottleneck. In *Cosmos and culture: Cultural evolution in a cosmic context*, ed. S. J. Dick, and M. L. Lupisella, 399–414. Washington, DC: NASA History Series.

Sinha, C. 2009. Language as a biocultural niche and social institution. In *New directions in cognitive linguistics*, ed. V. Evans, and S. Pourcel, 289–309. Amsterdam: Benjamins.

Sonesson, G. 2007. Preparation for discussing constructivism with a Martian. In *Les signes du monde: Interculturalité & Globalisation: Actes du 8ème congrès de l'Association International de sémiotique, Lyon, 7–12 juillet, 2004.* Lyon: Université de Lyon. http://jgalith.univ-lyon2.fr/Actes/articleAsPDF/SONESSON_(martiens)_pdf_20061114153124.

———. 2009. The view from Husserl's lectern: Considerations on the role of phenomenology in cognitive semiotics. *Cybernetics & Human Knowing* 16 (3–4): 107–48.

Tarter, J. 2001. The search for extraterrestrial intelligence (SETI). *Annual Review of Astronomy and Astrophysics* 39: 511–48.

Taylor, J. R. 2002. *Cognitive grammar.* Oxford: Oxford University Press.

———. 2003. *Linguistic categorization.* 3rd ed. Oxford: Oxford University Press.

Thompson, E., ed. 2001. *Between ourselves: Second-person issues in the study of consciousness.* Thorverton: Imprint Academic.

Tomasello, M. 1999. *The cultural origins of human cognition.* Cambridge: Harvard University Press.

———. 2005. Uniquely human cognition is a product of human culture. In *Evolution and culture: A Fryssen Foundation symposium,* ed. S. C. Levinson, and P. Jaisson, 203–17. Cambridge: MIT Press.

———. 2008. *Origins of human communication.* Cambridge: MIT Press.

Vakoch, D. A. 1998. Constructing messages to extraterrestrials: An exosemiotic perspective. *Acta Astronautica* 42 (10–12): 697–704.

———. 1999. The view from a distant star: Challenges of interstellar message-making. *Mercury,* March/April, 26–32.

———. 2009a. Anthropologial contributions to the search for extraterrestrial intelligence. In *Bioastronomy 2007: Molecules, microbes, and extraterrestrial life,* ed. K. J. Meech et al. ASP Conference Series, 421–27.

———. 2009b. Encoding our origins: Communicating the evolutionary epic in interstellar messages. In *Cosmos and culture: Cultural evolution in a cosmic context,* ed. S. J. Dick and M. L. Lupisella, 415–39. Washington, DC: NASA History Series.

Varela, F. J., E. Thompson, and E. Rosch. 1991. *The embodied mind: Cognitive science and human experience.* Cambridge: MIT Press.

Westin, A. 1987. Radio astronomy as epistemology: Some philosophical reflections on the contemporary search for extraterrestrial intelligence. *The Monist* 70: 88–100.

Workman, L., and W. Reader. 2004. *Evolutionary psychology: An introduction.* Cambridge: Cambridge University Press.

Wray, A. 1998. Protolanguage as a holistic system for social interaction. *Language & Communication* 18: 47–67.

Zlatev, J., T. P. Racine, C. Sinha, and E. Itkonen, ed. 2008. *The shared mind: Perspectives on intersubjectivity.* Amsterdam: Benjamins.

Culture, Meaning, and Interstellar Message Construction

John W. Traphagan

For she lived, or existed, or in the final analysis *hid* in the shadowed acreage behind the great tree where stood markers with names and dates peculiar to the Family.

—Ray Bradbury

1.0. Introduction

Culture is a concept that frequently arises in discussions related to communication with extraterrestrial intelligence (ETI). Less frequent are explorations of interstellar communication that problematize the nature of culture and unpack what we mean when we talk about culture—whether terrestrial or extraterrestrial. Indeed, as anthropologists have found repeatedly, "culture"—whatever that is—seems, like Bradbury's Angelina Marguerite, to lie in the shadows, just behind the tree or around the corner, despite the fact that identifiable markers seem to tell us it ought to be right there under our noses. In this paper, I am interested in considering some of the problems that arise when conceptualizing culture and applying notions of culture to thinking about both the process of interstellar message creation and the interpretation of any transmission we might receive. I am particularly interested in the problem of incommensurability between cultures in relation to communication, difficulties in reconciling the problem of cultural difference, cultural evolution, and challenges associated with interpretation of meaning cross-culturally, even among terrestrial cultures. Deep consideration of the nature, meaning, and construction of culture by scientists and others interested in the search for extraterrestrial intelligence (SETI) has the potential benefit of providing new and novel ways of thinking not only about ETI, but about human culture and social organization as well.

Central to my argument is the idea that directions for future thinking on interstellar message construction should involve not only research on the explicit message intended, but direct consideration of the implicit information that is being tacitly coded along with any explicit message that is sent (or received). Rather than only asking the questions, "What does ETI mean in a message we receive?" or "What information do we want to convey in a message from us to any possible ETI?" we should also be asking: "What are the implicit indicators and forms of information about ETI and ourselves that are tacitly coded in any message sent or received?" In many respects, initial focus on how to interpret implicit information may be more important than how to interpret the meaning of any explicit message, given the potential differences in culture and biology that may well exist between ourselves and an extraterrestrial other, as well as the inevitable differences in personal intentions and interpretations that will be fundamental parts of the context of contact on either side.

2.0. Defining Culture

Although used routinely in colloquial conversation, the term *culture* tends to lack definitional precision that allow it to be analytically useful for thinking about human behavior. In nonacademic circles—and even within those circles—the term is used in a variety of ways. For example, if we talk about a person being "cultured" we are using the term as a way to identify awareness and understanding of a narrow band of cultural production—that associated with intellectual elites. There is also a tendency within this usage to assume some degree of intellectual and even psychological superiority of the identified person, although the term can also be used derisively.

A second way in which the term is commonly used is as a label to identify a particular group of people in terms of seemingly essential characteristics that they appear to hold in common. Thus, we talk about "Japanese culture" or "French culture" with the notion that there is some level of uniformity in meaning that applies to members of those groups; or we talk about a society that is "multicultural" to suggest that it lacks the ability to be characterized uniformly, but that it consists of subgroupings of people who can be characterized in that way. Normally, those characteristics are left as tacitly understood, apparently lacking much need for explication. This tendency is not limited to the general public; academics also frequently use the culture concept in this way. Titles such as *The Japanese Mind: Understanding Contemporary Japanese Culture* are prime examples of approaches that treat a group of individuals as a collectivized unity able to be understood in terms of a set of common and uniform patterns of ideas and behaviors—in this book, we see a notion of the Japanese as having some sort of common mind that is reflected in their culture and can be both probed and talked about in generalized ways.

Anthropologists have recognized over the past twenty years or so that this is a very problematic way to think about human social organization or human minds and personalities. Rather than "The Japanese Mind," at the very least we need to talk about pluralities of Japanese minds that variously reflect, but also contest, tendencies that seem apparent among many Japanese people. Unfortunately, despite increasing sophistication—and often confusion—in thinking about culture over the past 150 years, sociocultural anthropologists have struggled to arrive at any widely agreed-upon definition of culture. In some ways, although we have developed sophisticated methods for collecting data and theorizing about culture and its influence on individuals and groups, we are not really any closer to a definition than E. B. Tylor was when he wrote in 1871 that culture is "that complex whole which includes knowledge, belief, art, morals, law, custom, and any other capabilities and habits acquired by man as a member of society" (Tylor 1924 [orig. 1871], 1).

This should not be interpreted as meaning that anthropologists have not learned anything over the past 150 years. We have an enormously rich and detailed catalogue of ethnographic data on human behavior and social organization.[1] We have also learned that the complexity associated with generalizing characteristics of any particular "culture" are sufficiently large that we probably should try to refrain from making those generalizations and we have recognized that the term *culture* itself tends to create a sense of boundedness to human groups that is not reflective of how they actually function. Usage of the term also inclines people toward assuming a deterministic understanding of human behavior—Tylor's idea of a set of ideas, capabilities, etc. "acquired by man as a member of society" leads to a notion of rather strong socialization into particular patterns of behavior and fails to capture human characteristics such as innovation, disagreement, intransigence, or overt opposition.

Contemporary anthropologists understand individuals as being both unified and divided by customs and beliefs, often simultaneously. For example, groups of people from different races may feel in one way unified by their support of a particular sports team, while having strong feelings of division in relation to other areas of their lives—such as ideas about race and social justice. They also may come together in times of crisis or triumph, while being largely divided otherwise. In short, people are not only unified, but also often divided by their customs and beliefs, even when they ostensibly identify themselves as being part of a common culture. For most anthropologists today, culture has come to be understood as an ongoing process through which individuals and groups invent, contest, and reinvent the customs, beliefs, and ideas that they use collectively, individually, and often strategically, to characterize both their own groups and the groups to which others belong and to manipulate their social and physical environments. If we can characterize culture as something, it is probably best to think about it as a moving context, framework, or environment of behaviors that people observe, mimic, and respond to; it is not a deterministic thing that generates a uniform set of beliefs or behaviors among a group of people. As a result, it is extremely difficult to predict the behaviors of individuals in terms of cultural patterns and we should avoid strongly characterizing any one individual psychologically as being representative of broader patterns in terms of thought and behavior.

Some anthropologists, myself included, have come to the conclusion that the term *culture*, either as an analytical concept or as a way of characterizing a group of people, does very limited theoretical work and actually tends to create more problems than it solves. But we seem to be stuck with it, so if we are going to use it we need to develop the idea with a great deal of nuance. The short form of this nuance for our purposes here means that "culture" should be seen as idiosyncratic and that when we talk about an alien civilization, we need to take a great deal of care in not assuming that the members of that civilization are going to think like us, or that they are even going to think like each other. If they are in any way like us, there will be patterns in common, but there will also be significant individual differences and they will likely argue about those differences. Furthermore, their "civilization" (itself a very loaded word) or their "culture" will, like ours, be a moving target, one that will change immediately upon contact with us.

3.0. Culture and Evolution

Change is perhaps the one thing that we need to pay the most attention to in contemplating the nature of either human societies or the societies of ETI. In his interesting and important article "The Postbiological Universe," Steven J. Dick (2008) notes that SETI researchers have not made significant strides to incorporate the idea of cultural evolution into the broader discourse speculating on the possible nature of extraterrestrial intelligence. Dick (2008, 502) takes an initial step in this direction by developing his ideas around what he refers to as the "Intelligence Principle," in which the driving force of cultural evolution is the "maintenance, improvement and perpetuation of knowledge and intelligence." Dick notes that his argument about the emergence of a postbiological university is based upon a set of assumptions and recognizes that several of these can be called into question. Here I want to focus on two of the assumptions from which Dick works: (1) that knowledge and intelligence improve and (2) that the driving force of cultural evolution is increasing intelligence or, more generally, that we should associate the notion of a driving force in any way with cultural change. Dick recognizes some of the problems with these assumptions; thus, my aim here is not a critique of his argument, but an extension and exploration of the problems associated with applying the idea of cultural evolution, particularly when it comes to SETI research.

What does it mean to say "culture evolves"? The answer to this question lies in clearly delineating what we mean by both the concepts of culture (explored above) and of evolution. When we talk about biological evolution, we are, as Vakoch notes, generally referring to an explanation of change in the natural environment through which constant laws of nature "are manifested through *transformations* of the stuff of the universe" (Vakoch 2009, 417). Among anthropologists, archaeologists, and evolutionary psychologists (to name a few) the meaning of the idea of evolution has taken on different connotations both across disciplines and in relation to the attempts to apply Darwinian and in some cases neo-Darwinian (e. g., Dawkins's [1976] notion of the meme) biological evolutionary ideas to social structures, ideational generation, and patterns of behavior. Influenced by Darwin, early anthropologists such as L. H. Morgan developed schemes of cultural evolution in which, broadly speaking, human societies moved from savagery to civilization as a result of both technological and ideational innovation and progress. The contemporary North Atlantic societies were viewed as the pinnacle of this process, exhibiting both advanced (interpreted as superior) technologies and social/moral structures—the latter being evident in the apparently more advanced condition of cultural features such as monotheistic religion (Morgan 1877).

Although this idea was panned by anthropologists in the early part of the twentieth century, the progress-centered approach to interpreting cultural change—in much more sophisticated and sometimes empirically accurate ways—has continued to be important, showing up in widely used theoretical frameworks such as modernization or globalization theories. In general, these notions work from an emphasis on a set of sequences or stages involving variables such as the scale of societies (population), levels of social integration, degrees in differentiation and specialization, etc., and can be recognized in classical anthropological divisions of social organization into types: bands, tribes, chiefdoms, and states (Shennan 2009, 2). These ideas also tend to work from an assumption about increasing complexity as being associated with progress, but do not sufficiently recognize the fact that complexity in biology is the result of a different processes from what occurs in cultural change, because cultural development is accumulative while biological evolution is substitutive (Kroeber 1948, 297; Williams 1966, 34). As Williams pointed out in the 1960s in relation to biological evolution, "There is nothing in the basic structure of the theory of natural selection that would suggest the idea of any kind of cumulative progress" in the development of biological organisms (Williams 1966, 34). This aspect of cultural evolutionary theory tends toward romanticized models of "civilization" as an improvement—due to increased complexity and capacity for technological innovation through specialization—over other forms of social organization, and these models, which are grounded in evolutionary models originating from the nineteenth century, tend to equate technological advancement with cultural advancement and also have normally affirmed the general superiority of North Atlantic cultures (Marcus and Fischer 1986, 128–29). This problem continues to lurk in discussions of cultural evolution, despite the fact that, as noted above, in the early twentieth century anthropologists empirically demonstrated the fundamental weaknesses of this approach to understanding cultural change (Steward 1955, 15).

While it seems clear that cultural change and increasing complexity are closely related to the accumulation of knowledge over time, we need to be careful not to equate this with progress in the sense of improvement. Although societies change over time, is there evidence that that change is *progressive* from a cultural perspective? I think that often when we answer "yes" to this question, we are equating technological innovation, development, and improvement through the accumulation of knowledge (progress) with culture itself. To be sure, technological progress is closely associated with cultural patterns and institutions; there are conditions that may arise in a given context that allow for increased technological innovation—the existence of the National Science Foundation or NASA are examples. But we need to be very careful in equating this sort of progress with more generalized cultural progress or with the idea that a specific culture or type of culture is more advanced than other types of cultures due to processes of evolutionary change that is directional. There is no reason to think that a tribal form of social organization is any worse, or any better, than a state-level form of social organization. One type of social structure works well in societies with mobile small populations; the other works better in sedentary societies with larger populations. One type of social organization is not better than the other, they are simply different and more appropriate for certain environmental (understood very broadly to include the social environment) conditions.

Additionally, while technology may improve, in the sense that we see increased efficiency and capabilities, there is little evidence to suggest that other areas of human behavior improve as well. For example, we might consider the atomic bomb as a significant technological development that *progressed* from a combination of knowledge related to previous types of bombs and nuclear physics. In this sense, it represents a change that involves increased elaboration of knowledge and even improved understanding of physics. However, from a moral standpoint, it might be argued that the atomic bomb represents regressive tendencies in human social capacities evident in our willingness in the twentieth century to engage in total warfare and indiscriminate attacks on noncombatants. If we think in terms of nondirectional and nondeterministic change, then perhaps we can talk about cultural evolution, but if we think in terms of progress, then it is difficult to see the atomic bomb as representing cultural progress, despite the fact that it represents technological progress based upon increased and better knowledge of physics.

The notion of cultural evolution is, rather than an objective product of empirical observation, a subjective reflection of a specific set of cultural values that tend to construct the world, and the human societies that inhabit it, in terms of a linear historical process that moves from primitive to increasingly advanced societies, or in more nuanced expressions from less to increasingly complex forms of social organization. While increased complexity and sophistication is obvious when it comes to technology and institutional structures (such as political institutions), it is far less obvious when one considers other aspects of culture such as moral values or notions about the nature of identity and self, both of which influence the flow of cultural change.

Additionally, to date no equivalent of Mendel has arisen for the study of culture (despite Dawkins's [1976] attempt to identify a unit of cultural selection with his notion of the meme) and, therefore, we have not determined a relatively nondiluting unit for the transmission of cultural inheritance that would serve as the basis for selection at the cultural level (Gaulin 2010). Of course, Dick is not writing about social structures, he is writing about the merging of the biological and nonbiological into a different kind of organism, but he is still working from a notion of directionality and an assumption that technological innovation can be equated with cultural advancement— "[t]he age and longevity of ETI is important for the overarching postbiological universe argument, since ETI somewhat older than humans is a necessity for more advanced cultural evolution" (Dick 2008, 500). In this perspective, cultural evolution is leading to something and that something seems inexorable, inevitable, and universal, not just for humans but for all technologically driven beings. The problem here is that there is no reason to assume a single path or pattern for cultural change; in fact, there is really no reason to think in terms of a path at all. If we use biological evolution as a model, then the idea of progress, as noted above, simply is not accurate. However, if we abandon the biological model in favor of some other way of thinking about evolution in relation to social organization and culture, what is the driving force? What determines that a particular path is likely as opposed to others? While it may be that homo sapiens are on a path toward integration between our biological selves and the machines we make—and this is a fairly large assumption that does not sufficiently account for significant contestation of technology among groups such as some fundamentalist religious movements—and a postbiological society, there is no reason based upon empirical observation of human societies and cultural change within human societies to think that this is an inevitable outcome of cultural change, because the one constant we seem to find when looking at culture is diversity and variation.

4.0. Culture, Evolution, and Human Universals

If there is a single thing that anthropologists have identified fairly clearly about humans it is that a comparison of processes of change in human cultures shows some consistency and enormous diversity. While there are a few commonalities in all human cultures—such as kinship and religion—there are enormous differences in how these human universals are constructed and interpreted. Rather than thinking specifically in terms of cultural evolution, it may be more useful to think in terms of characteristics that appear to be universal across human cultures without reference to concepts such as progress; in part this is due to the fact that even if we associate progress with cultural evolution, there is no reason to think that the particular path(s) of progress among humans should be held in common with other beings. Lemarchand and Lomberg (2009, 398) have approached this issue from a somewhat different perspective, looking at potential cognitive universals and note that there are significant differences in the cognitive maps that people in various societies use to interpret and structure their surroundings. Here, I am interested in thinking about universals of human social organization and considering these as another way of thinking about common elements of culture that cross enormously diverse ways of organizing human societies.

Religion and kinship are social structures found not only across currently extant human societies, but also appear to be evident in some form throughout the history of modern humans and may have existed among some other hominid species. Although there is long-standing debate about its origins, it seems that all modern human societies have shown some form of religious and ritual activity and that, perhaps, other hominid species such as *homo neanderthalensis* may have as well (Wunn 2000). And it is interesting if we think about cultural evolution to note that there is little or nothing empirically evident that would suggest that contemporary religions such as Christianity are much of an improvement over past religions such as the polytheism of the Greeks. And, in fact, we find both types of religious organization evident in technologically equivalent societies—for example, the United States is largely monotheistic and its religions tend to emphasize a god-being, while Japan is polytheistic and its religions tend to be animistic.

Here, I want to focus on the second of these universals—kinship—and discuss the extent to which we find variation within a common social structure and also point out that this variation, while certainly related to the evolution of cultures, is not directional. Kinship is particularly interesting from an evolutionary perspective as it relates to culture, because while it can be studied along genetic lines at the biological level, for sociocultural anthropologists kinship refers not to biology but to cultural constraints and interpretations of biological universals—such as genetic relatedness and human reproduction through sexual relations and birth—and the privileging of specific culturally defined relations over those biological universals (Parkin 1997, 3). People in the West tend to work from the assumption that our system of kinship is "natural," that is, constructed along biological lines—terminologies such as aunt or uncle seem opposed to father and mother seem to reflect biological relatedness (despite the fact that when these terms emerged there was little understanding of genetics). However, Western kinship systems are not based purely upon biology, as there is no reason, for example, to privilege patrilineal descent over matrilineal descent.

The first scholars to look carefully at kinship were heavily influenced by Spencerian theories of cultural evolution. Morgan, for example, argued that there were five successive forms of family structure evident in human societies: Consanguine (based upon group marriage within generations, such as brother/sister marriages), Punaluan (group marriage that forbade sibling marriage), pairing (an intermediate form between group marriage and monogamy), patriarchal (power within the family was located in the male head and polygyny was allowed), and monogonarian (monogamous marriage, female equality, and nuclear family structure) (Harris 1968, 181). Rather than the evolutionary schema he uses, which is clearly not empirically accurate, what is most important about Morgan's work is his recognition that kinship structure is not necessarily directly related to biology and that patterns of marriage and descent, as well as reckoning relatedness, are products of social and cultural variables.

Contemporary anthropologists recognize a variety of schemes for reckoning kinship, none of which can be neatly aligned to a lineal path of cultural evolution, although there is good evidence to support the idea of close interactions of factors such as innovations in subsistence like domestication, genetic and cultural diffusion, and demic expansion that exhibits connections to distributions of language groups, genetic frequencies, and to some extent kinship patterns (Jones 2003, 502). For the most part, kinship systems can be organized along two broad axes: gender and the cohesiveness of unilineality as shown in Figure 32.1 (Jones 2003, 503; Burton, et al. 1996).

Patrilineal and matrilineal systems of descent emphasize one parent at the expense of the other, in terms of tracing links of relatedness and often in determining rights and responsibilities related to variables such as inheritance, care of parents, and ritual obligations. All offspring born into a particular line belong to that line and trace ancestry to the originator of that line, but

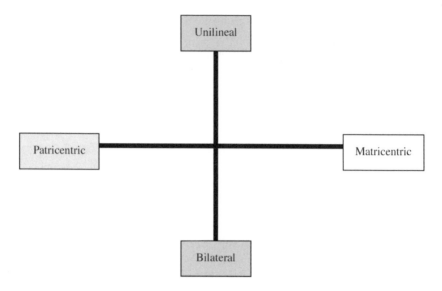

Figure 32.1. Types of kinship systems.

certain children will be emphasized—in the case of patrilineal descent boys and matrilineal descent girls—and certain relationships within the kinship system will also be emphasized, not necessarily in ways that reflect biological relatedness. For example, many societies downplay the relationship between the father and child in preference for emphasizing the relationship between the child and an uncle, such as the mother's brother or the mother's sister's husband (see Figure 32.2). This pattern of kinship is referred to as avunculate and occurs in a variety of forms that stipulate rights and obligations between children and affinal and consanguineal male relatives of a specific woman in ways quite different from what is expected in, for example, bilateral systems of descent (Lévi-Strauss 1963, 44).

In societies where there is no stress placed on gender in relation to descent, the system is called bilateral. The Eskimo pattern of kinship is an example of bilateral descent, because it distinguishes individuals not on the basis of gender, but on the basis of distance, with the conjugal family being emphasized and relatedness being associated with levels of prestige. Other kinship systems also do not distinguish between relatives in the way that Americans do. For example, the Iroquoian system, which is a matrilineal system, does not distinguish between aunts and uncles like the American system. One's father's brothers are all father and one's mother's sisters are all mother and those individuals have parental rights vis-à-vis the child in

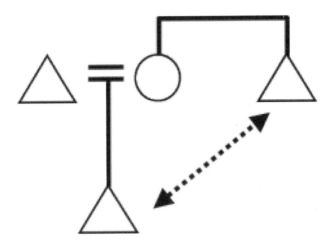

Figure 32.2. Avuncular kinship.

question. And in many societies marriage patterns are closely related to cultural emphases on gender or descent. In parts of Tibet, for example, marriages sometimes follow the pattern of fraternal polyandry, in which a single woman is married to two or more brothers. This does not mean that women are necessarily privileged in terms of power and authority; the eldest brother tends to have some degree of authority in domestic decision making and also has privileged mating access, thus allowing him to father the majority of children (Levine 1988, 5).

We could continue examining polygynous marriage patterns, or consider the relationship between descent and virilocal, matrilocal, or uxorilocal residence patterns at length, but the point is already clear. Despite the fact that kinship is a universal pattern in human social organization, the manner in which it is conceptualized and the ways in which it influences the construction of gender roles, power relationships, marriage patterns, residence patterns, economic activities, and so on, is remarkably complex.

Additionally, while it does seem that certain kinship patterns may be declining in frequency (polyandry may be one, although it was never all that common), there is no reason to think that the world is inexorably marching toward a single pattern for reckoning kinship. Rather, while kinship evolves, it evolves nondirectionally and new ways of thinking about kinship emerge that are related to other patterns operating within a given society. In contemporary American society, a variety of new approaches to kinship have emerged in recent years, the most notable being gay marriage. This represents simply a different, new, pattern of thinking about marriage that reflects other elements of contemporary American society. While it does represent a change, it does not represent a change that has a direction—it is just a change related to other changes in the social and political environment. And in contemporary rural Japan, the pattern of adopting adult males into lineages where no competent male offspring were available to take over responsibility for household headship that at one time was thought to be disappearing, has reemerged as a result of population pressures related to fewer children and consequent reduced marriage opportunities (Brown 1966; Traphagan 2000).

To reiterate, culture does not change with any specific directionality. A rather vast array of variables interact—including often relatively arbitrary social values, but also including variables such as population sizes and subsistence patterns—to generate novel approaches to organizing people as needed. This is evolution, to be sure, but it is not evolution that can be equated with other cultural patterns such as the drive toward technological innovation, elaboration of knowledge, and improved understanding of the world. Recognition of this point has major consequences for how we think about ETI. For example, Lemarchand and Lomberg (2009, 401) argue that there should be some universal evolutionary tendency in the formation of ethical ideas such that

> all civilizations should evolve ethically at the same time as they evolve technologically. When these civilizations reach their technological *adolescent* stage, they must perform an ethical societal mutation or become extinct. After learning how to reach a synergetic harmony among the individual members, their groups and their habitat, they would extend this praxis to all living beings, including their hypothetical galactic neighbors.

This passage harkens back to the cultural and social evolutionary schemes typical of nineteenth-century anthropologists such as Morgan, which assume unidirectionality in moral development and a specific type of moral behavior as universally good, evident in a romanticized "advanced" civilization (remember, for those anthropologists it was nineteenth-century Europe). Elsewhere (Traphagan in press) I have argued that altruism is difficult conceptualize from a cross-cultural perspective; what it means to be altruistic varies from one society to another. In one society it may be a supreme good, even an act of altruism, to hasten the death of elderly who are no longer viewed as contributors to society; in societies such as the United States, this is clearly not an accepted practice (Glascock 1990). It would be problematic, to say the least, and quite ethnocentric to argue that one of these patterns is more ethically advanced than the other. They are different, but there is no basis empirically (or otherwise, as far as I can tell) to determine which one is objectively more advanced. One might argue that death-hastening practices, as Glascock notes, are more evident in preindustrial societies, but the fact remains that all human societies accept killing of humans in some form, whether it be self-defense, war, capital punishment, euthanasia (a form of death hastening that is becoming more common in industrial societies), or suicide. On our own planet, there is no ethnographic (empirical) evidence of ethical values that are universally commensurable in all known cultural contexts present or historical, thus we need to take extreme caution in assuming an evolutionary model that equates technological advancement with moral advancement.[2]

While we may want to assume that ETI is technologically more advanced than are we, we cannot easily talk about cultural or moral advancement in relation to elements such as structures of social organization or values if we take seriously empirical ethnographic data related to our own planet—such a thing does not really exist, at least as far as humans go. Of course, technology is a product of culture, and the interaction between humans and their technology is producing new ways of living and new ways of being that integrate us in novel ways with the technologies we are creating. As Dick points out, this is certainly a possible outcome for other technological species, but it does not mean that there will be a corresponding direction in creativity among extraterrestrial societies. The picture we often see in writing about ETI as advanced is that it is advanced and stable. Humans are advanced and unstable; we have always been unstable because culture is not a static thing; it is a process of change, but a process of change that lacks direction. Thus, it is relatively unpredictable.

5.0. Culture and Contact

What does all of this mean in relation to culture and contact with ETI? First, we need to be aware of the fact that not only will the messages received be encoded with cultural elements from ETI, but the messages we send also will be encoded with implicit information, not about human civilization, but about the specific human culture in which the message is generated. And this is going to be something that is not stable, but will change. For example, should we encounter a signal from an ETI that is five hundred light-years from here, we are looking at one thousand years between sending and receipt of a message by those who sent it. Both cultures will have changed dramatically (and most likely the earthly one will have ceased to exist in any recognizable form, if we take history as a lesson). One question is whether or not we can interpret any meanings in what was sent by ETI, but the other question is, Will we be able to successfully interpret the meanings of what we sent five hundred or one thousand years in the past? In other words, we need to be very cognizant of the fact that our own cultures on earth are changing rapidly—not simply in terms of technology, but also in terms of forms of social organization that react to other changes such as technological innovation.

If we engage in Active SETI and send out transmissions from Earth, as Vakoch (2011) and others have encouraged, then we need to develop mechanisms for careful interpretation and exegesis of our *own* messages in the future. In other words, we must not only to be able to interpret what we receive, we need to be able to understand and interpret what we sent hundreds of years in the past. This is not an easy task, because culture itself is a moving target.

Second, the idea that there may well be an asymmetrical relationship in the ages of civilizations of humans and those of ETI does not mean that there is a necessarily correlated asymmetry in their respective cultural evolution. Nor does it mean that technologically sophistications such as the capacity to build radio telescopes, are equatable with cultural values that we—particularly we scientists—associate with being advanced. The fact that most scientists working in SETI probably think that global cultural interaction is good, that altruism is a positive thing and is associated with advanced intelligence, and that technological capabilities index advanced (and peaceful) forms of social organization, is more a product of our own values than it represents anything that is even remotely empirically evident in the broad range of human cultures—the very meaning of terms such as *altruism* varies significantly from one society/language to another (Traphagan in press). Human societies vary quite a bit in how they respond to potentialities such as warfare.

Take two very technologically advanced societies—the United States and Japan. Japan is notable for Article 9 of its constitution, which has normally been interpreted as a renouncement of war and even self-defense. Of course, Article 9 was imposed by the United States, but most—not all—Japanese have bought into the idea of Japan as a pacifist country. This is despite the fact that Japan has a relatively sophisticated military and is the seventh-largest spender on the military in the world, representing 3 percent of the total international expenditures. By contrast, the United States, also a technologically sophisticated place, has not had a great deal of interest in renouncing war, and seems comparatively comfortable with representing 41 percent of all money spent on the military on Earth. Structurally, there are numerous similarities between the United States and Japan, but historically and culturally there are differences that have generated very different responses to the same problem of self-defense. Furthermore, these approaches did not arise in a vacuum; Americans significantly influenced the Japanese mindset in relation to war and the military, but pro and con.

Would an extraterrestrial civilization be living in complete isolation from other such civilizations? Speculation about an answer to this is of course closely related to how we calculate the Drake equation and how we think about the Fermi paradox or whether we accept the Rare Earth Hypothesis (Ward and Brownlee 2000). We can't really answer this question, but we can speculate that an extraterrestrial "civilization" might, in fact, be many civilizations, rather than a romanticized unified society in which internal conflict and dissent is limited. If humans are any kind of model at all, then that may not be likely, whether that extraterrestrial civilization is on one world or many worlds, and whether or not it is postbiological. Indeed, given our own lack of understanding of how we are moving toward, or even if we are moving toward, a postbiological world, it is difficult to imagine what postbiological would mean. But if we drop the notion of cultural evolution as directional, then there is no reason to think that a postbiological universe will be any more unified, or any less culturally diverse, than our own little universe right here on earth.

Finally, it is important to recognize that should contact be made, if the aliens are like us, then it will not be with another civilization nor with another culture, it will be with an individual or a group of individuals, unless we run into beings akin to Star Trek's Borg who think and act collectively—maybe the postbiological outcome of which Dick writes. Perhaps their world is more unified than ours, perhaps it is not. However their cultures have evolved, it would be a mistake to assume that even if they have technological superiority, they also have cultural superiority—this is not something that has been proven even to exist on Earth.

Notes

1. The Human Relations Area Files are an excellent example of the accumulation of empirical data on a very large array of societies.

2. It is important to point out that within moral philosophy there is a long-standing debate about the extent to which we can identify objective and universal values as opposed to having to accept a position of moral relativism. Furthermore, not all societies accept the idea of ethical universals—Japanese, for example, are much more inclined toward highly situational approaches to ethics. See Rorty 1979 and Traphagan in press.

Works Cited

Bradbury, R. 2001. *From the dust returned.* New York: Avon Books.
Brown, K. 1966. Dōzoku and descent ideology in Japan. *American Anthropologist* 68: 1129–51.

Burton, M. L., C. C. Moore, J. W. M. Whiting, and A. K. Romney. 1996. Regions based on social structure. *Current Anthropology* 37: 87–123.

Dawkins, R. 1976. *The selfish gene.* New York: Oxford University Press.

Dick, S. J. 2003. Cultural evolution, the postbiological universe and SETI. *International Journal of Astrobiology* 2: 65–74.

———. 2008. The postbiological universe. *Acta Astronautica* 62: 499–504.

Gaulin, S. 2010. Personal communication.

Glascock, A. P. 1990. By any other name, it is still killing: A comparison of the treatment of the elderly in American and other societies. In *The cultural context of aging: Worldwide perspectives*, ed. Jay Sokolovsky, 43–56. Westport: Bergin and Garvey.

Harris, M. 1968. *The rise of anthropological theory.* New York: Harper and Row.

Jones, D. 2003. Kinship and deep history: Exploring connections between culture areas, genes, and languages. *American Anthropologist* 105 (3): 501–14.

Kroeber, A. 1948. *Anthropology.* New York: Harcourt, Brace.

Lemarchand, G. A., and J. Lomberg. 2009. Universal cognitive maps and the search for intelligent life in the universe. *Leonardo* 42 (5): 396–402.

Levine, N. E. 1988. *The dynamics of polyandry: Kinship, domesticity, and population on the Tibetan border.* Chicago: University of Chicago Press.

Lévi-Strauss, C. 1963. *Structural anthropology.* New York: Basic Books.

Marcus, G. E., and M. M. J. Fischer. 1986. *Anthropology as cultural critique: An experimental moment in the human sciences.* University of Chicago Press, Chicago.

Morgan, L. H. 1877. *Ancient society: Or, researches in the line of human progress from savagery through barbarism to civilization.* New York: H. Holt and Co.

Parkin, R. 1997. Kinship: An introduction to the basic concepts. Malden, MA: Blackwell.

Rorty, R. 1979. *Philosophy and the mirror of nature.* Princeton: Princeton University Press.

Shennan, S. 2009. Introduction. In *Pattern and process in cultural evolution: An introduction*, ed. Stephen Shennan, 1–18. Berkeley: University of California Press.

Steward, J. H. 1955. *Theory of culture change: The methodology of multilinear evolution.* Urbana: University of Illinois Press.

Traphagan, J. W. 2004. *The practice of concern: Ritual, well-being, and aging in rural Japan.* Durham, NC: Carolina Academic Press.

———. 2008. Embodiment, ritual incorporation, and cannibalism among the Iroquoians after 1300 C.E. *Journal of Ritual Studies* 22 (2): 1–12.

———. In press. Altruism, pathology, and culture. In *Pathological Altruism*, ed. Barbara Oakley, Ariel Knafo, Guruprasad Madhavan, and David Sloan Wilson. New York: Oxford University Press.

Tylor, E. B. 1924 [orig. 1871]. *Primitive culture.* 2 vols. 7th ed. New York: Brentano's.

Vakoch, D. A. 2009. Encoding our origins: Communicating the evolutionary epic in interstellar messages. In *Cosmos and culture: Cultural evolution in a cosmic context*, ed. Steven J. Dick and Mark L. Lupisella, 415–39. Washington, DC: NASA History Series.

———. 2011. Asymmetry in Active SETI: A case for transmissions from Earth. *Acta Astronautica* 68 (3–4): 476–88. doi:10.1016/j.actaastro.2010.03.008.

Ward, P. D., and D. Brownlee. 2000. *Rare Earth: Why complex life is uncommon in the universe.* New York: Copernicus Books.

Williams, G. C. 1966. *Adaptation and natural selection.* Princeton: Princeton University Press.

Wunn, I. 2000. Beginning of religion. *Numen* 47 (4): 417–52.

Index